Advanced
NANOELECTRONICS

T0225623

Nano and Energy

Series Editor: Sohail Anwar

PUBLISHED TITLES

Advanced Nanoelectronics
Razali Ismail, Mohammad Taghi Ahmadi, and Sohail Anwar

Computational Nanotechnology: Modeling and Applications with MATLAB®
Sarhan M. Musa

Nanotechnology: Business Applications and Commercialization
Sherron Sparks

Nanotechnology: Ethical and Social Implications
Ahmed S. Khan

Advanced
NANOELECTRONICS

Edited by
Razali Ismail
Mohammad Taghi Ahmadi
Sohail Anwar

CRC Press
Taylor & Francis Group
Boca Raton London New York

CRC Press is an imprint of the
Taylor & Francis Group, an **informa** business

MATLAB® and Simulink® are trademarks of The MathWorks, Inc. and are used with permission. The Math-Works does not warrant the accuracy of the text or exercises in this book. This book's use or discussion of MATLAB® and Simulink® software or related products does not constitute endorsement or sponsorship by The MathWorks of a particular pedagogical approach or particular use of the MATLAB® and Simulink® software.

CRC Press
Taylor & Francis Group
6000 Broken Sound Parkway NW, Suite 300
Boca Raton, FL 33487-2742

First issued in paperback 2017

© 2013 by Taylor & Francis Group, LLC
CRC Press is an imprint of Taylor & Francis Group, an Informa business

No claim to original U.S. Government works

ISBN-13: 978-1-4398-5680-2 (hbk)
ISBN-13: 978-1-138-07287-9 (pbk)

This book contains information obtained from authentic and highly regarded sources. Reasonable efforts have been made to publish reliable data and information, but the author and publisher cannot assume responsibility for the validity of all materials or the consequences of their use. The authors and publishers have attempted to trace the copyright holders of all material reproduced in this publication and apologize to copyright holders if permission to publish in this form has not been obtained. If any copyright material has not been acknowledged please write and let us know so we may rectify in any future reprint.

Except as permitted under U.S. Copyright Law, no part of this book may be reprinted, reproduced, transmit-ted, or utilized in any form by any electronic, mechanical, or other means, now known or hereafter invented, including photocopying, microfilming, and recording, or in any information storage or retrieval system, without written permission from the publishers.

For permission to photocopy or use material electronically from this work, please access www.copyright.com (http://www.copyright.com/) or contact the Copyright Clearance Center, Inc. (CCC), 222 Rosewood Drive, Danvers, MA 01923, 978-750-8400. CCC is a not-for-profit organization that provides licenses and registration for a variety of users. For organizations that have been granted a photocopy license by the CCC, a separate system of payment has been arranged.

Trademark Notice: Product or corporate names may be trademarks or registered trademarks, and are used only for identification and explanation without intent to infringe.

Library of Congress Cataloging-in-Publication Data

Advanced nanoelectronics / edited by Razali Bin Ismail, Mohammad Taghi Ahmadi, and
  Sohail Anwar.
    p. cm. -- (Nano and energy)
  Includes bibliographical references and index.
  ISBN 978-1-4398-5680-2 (hardback)
  1. Nanoelectronics. I. Bin Ismail, Razali. II. Ahmadi, Mohammad Taghi. III. Anwar,
Sohail.

TK7874.84.A38 2012
621.381--dc23   2012021701

Visit the Taylor & Francis Web site at
http://www.taylorandfrancis.com

and the CRC Press Web site at
http://www.crcpress.com

Contents

v

Preface

Nanoelectronics refers to the technology of electronic devices, especially transistors, whose dimensions range from atoms to 100 nm. Such transistors are so small that the interatomic interactions and quantum mechanical properties need to be studied extensively. This book provides research information regarding advanced nanoelectronics concepts, focusing primarily on the aspects of modeling and simulation. It develops and applies numerical algorithms to investigate nanodevices. While theories based on classical physics have been very successful in helping experimentalists design microelectronic devices, new approaches based on quantum mechanics are required to accurately model nanoscale transistors and to predict their characteristics even before they are fabricated.

The book is organized into 14 chapters. Chapter 1 introduces the basic ideas related to quantum theory required for understanding nanoscale structures found in nanoelectronics. The key principles of quantum theory are briefly explained, followed by an explanation of the basic quantum theory of electrons relevant to nanoelectronics. A brief outline of standard solid state physics is presented and how this has to be extended to account for the electronic properties of nanoscale systems is also shown. At the end of the chapter, these fundamental theoretical ideas are applied to nanostructures such as graphenes, carbon nanotubes, quantum wells, quantum dots, and quantum wires.

Chapter 2 highlights some of the key concepts required to understand nanotransistors. The quantum theory of solids and the Fermi–Dirac distribution function are introduced, followed by the application of these concepts to three-dimensional materials in the nondegenerate and degenerate limits. Quasi two-dimensional and one-dimensional devices are then considered. These concepts are applied to the carbon nanotube (CNT). A comprehensive study of the CNT is presented.

Chapter 3 describes the carbon nanotube field effect transistor (CNTFET) model in detail. The key concepts relevant to quantum electronic and semiconductor physics, which were presented in Chapters 1 and 2, provide the background for this chapter. The types of CNTFET are first discussed, followed by the CNTFET design. Next, the CNTFET models used are reviewed. The chapter goes on to describe how CNTFET models are developed. Carrier statistics are formulated and results presented.

Circuit theory in electronic textbooks relies heavily on Ohm's law. Ohm's law enjoyed its superiority in the performance assessment of all conducting materials until it was discovered that the carrier velocity cannot increase indefinitely with the increase of electric field, and eventually saturates to a value that leads to current saturation. Chapter 4 discusses the application of nonohmic law to CNTFET circuits. The ultimate goal of this chapter is to verify the application of nonohmic law to the CNTFET circuit by comparing the theoretical results with the Hspice simulation results. This chapter concludes with a discussion of two quality measures, which indeed examine the overall performance of CNTFET in logic gates, comparing it with that of the logic gates made of state-of-the-art metal–oxide–semiconductor field effect transistor (MOSFET).

Chapters 5 through 9 focus on graphene. This unzipped form of CNT is the recently discovered allotrope of carbon that has gained a tremendous amount of scientific and technological interest. Prototype structures showing good performance for transistors, interconnects, electromechanical switches, infrared emitters, and biosensors have been demonstrated. The graphene nanoribbon (GNR) MOSFET (GNRFET) has been developed and used as a possible replacement to overcome the CNT chirality challenge. In recent years, single-layer graphene (SLG) has attracted a great deal of interest. Chapter 5 reviews and discusses the theoretical physics related to GNR, SLG, and GNRFET. Chapter 6 describes bilayer graphene nanoribbon (BGN). The carrier density and temperature effects on mobility of carriers in BGNs are explored. Further, variation of gate voltage during the device operation and its effect on mobility are discussed, and a numerical mobility model is presented based on channel conductance. Chapter 7 continues with the discussion on BGN models. The BGN carrier statistics and ballistic conductance in the nondegenerate and the degenerate limits are presented. The proposed model shows good agreement with experimental data. Since a BGN field effect transistor (BGNFET) can be shaped by using graphene bilayers with an external controllable voltage that is perpendicular to the layers, its application as a future field effect transistor channel is expected to be widespread.

Chapter 8 presents trilayer graphene nanoribbon (TGN). In this chapter, a tight-binding method for the band structure of TGN in the presence of a perpendicular electric field is employed. An analytical model of ABA-stacked TGN carrier statistics incorporated with a numerical solution in the degenerate and nondegenerate regimes is presented. Simulated results based on the presented model indicate that this model can be approximated by degenerate and nondegenerate approximation in some numbers of normalized Fermi energy.

Chapter 9 provides additional GNR transistor modeling concepts. Analytical models for surface potential, lateral electric field, and length of saturation in the saturation region are presented. The behavior of the GNR transistor in the saturation region is also studied.

In order to achieve downscaling of devices, the use of new material or structure should be explored. Nanowire is one such candidate. A silicon nanowire (SiNW) is an elongated crystalline or amorphous silicon with the diameter ranging from ten to hundred nanometers and a length of several micrometers. Researchers have focused on SiNWs because of their unique properties being significantly different from those of bulk silicon. The electronic band gap of SiNW is adjustable with the nanowire diameter. Chapters 10 through 12 focus on SiNW. Chapter 10 describes the modeling and simulation of SiNW, while Chapter 11 describes its properties and growth techniques. Experimental results obtained are also discussed. ZnO nanowires are generally fabricated using bottom-up technology and require the use of tedious pick and place method and electron beam lithography, which limits their use in large area and low-cost applications. Chapter 12 provides a new perspective of top-down fabrication technique to produce highly oriented and reproducible nanowires with different channel lengths in defined locations on a larger processing scale.

One of the most popular new material technologies is strained technology. Strained technology changes the properties of device materials rather than changing the device geometry. Chapter 13 presents a comprehensive knowledge of the stress technology using $Si/Si_{1-x}Ge_x$ materials. In addition, it provides a useful threshold voltage model using the quantum mechanical effect approach.

Although nanoelectronics have the potential for numerous applications, they face technical and economic challenges. A major obstacle is the long time period from research to commercialization. This gap must be addressed by industry, government, and academia. Chapter 14 discusses the triple helix models involving universities, industries, and government, which can help bridge this gap and lower the barriers to nanoelectronics commercialization.

Finally, the book concludes with an Appendix containing the MATLAB® codes used to generate some of the figures, followed by a Glossary of terms used in the chapters concerning nanoelectronics. The Glossary is not comprehensive and does not list every technical term used to describe advanced nanoelectronics. However, effort has been made to include important terms.

It is hoped that this book will serve as a useful source of technical and scientific information for professionals, researchers, and scientists who want to know about topics that deal with advanced nanoelectronics.

Razali Ismail

MATLAB® and Simulink® are registered trademarks of The MathWorks, Inc. For product information, please contact:

The MathWorks, Inc.
3 Apple Hill Drive
Natick, MA 01760-2098 USA
Tel: 508 647 7000
Fax: 508-647-7001
E-mail: info@mathworks.com
Web: www.mathworks.com

Acknowledgments

First of all, I express my sincere appreciation to the authors, who provided wonderful contributions and were willing to accept request for modifications where necessary. My special thanks to my coeditor, Dr. Sohail Anwar, who initiated the book and provided advice and comments during the preparation of the manuscripts. Also special thanks to my other coeditor, Dr. Mohammad Taghi Ahmadi, who has been instrumental in the successful completion of this book. The following group members are especially thanked for assisting in the editing: Meghdad Ahmadi, Fatimah Khairiah Abd Hamid, and Muhammad Afiq Nurudin Hamzah. Finally, I would like to gratefully acknowledge the Research Management Centre (RMC) of Universiti Teknologi Malaysia (UTM) for providing excellent research environment and support in which to complete a significant portion of the work in this book.

Editors

Razali Ismail received his BSc and MSc in electrical and electronic engineering from the University of Nottingham, Nottingham, UK in 1980 and 1983, respectively, and his PhD from Cambridge University, Cambridge, UK, in 1989.

In 1984, he joined the Faculty of Electrical Engineering, Universiti Teknologi Malaysia, as a lecturer in electrical and electronic engineering. He has held various faculty positions including department head and chief editor of the university journal. In 1985, he began his PhD work at the Department of Electronics and Computer Science, University of Southampton, Southampton, UK. From 1987 to 1989, he continued his research work at the Department of Engineering, Cambridge University, Cambridge, UK, where he completed his PhD in microelectronics.

His main research interest is in the field of microelectronics, which includes the modeling and simulation of IC fabrication process and modeling of semiconductor devices. He has worked for more than 20 years in this research area and has published various articles on the subject. His current research interest is in the emerging area of nanoelectronics devices focusing on the use of carbon-based materials and novel device structure. He is presently with the Universiti Teknologi Malaysia as a professor and head of the Computational Nanoelectronics Research Group. He is a member of the IEEE Electron Devices Society (EDS).

Mohammad Taghi Ahmadi received his BSc in 1997, MSc in 2006 in solid state physics, and PhD in electrical engineering in 2009. Since 2004, he has been active in research related to carbon-based devices and graphene-based transistor modeling. He completed a postdoctoral program with the Computational Nanoelectronic Research Group, Universiti Teknologi Malaysia, in 2010, after which he joined the Faculty of Electrical Engineering of the same university as a senior lecturer.

His main research interests are in nanoscale device modeling, simulation, and characterization. His research has resulted in a number of publications in high-impact journals for which he has been awarded the UTM Chancellor Award (2010), academic excellence awards, and the Best Student Award (International Student Center Iranian Student Society). He is an IEEE and American Nano Society member. He is also listed in *Who's Who in the World* (2010) and IBC Top 100 Engineers (2012).

Sohail Anwar is an associate professor of engineering at the Altoona campus of Pennsylvania State University. In addition, he is a professional associate of Management Development Programs and Services at Pennsylvania State University, University Park.

Dr. Anwar has served as the editor in chief of the *Journal of Engineering Technology*. He is currently serving as the editor in chief of the *International Journal of Engineering Research and Innovation*, executive editor of the *International Journal of Modern Engineering*, and an associate editor of the *Journal of The Pennsylvania*

Academy of Science. In addition, he is the series editor of the *Nanotechnology and Energy Series*, Taylor & Francis Group/CRC Press.

Dr. Anwar recently edited *Nanotechnology for Telecommunications* published by Taylor & Francis Group/CRC Press in June 2010. He is also editing the *Handbook of Research on Solar Energy Systems and Technologies* to be published by IGI Global Press in 2012. He is the editor-in-chief of the *Encyclopedia of Energy Engineering and Technology* published by Taylor & Francis Group.

Dr. Anwar is a senior member of the IEEE, and a member of the ASEE, ATMAE, and PAS. He recently served as a member of the IEEE Committee on Technology Accreditation Activities (CTAA). In addition, he is a commissioner of the Technology Accreditation Commission (TAC) of ABET.

Contributors

Mohammad Taghi Ahmadi
Faculty of Electrical Engineering
Universiti Teknologi Malaysia
Johor Bahru, Malaysia

Noraliah Aziziah Amin
Faculty of Electrical Engineering
Universiti Teknologi Malaysia
Johor Bahru, Malaysia

Sohail Anwar
Division of Business and
    Engineering
Pennsylvania State University
Altoona, Pennsylvania

Peter Ashburn
School of Electronics and Computer
    Science
University of Southampton
Southampton, United Kingdom

Desmond C. Y. Chek
Faculty of Electrical Engineering
Universiti Teknologi Malaysia
Johor Bahru, Malaysia

Harold M. H. Chong
School of Electronics and Computer
    Science
University of Southampton
Southampton, United Kingdom

Amir Hossein Fallahpour
Department of Microelectronics
    and Telecommunication
University of Rome "Tor Vergata"
Rome, Italy

Mahdiar Ghadiry
School of Electrical and Electronic
    Engineering
Universiti Sains Malaysia
Nibong Tebal, Malaysia

Habib Hamidinezhad
Physics Department
Universiti Teknologi Malaysia
Johor Bahru, Malaysia

Razali Ismail
Faculty of Electrical Engineering
Universiti Teknologi Malaysia
Johor Bahru, Malaysia

Asrulnizam Abd Manaf
School of Electrical and Electronic
    Engineering
Universiti Sains Malaysia
Nibong Tebal, Malaysia

Seyed Mahdi Mousavi
Faculty of Electrical Engineering
Universiti Teknologi Malaysia
Johor Bahru, Malaysia

Mahdieh Nadi
Department of Computer Engineering
Islamic Azad University
Ashtian, Iran

Moones Rahmandoust
Faculty of Mechanical
    Engineering
Universiti Teknologi Malaysia
Johor Bahru, Malaysia

Meisam Rahmani
Faculty of Electrical Engineering
Universiti Teknologi Malaysia
Johor Bahru, Malaysia

Hatef Sadeghi
Faculty of Electrical Engineering
Universiti Teknologi Malaysia
Johor Bahru, Malaysia

Kang Eng Siew
Faculty of Electrical Engineering
Universiti Teknologi Malaysia
Johor Bahru, Malaysia

Suhana Mohamed Sultan
Faculty of Electrical Engineering
Universiti Teknologi Malaysia
Johor Bahru, Malaysia
and
School of Electronics and Computer
    Science
University of Southampton
Southampton, United Kingdom

Yussof Wahab
Physics Department
Universiti Teknologi Malaysia
Johor Bahru, Malaysia

Jeffrey Frank Webb
School of Engineering, Computing
    and Science
Swinburne University of
    Technology
Kuching, Malaysia

1 Fundamentals of Quantum Nanoelectronics

Jeffrey Frank Webb and Mohammad Taghi Ahmadi

CONTENTS

1.1   INTRODUCTION

The size scales in microelectronics are approaching nanoscale dimensions. Moore's law (Moore, 1965) predicts that transistors will be manufactured with 16 nm features within the next few years. Fabrication at this scale is already possible for research purposes. With these size reductions, microelectronics is starting to be superseded by nanoelectronics. A key feature is that at this scale truly quantum mechanical effects start to become much important than hitherto.

Microelectronics has already brought about a tremendous reduction in size from the several centimeters of the first transistor invented at Bell Laboratories in 1947 (Riordan and Hoddeson, 1998) to the 50 nm point that we have currently reached on the semiconductor road map. Hence we are reaching the size scale at which quantum effects become much more important. Before this, even for transistor sizes of about 100 nm, although quantum effects were apparent as reflected in semiconductor band theory and the concept of effective mass, an adequate approach was semiclassical—classical physics modified by some quantum principles. It was not necessary to completely replace the classical ideas with fully quantum mechanical concepts.

For sizes below about 100 nm, the semiclassical treatment starts to break down and a more quantum mechanical approach is required. Importantly, this is not just a scale change; it is also a change in qualitative behavior. Following are some examples of this qualitative change:

- Energy levels within bands are discrete rather than quasi-continuous, as encountered in microelectronics.
- Excitons are more influential.
- Quantum tunneling becomes much more likely, as the gate oxide in metal oxide semiconductor field effect transistors (MOSFETs) decreases to a few nanometers, for example.

In this chapter, the main principles of quantum theory and how it arose will be discussed and some applications to the modeling of charge carriers in semiconductor nanostructures will be considered. In particular, it will be shown that the confinement of charge carriers on scales of the order of the de Broglie wavelength will produce discrete quantum effects rather than the continuous (within bands) semiclassical behavior observed for scales much larger than the de Broglie wavelength. At least one dimension has to be confined in this way for quantum effects to be apparent. Three cases will be considered: (1) Quantum wells in which one dimension is small enough to create quantum confinement. Geometrically, this could be a thin film structure. Two dimensions are free in the sense that the quantum confinement is not apparent in these dimensions. (2) Quantum wires confined in two dimensions; one dimension is free. (3) Quantum dots confined in all three dimensions; no dimensions are free. In this sense, a quantum dot is zero dimensional. The main ideas from conventional condensed matter physics such as band theory and effective mass will be used with only brief introductory explanations, with the assumption that the reader has some background in condensed matter (of the sort covered in many undergraduate-level textbooks).

1.2   QUANTUM PHYSICS

Classical or Newtonian physics held sway, unchallenged, for several hundred years. Around the beginning of the twentieth century, however, the results of a number of experimental studies could not be explained by classical physics. This heralded the need for a new theory and thus begun the development by many scientists, including some great minds, of quantum theory or quantum mechanics. Over the ensuing 25 years or so the theory was developed. Among the first experiments that could not be explained in the classical framework concerned blackbody radiation and the photoelectric effect. These will now be explained briefly—a more detailed account is given in Bohm (1951) and Hund (1974).

1.2.1 EXPERIMENTAL RESULTS BEYOND CLASSICAL PHYSICS

1.2.1.1 Blackbody Radiation

All materials emit electromagnetic radiation over a wide frequency range. Thus, a cavity will be filled with radiation, and in equilibrium the amount emitted from inner walls of the cavity is equal to that absorbed by them. It can be measured by making a small hole in the cavity. The hole is close to a perfect absorber (an emitter) of radiation, hence the term "blackbody radiation."

Experiments at the end of the nineteenth century showed that the energy distribution of electromagnetic radiation in a cavity in equilibrium at constant temperature increases with frequency until a peak is reached after which it decreases, and at high enough frequencies it decays exponentially. However, a theoretical analysis of the situation using classical electromagnetic theory and Maxwell–Boltzmann statistical mechanics (also classical) predicted that the amount of energy in the frequency range between v and $v + dv$ was $U(v) \sim k_B T v^2$, where k_B is Boltzmann's constant and T is the absolute temperature. Although this is in agreement with the experimental values at low frequencies, it gives too much radiation for high frequencies and an attempt to integrate over all frequencies to find the total energy results in divergence that leads to an absurd conclusion that the cavity contains an infinite amount of energy (Bohm, 1951). This signaled that there was something wrong with classical physics.

In 1900, Max Planck was able to resolve this problem by making an assumption equivalent to the following: All components of the electromagnetic radiation in the cavity of frequency v can only have an energy given by an integral multiple of hv, where h is now known as Planck's constant.

1.2.1.2 Photoelectric Effect

The photoelectric effect was first observed during experiments in 1897 by Hertz who showed that electrons are emitted from a metal surface that is irradiated with light or ultraviolet rays. Furthermore, the kinetic energy of the electrons is independent of the intensity of the radiation, but depends only on the frequency. Classically, this did not make sense because the kinetic energy should depend on the amplitude of the radiation rather than on the frequency. In 1905, Einstein explained this by supposing that light is composed of particles—called photons—each of energy hv, consistent with the energy relation proposed by Planck to explain blackbody radiation; now the new feature of light as particles is also present.

1.2.2 THE CONSEQUENCES: WAVE–PARTICLE DUALITY

By understanding the aforementioned experiments, it appears that a light wave with frequency v can behave as a stream of particles or photons. Each photon has energy $E = hv$, where h is Planck's constant; this means that the energy of a given light wave is fixed, or quantized, according to $E = hf$, rather than varying continuously with the wave amplitude. It is true that the wavelength can vary, but in confined systems the wavelengths are often restricted to discrete rather than continuous values analogous to waves on a stretched string. The particle-like behavior and energy quantization are both significant departures from the classical description of light of frequency v: a wave whose energy could be varied continuously.

However, this does not mean that light always behaves as a particle. Rather, it depends on the context—the experiment being done. So the behavior is described as wave–particle duality. This was not very agreeable to many scientists at the time, but was more or less forced on them by the experimental evidence.

1.2.2.1 Young's Double Slit Experiment: Wave Interference

An example of an experiment in which light behaves as a wave is the classic Young's double slit experiment (Young, 1804; Halliday et al., 2008). It is a demonstration of interference that occurs for waves but not for particles. The experiment is illustrated in Figure 1.1. Light incident on two slits will, if it is a wave, diffract as it passes through them; the two diffracted beams emerging from the slits will interfere producing a pattern of constructive interference peaks and destructive interference troughs on the screen. For the effect to be noticeable, the slit width has to be of the order of the wavelength of the incident light.

1.2.2.2 de Broglie Waves

In 1923, the wave–particle duality of light led Louis de Broglie to suggest that particles, such as electrons, may also, under certain experimental conditions, behave as waves (his ideas appear in his PhD thesis published as a long journal paper [de Broglie, 1925]). If this had been suggested before the experiments leading to the introduction of photons and quantization for light, it would have seemed absurd. However, de Broglie's suggestion was shown to be true through the experimental work of Davisson and Germer (1927) involving crystal diffraction. The idea here is that the regularly repeating crystal structure would diffract waves with a wavelength of the order of the lattice spacing. In fact, x-ray diffraction from such crystals was already known. However, since x-rays are electromagnetic waves, this is not a test of wave–particle duality. Davisson and Germer rather discovered that beams of

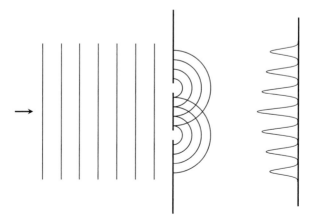

FIGURE 1.1   Young's double slit experiment. A plane wave incident on the slits from the left is diffracted at each slit leading to the interference pattern on the screen.

electrons were also diffracted by the crystal, as though they were waves rather than particles. A diffraction pattern on a screen sensitive to electrons was observed consistent with the electrons behaving like waves.* Potentially, all particles can exhibit wave-like qualities. However, on the scale of everyday objects—billiard balls for example—the associated wavelength is far too small to be noticeable, as will be shown next.

Qualitatively, de Broglie proposed that each particle having a momentum p has a wave behavior associated with a wave of wavelength, which can be referred to as the de Broglie wavelength,

$$\lambda = \frac{h}{p}, \tag{1.1}$$

where $h = 6.6 \times 10^{-34}$ J s is Planck's constant. For a billiard ball of mass 0.2 kg, this formula gives $\lambda = 3.3 \times 10^{-33}$ m—too small to be noticeable. If the particle were an electron, though, with rest mass $m_e = 10 \times 10^{-34}$ m, the wavelength would be much greater, 6.6×10^{-3} m. In electron diffraction experiments, the speed of an electron is greater than in this example and the resulting wavelength is of the order of the crystal lattice spacing; if it was not of this order, the diffraction effect would be very small, and crystal diffraction would not be a good way of demonstrating the wave nature of electrons.

1.2.2.3   Young's Double Slit Experiment with Electrons: A Strange Feature of Quantum Mechanics

In addition to the crystal diffraction techniques, it is possible to use Young's double slit experiment to show that electrons behave as waves: A beam of electrons passing through two slits will form a diffraction pattern just as would be expected if the electrons behaved as waves. To illustrate just how strange quantum mechanics seems compared with classical physics, consider the case where a single electron passes through the slits at a time (Tonomura et al., 1989). Even then, the diffraction pattern builds up after enough electrons have passed through. Strikingly, if one slit is closed and the experiment repeated with one electron at a time (Tonomura et al., 1989), the diffraction pattern disappears as there are no longer two slits to create waves from each slit that interfere. Although this is expected for waves, the question often asked when pondering on this single electron experiment is this: How does a single electron know when it passes through a slit whether or not the other is open? In this experiment, it is the wave nature of the electron that is in control, even if it is the wave nature of a single electron. It is hard to give up the idea that an electron is intrinsically a particle. This experiment is showing us that this is not an intrinsic property of the electron; rather, the particle quality depends on the context, that is, on the experiment. In the double slit experiment, the electron does not have the quality of a particle; it is truly behaving as a wave.

---

* Since then, some highly accurate experiments confirming these findings have been carried out, such as those of Tonomura et al. (1989), who used a Young's double slit setup rather than crystal diffraction.

1.2.2.4 Spin: Quantum or Classical?

It was demonstrated experimentally that electrons in a magnetic field behave in a way that would be expected if they had angular momentum (Weinert, 1995). At first this was thought to be due to the electrons spinning about their axes—hence the term spin. However, a closer examination revealed that there was no spin akin to classical spin. Instead, an intrinsic and quantized property was evident (Tumulka, 2009); nonetheless, the name spin was retained. Classically, particles can possess spin due to rotation about an axis through it; but there are no restrictions on the orientation of the axis and the angular rotation is a continuous quantity. What's more, this spin is not intrinsic in the sense that a particle may have no spin (but could still undergo translational motion). In contrast, the quantum spin is intrinsic: the particle and its spin are a whole, and the particle always possesses spin. Some particles, photons for example, are quantized in integral multiples of $\hbar$ and are called bosons; others, called fermions, are quantized with half-integral spin. An electron is an example of a fermion; it can posses a quantum spin number of either $1/2$ or $-1/2$, sometimes referred to as spin up or spin down, respectively.

1.2.2.5 Pauli Exclusion Principle

Another property with no classical analogue, and which is related to the concept of spin, is the Pauli exclusion principle proposed by Wolfgang Pauli (Massimi, 2009). It states that two or more identical fermions cannot be in the same quantum state. This principle is very important in understanding how quantum energy levels in confined structures, such as atoms, fill up. For example, a particular energy state in an atom can only be occupied by two electrons at most—one with spin up and the other with spin down.

1.2.3 INTRODUCTORY THEORETICAL CONCEPTS IN QUANTUM PHYSICS

1.2.3.1 Schrödinger's Wave Equation

A wavefunction Ψ proposed by Erwin Schrödinger, soon after the de Broglie wavelength idea had been established, became the fundamental equation for quantum mechanics replacing the Newtonian or classical mechanics described for a single particle of mass m by $\mathbf{F} = m \cdot d^2\mathbf{r}/dt^2$, where $\mathbf{F}$ is the net force on the particle at position $\mathbf{r}$, $d^2\mathbf{r}/dt^2$ is the acceleration, and $d\mathbf{r}/dt$ is the velocity. However, the wavefunction and its relation to the physical properties of quantum mechanics are much more abstract than for Newtonian mechanics. For larger sizes compared to h, the quantum effects are negligibly small and Newtonian mechanics is sufficient. But at small enough scales, such as the atomic scale, quantum mechanics must be used.

There is an alternative formulation in terms of matrices formulated by Heisenberg (Born and Jordan, 1925; Heisenberg, 1925; Born et al., 1926). Here, we consider only the Schrödinger wave equation approach. Mathematically, it was shown that the two alternatives are equivalent (Schrödinger, 1926; von Neuman, 1955).

1.2.3.2 Heisenberg Uncertainty Principle

An important implication of the mathematical description of quantum behavior is the Heisenberg uncertainty principle, formulated in 1927 (Heisenberg, 1927).

It states that there is a limit to the accuracy with which the position and momentum of a particle* can be simultaneously measured. The principle is expressed mathematically as

$$\Delta p \Delta x \geq \frac{\hbar}{2}. \tag{1.2}$$

This means that if the accuracy to which x is known is high so that Δx is small, then Δp, the uncertainty in momentum, is large; on the other hand, if Δp is small, then Δx is large. Again, this is very different to classical physics where in principle it is possible to know both p and x with arbitrarily high accuracy, because there is no inequality of the form of Equation 1.2 to restrict Δx or Δp. At the macroscopic scale, even high accuracy could be achieved with values of Δx and Δp greater than the order set by $\hbar$. Hence, the effect of the uncertainty principle is hidden from the macroscopic-scale phenomena that dominate our everyday experiences.

We have just considered a one-dimensional situation for which $p = p_x$. In three dimensions, $p^2 = p_x^2 + p_y^2 + p_z^2$ and Equation 1.2 generalizes to

$$\Delta p_x \Delta x \geq \frac{\hbar}{2}, \quad \Delta p_y \Delta y \geq \frac{\hbar}{2}, \quad \Delta p_z \Delta z \geq \frac{\hbar}{2}. \tag{1.3}$$

Therefore, the uncertainty principle only prevents simultaneous measurements of position and momentum along a common axis, so that combinations such as x and p_z can be measured simultaneously.

1.2.3.3 Wave Packets

Some insight into the uncertainty principle and how it is related to waves can be appreciated by considering wave packets. To start with, assume a quantum object localized over a small region of space. It can be represented by a superposition of plane waves $a(\mathbf{k})e^{-i[\omega(\mathbf{k})t-\mathbf{k}\cdot\mathbf{r}]}$ of the form

$$\Psi(x,t) = \int_{\text{over } k\text{-space}} a(\mathbf{k})e^{-i[\omega(\mathbf{k})t-\mathbf{k}\cdot\mathbf{r}]}\mathrm{d}V_k. \tag{1.4}$$

To illustrate, we consider one-dimensional space and expand the $\omega(k)$ in a Taylor series about a wave number k_0. Considering only the linear terms of the expansion gives a dispersion relation

$$\omega(k) = \omega(k_0) + \frac{\partial\omega}{\partial k}\bigg|_{k=0}(k-k_0) = \omega_0 + \alpha(k-k_0). \tag{1.5}$$

---

* Since we are discussing quantum particles, they are affected by wave–particle duality, and in fact whether or not they are particles depends on the context. The word "particle" will be retained but this context dependence should always be borne in mind when considering the so-called quantum particles.

Now, if $a(k)$ is defined by

$$a(k) = \begin{cases} 1 & \text{if } k_0 - \Delta k \le k \le k_0 + \Delta k, \\ 0 & \text{otherwise,} \end{cases} \qquad (1.6)$$

Equation 1.4 defines a wave packet localized in space with a k range centered on k_0:

$$\Psi(x,t) = e^{-ik_0(v_p t - x)} \int_{k_0 - \Delta k}^{k_0 + \Delta k} a(k) e^{-i(k-k_0)(v_g t - x)} dk = 2 e^{-ik_0(v_p t - x)} \Delta k \frac{\sin[\Delta k(v_g t - x)]}{\Delta k(v_g t - x)}, \quad (1.7)$$

where $v_p = \omega_0/k_0$ is the phase velocity of the packet, and the envelope has a velocity, that in general is different from this and given by

$$v_g = \alpha = \left. \frac{\partial \omega}{\partial k} \right|_{k=0}, \qquad (1.8)$$

called the group velocity. The wave packet described by Equation 1.7 is plotted in Figure 1.2 at time $t = 0$.

The wave packet in this model moves through space and time as a bundle. The width of the region over which most of the packet is concentrated can be expressed by

$$\Delta k(v_g t - x) = \frac{\pi}{2} \qquad (1.9)$$

centered at $(v_g t - x) = 0$, as illustrated in Figure 1.3. As a whole, the packet moves with the velocity of the envelope, which is the group velocity v_g; the movement inside the envelope is according to the phase velocity v_p.*

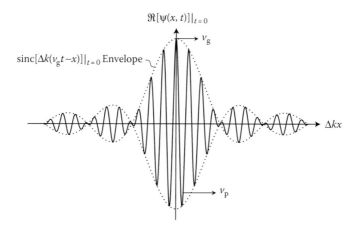

FIGURE 1.2 Wave packet at $t = 0$.

---

* It can turn out that v_p is greater than the speed of light in a vacuum, c. However, physically, the overall motion is determined by v_g, which does not exceed c, as required by Einstein's relativity theory that limits particles of finite mass such as electrons from exceeding c.

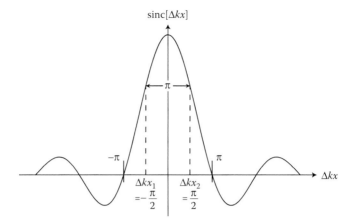

FIGURE 1.3   A measure of the width of a wave packet.

Insight into the Heisenberg uncertainty principle can be gained from the wave packet expression (see Equation 1.7). Consider the time $t = 0$; the wave packet envelope is centered at $x = 0$ and is given by

$$\frac{\sin \Delta k(-x)}{\Delta k(-x)} = \frac{\sin \Delta kx}{\Delta kx} = \text{sinc } \Delta kx. \tag{1.10}$$

If the wave number range Δk is small, then the momentum $p = \hbar k$ varies over a small range: Δp is small and p is well defined. Since the sinc function argument is Δkx, this means that the spatial extent of the wave packet as measured by Equation 1.9, as can be seen from Figure 1.3, is

$$\Delta kx_2 - \Delta k_1 x_1 = \Delta k \Delta x = \pi. \tag{1.11}$$

The de Broglie relation in Equation 1.1 can be written as

$$p = \hbar k, \tag{1.12}$$

where $k = 2\pi/\lambda$. Thus, Equation 1.11 becomes

$$\Delta p \Delta x = \hbar \pi, \tag{1.13}$$

which is similar to the Heisenberg uncertainty principle of Equation 1.2 and in essence expresses the same idea: a small Δp (momentum known to a high degree of accuracy) implies a large Δx (the quantum object is not very localized in space) and vice versa, so that a well-defined momentum and a high degree of localization—a well-defined position—cannot occur simultaneously. Note that the factor of a half difference between Equations 1.2 and 1.13 is because the model used to arrive at the latter used a simple form for $a(k)$. Although this is sufficient for illustration, a Gaussian form would be more realistic and lead to Equation 1.2 more readily, but with an increase in the complexity of the calculations.

1.2.4 Schrödinger's Equation and the Main Principles of Quantum Theory

Schrödinger wave equation has been very successful with regard to being consistent with experimental measurements on quantum systems that could not be dealt with by classical physics. It is in this sense that the equation is valid; it cannot be derived from more fundamental principles. In many cases, Schrödinger's equation cannot be solved analytically. However, it is sometimes possible to get useful results by employing simple geometry and approximations that result in analytical solutions, as will be seen later. Numerical techniques and perturbation analysis are also important ways of obtaining solutions, but will not be used here.

After covering Schrödinger's equation, this section describes the general principles according to the standard interpretation of quantum theory; this is an attempt to show how the theory connects to reality. In terms of experimental predictions, the standard interpretation is adequate; however, it is an abstract mathematical algorithm for calculating results (Bohm and Hiley, 1993) with little ontology to explain or make it understandable beyond this. As a result, there are various other interpretations that try to make quantum theory more understandable; this issue will also be discussed here briefly.

1.2.4.1 Schrödinger's Equation for a Single Particle

Schrödinger's equation for a single particle is

$$i\hbar \frac{\partial \Psi(\mathbf{r},t)}{\partial t} = \left(-\frac{\hbar^2}{2m}\nabla^2 + V(\mathbf{r},t) \right)\Psi(\mathbf{r},t), \qquad (1.14)$$

where

$\Psi(\mathbf{r}, t)$ is the wavefunction that depends on position $\mathbf{r}$ and time t

∇^2 is the Laplacian given in Cartesian coordinates by

$$\nabla^2 \equiv \frac{\partial^2}{\partial x^2} + \frac{\partial^2}{\partial y^2} + \frac{\partial^2}{\partial z^2}; \qquad (1.15)$$

$V(\mathbf{r}, t)$ is the potential energy of the quantum system. Schrödinger's equation can also be written using the Hamiltonian energy operator:

$$H = -\frac{\hbar^2}{2m}\nabla^2 + V(\mathbf{r},t). \qquad (1.16)$$

This operator will be discussed in the following.

The boundary conditions are that Ψ and its derivative are continuous at boundaries. This can be expressed by

$$\Psi_{s_{1,i}}(\mathbf{r}_{b_i},t) = \Psi_{s_{2,i}}(\mathbf{r}_{b_i},t), \tag{1.17}$$

$$\nabla\Psi'_{s_{1,i}}(\mathbf{r}_{b_i},t)\cdot\hat{\mathbf{n}}(\mathbf{r}_{b_i}) = \nabla\Psi'_{s_{2,i}}(\mathbf{r}_{b_i},t)\cdot\hat{\mathbf{n}}(\mathbf{r}_{b_i}), \tag{1.18}$$

$$\forall \mathbf{r}_{b_i} \text{ on } b_i, \quad i = 1,2,\ldots,N_b,$$

where
  N_b are the number of boundaries
  $\mathbf{r}_{b_i}$ is a point on boundary b_i
  $\Psi_{s_{1,i}}$ is the wavefunction on one side of b_i
  $\Psi_{s_{2,i}}$ is the wavefunction on the other side of b_i
  $\hat{\mathbf{n}}(\mathbf{r}_{b_i})$ is a unit vector normal to the tangent plane at $\mathbf{r}_{b_i}$

Further, the wavefunction is required to be finite everywhere.

1.2.4.2 Schrödinger's Time-Independent Equation

A time-independent form is possible if the potential is independent of time ($V(\mathbf{r}, t) = V(\mathbf{r})$), and the wavefunction is expressible as a product of spatial-only and time-only functions:

$$\Psi(\mathbf{r},t) = \psi(\mathbf{r})g(t). \tag{1.19}$$

Under these conditions, it is easy to show (Bohm, 1951) that

$$\Psi(\mathbf{r},t) = \psi(\mathbf{r})e^{-iEt/\hbar}, \tag{1.20}$$

where
  E, as can also be shown, is the total energy
  $\psi(\mathbf{r})$ is given by Schrödinger's time-independent equation

$$\left[-\frac{\hbar^2}{2m}\nabla^2 + V(\mathbf{r})\right]\psi(\mathbf{r}) = H\psi(\mathbf{r}) = E\psi(\mathbf{r}), \tag{1.21}$$

in which the Hamiltonian now takes the form

$$H = -\frac{\hbar^2}{2m}\nabla^2 + V(\mathbf{r}). \tag{1.22}$$

Equation 1.21 is an eigen equation and in general will have solutions corresponding to eigenfunction, eigenvalue pairs ($\psi_n(\mathbf{r})$, E_n). Since Schrödinger's equation is a linear homogeneous equation, the most general solution for the time-dependent equation can be written as an expansion in terms of eigenfunctions and eigenvalues as

$$\Psi(\mathbf{r},t) = \sum_n a_n\psi_n(\mathbf{r})e^{-iEt/\hbar}, \tag{1.23}$$

where the a_n are suitable constants that can be found from ψ_n and E_n, as will be seen later.

1.2.4.3 Schrödinger's Equation for a System of Particles

For an N particle system with a particle at each of the positions $\mathbf{r}_1, \mathbf{r}_2, \ldots, \mathbf{r}_N$, the wave equation becomes $\Psi(\mathbf{r}_1, \mathbf{r}_2, \ldots, \mathbf{r}_N, t)$, and the Schrödinger's equation is

$$i\hbar \frac{\partial \Psi(\mathbf{r}_1, \mathbf{r}_2, \ldots, \mathbf{r}_N, t)}{\partial t} = H\Psi(\mathbf{r}_1, \mathbf{r}_2, \ldots, \mathbf{r}_N, t), \tag{1.24}$$

where the Hamiltonian is given by

$$H = \sum_{n=1}^{N} H_n + H_{1,2,\ldots,N}, \tag{1.25}$$

where

H_n is the Hamiltonian that the nth particle would have if there were no other particles

$H_{1, 2, \ldots, N}$ represents the interactions between the particles

In general, this is a very difficult problem to solve. However, in nanoelectronics it is often possible, fortunately, to treat the N particles as if they do not interact. In this case, $H_{1, 2, \ldots, N} = 0$ (Chazalviel, 1999) and Schrödinger's equation separates into N independent single-particle equations,

$$i\hbar \frac{\partial \Psi_j(\mathbf{r}_j, t)}{\partial t} = H_j \Psi_j(\mathbf{r}_j), \quad j = 1, 2, \ldots, N. \tag{1.26}$$

1.2.4.4 Principles of the Standard Interpretation of Quantum Theory

The standard interpretation, also known as the Copenhagen interpretation, is the most commonly used set of ideas and principles; it allows quantum theory to be applied to the calculation of properties at the quantum level. A postulational approach (Zettili, 2009) could be taken here; but instead we outline the main points and associated ideas inherent in what could be called postulates without referring to them as such.

1.2.4.4.1 Significance of the Wavefunction

The aforementioned wavefunction is also known as a state function. Its significance is discussed in the following.

A quantum system is completely described by a state function $\Psi(\mathbf{r}, t)$ that contains everything that can be known about the system. Ψ is a continuous-valued, finite function having a single complex value at each point in its domain. It is related to physical observables through probability via a probability density for finding the particle at $(\mathbf{r}, t)$. Thus, the probability of finding a particle in a small volume dV centered at $\mathbf{r}$ is $\rho(\mathbf{r}, t) = |\Psi(\mathbf{r}, t)|^2 dV$, and the probability of finding it in a region $\mathcal{R}$ is

$$P = \int_{\mathcal{R}} |\Psi(\mathbf{r}, t)|^2 \, dV = \int_{\mathcal{R}} \Psi^*(\mathbf{r}, t) \Psi(\mathbf{r}, t) \, dV, \tag{1.27}$$

where * denotes the complex conjugate. Since the existing particle must be somewhere in space, we also have

$$\int_{\mathcal{R}=\infty} \Psi^*(\mathbf{r},t)\Psi(\mathbf{r},t)\,dV = 1. \tag{1.28}$$

This is used to normalize the solutions to Schrödinger's equation.

Note that this is very different from classical physics where the equations usually give values of the physical observables (even if they are statistical averages), rather than the probability of finding a particular value. It is true that in classical measurements there may be inaccuracies leading to uncertainties in the obtained values. However, in principle, according to classical physics, a precise value can be found. The probabilistic nature of Ψ, on the other hand, is taken to imply that there is an irreducible uncertainty at the quantum level that cannot be overcome—as though it were an inherent property of nature at this level.

There is no contradiction, though, with the experiences at the larger scales where classical physics is valid, because the quantum uncertainties average out to give well-defined values on these scales (Bohm, 1951) whereas experiments at the quantum level do show randomness. For example, a Geiger counter that measures radioactive decays—quantum events—registers decays that occur randomly; the precise time that a decay will be registered is not predictable. Another example is that the buildup of the diffraction pattern in the single-electron double slit experiment starts of as a random buildup in time and space. Many particles have passed through the slits before the diffraction pattern is obvious.

1.2.4.4.2   Observables and Operators

The key principles here are as follows:

- Every physical observable (such as position, momentum, and energy) has a corresponding Hermitian operator $\hat{O}$, which corresponds to an eigenvalue problem

$$\hat{O}\psi_n = \lambda_n\psi_n. \tag{1.29}$$

- The result of a measurement of an observable is one of the eigenvalues λ_n of $\hat{O}$. But which one will be obtained is probabilistic as described next.
- If a system is in an initial state Ψ, measurement of an observable will result in an eigenvalue of $\hat{O}$ with probability

$$P(\lambda_n) = \left|\int \Psi(\mathbf{r},t)\psi_n^*\,dV\right|^2, \tag{1.30}$$

and the system will change from state Ψ to ψ_n.

1.2.4.4.3   Hermitian Operators

The important operators in quantum theory are usually formulated as Hermitian operators. These are a class of operators having real eigenvalues, and the eigenfunctions form a complete, orthogonal set of functions such that

$$\int \psi_n^* \psi_n \, dV = \begin{cases} C & \text{if } m = n, \\ 0 & \text{if } m \neq n, \end{cases} \tag{1.31}$$

where C is a real constant. The set is orthonormal if $C = 1$. The completeness of the set means that any function Ψ can be expanded in terms of the eigenfunctions, that is,

$$\Psi = \sum_n a_n \psi_n. \tag{1.32}$$

For an orthonormal set it is easy to show that the a_n are given by

$$a_n = \sum_m a_m \int \psi_n^* \psi_m \, dV = \int \psi_n^* \Psi \, dV. \tag{1.33}$$

Following are some examples of quantum mechanical operators:

$$\mathbf{p} = -i\hbar \nabla \quad \text{(momentum operator)}; \tag{1.34}$$

$$\hat{E} = i\hbar \frac{\partial}{\partial t} \quad \text{(energy operator)}; \tag{1.35}$$

$$\mathbf{r} \quad \text{(position operator)}. \tag{1.36}$$

1.2.4.4.4   Operator Commutation Relations and Heisenberg Uncertainty

Consider two observables, O_1 with operator $\hat{O}_1$ and O_2 with operator $\hat{O}_2$. The Heisenberg uncertainty principle implies that not all observables can simultaneously be measured to an arbitrary degree of accuracy. To find out if O_1 and O_2 can both be measured accurately, we proceed as follows. If they can be measured, the measurement of one will not influence the other. If this is true, which one is measured first will not matter and we can write

$$\hat{O}_1 \hat{O}_2 \Psi = \hat{O}_2 \hat{O}_1 \Psi \Rightarrow \left(\hat{O}_1 \hat{O}_2 - \hat{O}_2 \hat{O}_1 \right) \Psi = 0,$$

which can be expressed as

$$\left(\hat{O}_1 \hat{O}_2 - \hat{O}_2 \hat{O}_1 \right) = \left[\hat{O}_1, \hat{O}_2 \right] = 0. \tag{1.37}$$

The square bracket notation expressed for the difference operator is the commutator.

It follows that if $[\hat{O}_1, \hat{O}_2] = 0$, the operators commute and the corresponding observables can be simultaneously measured to arbitrary precision. On the other hand, if $[\hat{O}_1, \hat{O}_2] \neq 0$, the operators do not commute and the observables cannot possess well-defined values simultaneously. For example, it can be shown that $[\hat{x}, \hat{p}_x] \neq 0$ and $[\hat{x}, \hat{p}_z] = 0$, consistent with the uncertainty relations in Equation 1.3.

1.2.4.4.5   More on Probability

Equation 1.30 indicates how to find the probability that a measurement will yield a particular value λ_n for an observable. We have seen that the corresponding eigenfunction Ψ_i can be expressed using Equation 1.32. From Equations 1.30 and 1.32, the probability of obtaining λ_n in a measurement is

$$P(\lambda_n) = \left| \int \Psi \psi_n^* dV \right|^2 = P(\lambda_n) = \left| \int \left(\sum a_m \psi_m \right) \psi_n^* dV \right|^2. \tag{1.38}$$

Invoking orthonormality yields from this the simple result

$$P(\lambda_n) = |a_n|^2. \tag{1.39}$$

This expresses the algorithmic relation of quantum theory to reality: we end up with only a probability that a value (the reality) will be measured in an experiment. The wavefunction in the standard interpretation only corresponds to physical reality in this way. Contrast this with the wave equations for a classical electromagnetic field derived from Maxwell's equations. Here the field terms $\mathbf{E}$ and $\mathbf{H}$ represent the actual values that would (in principle with arbitrary accuracy) be observed—not to the probability that they would be observed.

1.2.4.4.6 Quantum Measurements: Collapse of the Wavefunction

It has been shown that before a measurement, Ψ can be expressed as an expansion in terms of the eigenfunctions of the observables concerned. Note that although different observables have different eigenfunctions, any set of eigenfunctions for a given observable can be used to express Ψ as a series of the form of Equation 1.32. Consider a measurement of momentum, for example, before Ψ can be expressed as an expansion in terms of the eigenfunctions of the momentum operator. If subsequently a momentum measurement is made, then Ψ will become equal to one of the eigenfunctions of the expansion and the eigenvalue for that eigenfunction will be the measured value of the momentum. The process of moving from Ψ as a series in terms of the eigenfunctions, a superposition of states, to a particular eigenfunction is known as the collapse of the wavefunction. Going back to the momentum measurement, shortly after the measurement and collapse of the wavefunction, if there are no further interactions between the quantum system and its surroundings, another momentum measurement would yield the same eigenvalue. However, external influences may put Ψ back into a superposition of states such that the outcome of a subsequent measurement would not be known definitely; only the probability of obtaining a certain value in accordance with Equation 1.39 would be known.

If Ψ is collapsed by measuring the momentum, a measurement of position would not be precisely known. With respect to position, the wavefunction is in a superposition of position states after the momentum measurement and only the probability of obtaining a particular position measurement would be known. Since the momentum has just been set by the momentum measurement, all values of position measurements are equally likely in correspondence with the Heisenberg uncertainty principle.

1.2.4.4.7 Expectation Values

Another important principle concerned with mean values is as follows:

The mean value of an observable is the expectation value of the corresponding operator.

Mathematically, this means that for any observable O, with operator $\hat{O}$, the mean value of many individual measurements under identical conditions is

$$\langle O \rangle = \int \Psi^* \hat{O} \Psi \, dV, \qquad (1.40)$$

where Ψ is a normalized wavefunction.

This gives some insight into how the macroscopic, everyday, scale that we are familiar with does not strongly manifest the probabilistic nature found at the quantum level. There are a very large number of quantum particles that constitute macroscopic systems. The mean for such large numbers is well defined, even though this in not so for an individual particle. An analogy to this given by Bohm (1951) is insurance statistics. A life insurance company is able to predict its earnings fairly well because the average age of death in a large population does not fluctuate much over a period of a few years. However, the age at which a particular individual will die (analogous to the quantum level) cannot be predicted well.

1.2.4.5 Difficulties with the Standard Interpretation and Possible Alternatives

Because of the outstanding success that quantum theory has had in predicting the results of experiments without the need for any alternative to the standard interpretation, many scientists feel that this interpretation is perfectly adequate and do not see any need to discuss the matter further or to look more deeply into what quantum theory might be telling us about the nature of reality. However, there are some scientists, mainly physicists, who are not satisfied with the standard interpretation because it does not really explain or make understandable the strange and seemingly paradoxical features of quantum theory, such as wave–particle duality and nonlocality. Nonlocality refers to a connection between distant particles that is instantaneous—as though in some way they are not really separate. Although this cannot be used to transmit information, which would violate relativity, it is rather disturbing with no parallel in classical physics where any connection between distant particles takes a finite time to occur (and so is not instantaneous) (Bohm, 1951; Davies and Brown, 1993).

The standard interpretation does not really offer any kind of ontology* and does not allow one to form concepts of what is actually going on at the quantum level, beyond giving mathematical apparatus to calculate the probability of obtaining results. In support of this statement, one of the founding fathers of quantum theory, Niels Bohr, felt that quantum theory could only be understood as an algorithm for calculating the possible outcomes of experiments. In fact, he went further than that and said that it was not even possible to form a concept of what was going on beyond this (Kalckar, 1985, 1996). Many scientists, even physicists, are not very aware of this and do not realize how radical Bohr's attitude was. The standard interpretation, though, arose from a variety of ideas as it was developed, not just from Bohr's (Bohm and Hiley, 1993).

It is well worth being aware of these issues in the foundations of quantum theory that to this day have not been fully resolved. Of the various other interpretations that have arisen, one in particular, the de Broglie–Bohm pilot wave interpretation developed by David Bohm and his colleague Basil Hiley (with an initial idea from earlier work by de Broglie), tries very hard to make quantum theory more intelligible and pays attention to the development of a coherent ontology for quantum theory;

---

* Ontology is concerned with discovering the nature or essence of things.

a major feature brought out by Bohm and Hiley is the importance of wholeness, which is not very often discussed in standard textbooks on quantum theory. More details can be found in Bohm and Hiley (1993). Other interpretations include the many worlds interpretation originally developed by Everret (DeWitt and Graham, 1973); a survey of various interpretations can be found in Rae (2004).

One of the problems with interpretations is that experimentally there is usually no way to distinguish them and so many people continue to work with the standard interpretation despite its abstract and rather unintelligible nature. With the advent of new technologies, including nanoelectronics, we can also hope that experiments may help clear up some of the long-standing difficulties.

1.3 QUANTUM THEORY OF ELECTRONS IN EMPTY REGIONS

Before going on to discuss electrons in nanostructures, it is instructive to consider the solutions to Schrödinger's equation for an electron in empty space. In the regions considered, the potential is taken to be constant. We look at only one electron and therefore the single-particle Schrödinger's equation will be sufficient. However, this can still be valid for many electrons, as will be briefly discussed next.

1.3.1 VALIDITY OF THE SINGLE-PARTICLE WAVE EQUATION TREATMENT OF MANY ELECTRON SYSTEMS

Since electronic materials contain many carriers, the use of the single-particle wave equation may seem inappropriate. If the electrons are assumed to be very far apart, the interactions can be neglected because the Coulomb potential between them varies as the reciprocal distance. In this case, the single-particle Schrödinger's equation can be solved for any particle and that solution applies to each electron (the wavefunctions obtained are called orbitals). In solid materials in which electron densities are of the order of 10^{22} electrons per cm^3, the separating distances are clearly not large enough. However, in many materials, the noninteraction assumption is still rather good due to the screening effect in which electrons shield themselves. This effect is related to the Pauli exclusion principle: Each electron repels the other electrons due to both the Coulomb interaction between them and the exclusion principle that implies that electrons with the same spin tend to stay apart. Since no two electrons can be in the same state, those with the same spin will have a low probability of being close together. Hence to some extent, electrons move through a material somewhat independently of each other.

In a solid, the presence of the crystal lattice is accounted for by band theory and together with the effective mass this still allows the single-particle Schrödinger's equation to be used as an approximation (Hanson, 2008).

However, to some extent, electron–electron interactions in materials can be ignored as an approximation. This is because it is often reasonable to assume that the electrons are far enough apart for the Coulomb potential between them, which varies as the reciprocal of the distance between electrons, to be small enough to make the interactions negligible (Hanson, 2008).

1.3.2 ELECTRONS IN UNBOUNDED SPACE

We start with an unbounded region, which is also valid for a region of space far from any boundaries.

1.3.2.1 One-Dimensional Problem

In a one-dimensional problem, Schrödinger's time-independent equation, Equation 1.21, for a constant potential V_0 reduces to

$$\left(-\frac{\hbar^2}{2m}\frac{d^2}{dx^2} + V_0\right)\psi(x) = E\psi(x). \tag{1.41}$$

Starting with an ansatz of the form e^{ikx}, it is easy to show that the general solution is given by

$$\psi(x) = Ae^{ikx} + Be^{-ikx}, \tag{1.42}$$

where
   A and B are constants
   the wave number k is given by

$$k^2 = \frac{2m_e(E - V_0)}{\hbar^2}. \tag{1.43}$$

Note that this produces a parabolic E–k relation, as shown in Figure 1.4.

The solution can be interpreted as the sum of plane waves Ae^{ikx}, moving along positive x, and Be^{-ikx}, moving in the opposite direction. The inclusion of the time-dependent factor from Equation 1.20 makes this more explicit:

$$\Psi(x,t) = (Ae^{ikx} + Be^{-ikx})e^{-iEt/\hbar}. \tag{1.44}$$

Since the electron is free, it will be moving in either of the directions, and it is permissible to set $B = 0$ because the direction of the motion can be set by the sign of k. Thus, it is sufficient to consider the solution

$$\Psi(x,t) = Ae^{ikx}e^{-iEt/\hbar}. \tag{1.45}$$

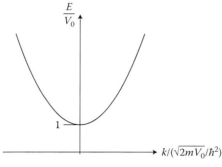

FIGURE 1.4 E–k plot for $k^2 = 2m_e(E - V_0)/\hbar^2$, a parabola.

The absence of confinement has led to a continuous range of allowed values for the energy, as is implicit in Equation 1.43. It is the confinement (of an electron in a hydrogen atom, for example) that leads to discrete energy levels.

1.3.2.2 Three-Dimensional Problem

In three dimensions, Schrödinger, Equation 1.21, in Cartesian coordinates (x, y, z), takes the form

$$\left[-\frac{\hbar^2}{2m}\left(\frac{\partial^2}{\partial x^2} + \frac{\partial^2}{\partial y^2} + \frac{\partial^2}{\partial z^2}\right) + V_0 - E\right]\psi(x, y, z) = 0. \qquad (1.46)$$

It is straightforward to solve this using separation of variables by assuming a solution of the form

$$\psi(x, y, z) = \psi_x(x)\psi_y(y)\psi_z(z). \qquad (1.47)$$

Substituting this into Equation 1.46 shows that the terms

$$\frac{1}{\psi_\alpha}\frac{\partial^2}{\partial \alpha^2}\psi_\alpha \quad \alpha = x, y, z$$

must equal a constant for each α. It is convenient to write the constants as $-k_\alpha^2$, $\alpha = x, y, z$. Then,

$$\frac{1}{\psi_\alpha}\frac{\partial^2}{\partial \alpha^2}\psi_\alpha = -k_\alpha^2 \rightarrow \psi_\alpha(x) = A_\alpha e^{ik_\alpha\alpha} + B_\alpha e^{-ik_\alpha\alpha}, \quad \alpha = x, y, z, \qquad (1.48)$$

and the constants satisfy

$$k_x^2 + k_y^2 + k_z^2 = \frac{2m_e}{\hbar^2}(E - V_0), \qquad (1.49)$$

which gives the three-dimensional form of the E–k relation for a free electron—it is still basically a parabolic relationship and the energy varies continuously. The wave vector has components k_x, k_y, k_z. By a similar argument to that used for the one-dimensional case, it is possible to set $B_\alpha = 0, \forall \; \alpha = x, y, z$, and the product solution can then be expressed as

$$\psi(\mathbf{r}) = A_x A_y A_z e^{ik_y y} e^{ik_x x} e^{ik_z z} = A_0 e^{i\mathbf{k}\cdot\mathbf{r}}. \qquad (1.50)$$

Here, the product of constants has been absorbed into A_0, $\mathbf{r} = x\hat{\mathbf{a}}_x + y\hat{\mathbf{a}}_y + z\hat{\mathbf{a}}_z$, $\mathbf{k} = k_x\hat{\mathbf{a}}_x + k_y\hat{\mathbf{a}}_y + k_z\hat{\mathbf{a}}_z$, and $k =| \mathbf{k}| = k_x^2 + k_y^2 + k_z^2$.

We conclude this section with the following comments:

- The E–k relations for free electrons are more useful for understanding semiconductors than they may seem at first sight because, as mentioned in Section 1.4.1, near the band edges where the most influential charge carriers lie, the structure is often well approximated by parabolic shapes.

- The solutions to Schrödinger's equation obtained for the unbounded region is general. A wave packet can be formed from an integral superposition (see Equation 1.4) of the waves of the general solution. The wavepacket represents the electron in a superposition of states.
- The solutions, as is well known, can be used to describe a free-electron gas, which can explain various features of the behavior of electrons in metals (Ashcroft and Mermin, 1976; Mizutani, 2001).

1.3.3 ELECTRONS IN SPACE BOUNDED BY INFINITE POTENTIALS

The effects of the confinement of empty space will be considered next. In this section, only one dimension will be considered; the most important concepts can be illustrated in a single dimension. It is easy to generalize the treatment to three dimensions, and the way this can be done will be clear from the later discussion of semiconductor structures.

An infinite well is chosen as the simplest way to consider the effects of confinement which, again, allows some of the important points to be discussed without the need of extra mathematical details to deal with finite wells. However, Section 1.6 on tunneling discusses some of the important features of finite wells.

1.3.4 ONE-DIMENSIONAL SPACE BOUNDED BY AN INFINITE WELL

An infinite potential well in one dimension that confines the space within it can be expressed by

$$V(x) = \begin{cases} 0 & \text{if } 0 \leq x \leq L, \\ \infty & \text{if } x < 0, \text{ or } x > L. \end{cases} \tag{1.51}$$

This potential completely confines the electron to the region $0 \leq x \leq L$; it is depicted in Figure 1.5.

As it did for the free electron in one dimensional, Schrödinger's equation takes the form of Equation 1.41, but now $V_0 = 0$, and it is for the region $0 \leq x \leq L$. So we have

$$\left(-\frac{\hbar^2}{2m} \frac{d^2}{dx^2} \right) \psi(x) = E\psi(x), \quad 0 \leq x \leq L. \tag{1.52}$$

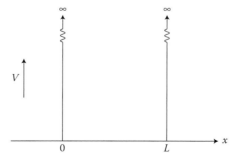

FIGURE 1.5 Infinite square well potential defined by Equation 1.51.

The general solution is still given by Equation 1.42, and from Equation 1.43 with $V_0 = 0$, we obtain the E–k relation

$$k^2 = \frac{2m_e E}{\hbar^2}. \tag{1.53}$$

For the infinite well, there is zero probability that the electron will be found outside of the region $0 \le x \le L$. Hence, $\psi(x) = 0$, if $x < 0$, or $x > L$, and continuity at the boundaries implies that

$$\psi(x) = \psi(L) = 0.$$

From these boundary conditions it follows that $A = -B$, and k is restricted to*

$$k = \frac{n\pi}{L}, \quad n = 1, 2, \ldots \tag{1.54}$$

and so the wavefunction is

$$\psi_n(x) = 2Ai \sin \frac{n\pi}{L} x. \tag{1.55}$$

Since the electron is completely confined to the well, the probability of it being somewhere on $0 \le x \le L$ is 1 and a value of A that normalizes the wavefunction can be found, following Equation 1.28, from

$$\int_0^L | \psi(x) |^2 \, dx = 4A^2 \int_0^L \sin^2 \frac{n\pi}{L} x \, dx = 1, \tag{1.56}$$

which leads to $A = 1/\sqrt{2L}$, and the normalized wavefunction is

$$\psi_n(x) = i\sqrt{\frac{2}{L}} \sin \frac{n\pi}{L} x. \tag{1.57}$$

In fact, physically, the phase factor i is unimportant[†] and it can be dropped to give

$$\psi_n(x) = \sqrt{\frac{2}{L}} \sin \frac{n\pi}{L} x. \tag{1.58}$$

Since k is discrete, the energy, from Equation 1.53, is quantized according to

$$E = E_n = \frac{\hbar^2 k^2}{2m_e} = \frac{\hbar^2}{2m_e} \left(\frac{n\pi}{L} \right)^2. \tag{1.59}$$

---

* The value $n = 0$ is not allowed because it would correspond to an eigenfunction that is identically zero and the energy would be zero. However, a zero energy would imply that the electron is not moving in which case its position could be measured precisely; also, its momentum would be zero and so the position and the momentum would both be known in violation of the Heisenberg uncertainty principle. Also, the general solution in Equation 1.42 implies that k there is positive, so n does not take negative values.

† This is because of the probabilistic connection between the wavefunction and experimental results given by Equation 1.30. It is clear from this that the probability involves taking a modulus squared; by doing this the phase factor i disappears.

Hence the confinement causes the energy to be quantized. The energy levels depend on the size L of the confinement. The corresponding wavefunction states are standing waves rather than the traveling waves found in unbounded space. The E–k relation is still parabolic, but now only points on a parabolic curve, given by Equation 1.59, are allowed.

Note that the electron can be either in one of the allowed states $\Psi_n(x,t) = \psi_n(x)e^{-iE_nt/\hbar}$, just after a measurement, or in a superposition of states $\Psi(x,t) = \sum_n a_n \Psi_n(x,t)$, with the probability that a measurement will yield a state $\Psi_n(x, t)$ given by $|a_n|^2$.

The de Broglie wavelength for the electron is an important parameter for determining the degree to which it is affected by the confinement. Since the energy is the sum of kinetic ($m_e v^2/2 = p^2/(2m_e)$) and potential energies, we can write

$$E = \frac{p^2}{2m_e} + V \rightarrow p = \sqrt{2m_e(E - V)}. \tag{1.60}$$

This expression for p, together with Equation 1.1, and substituting $V = 0$ leads, here, to the de Broglie wavelength

$$\lambda_n = \frac{h}{\sqrt{2m_e E_n}}. \tag{1.61}$$

Substituting for E_n using Equation 1.59 gives

$$L = n\left(\frac{\lambda_n}{2}\right), \tag{1.62}$$

showing that an integral number of de Broglie wavelengths fit into the confined region.

Whether the confinement effect is significant or not depends on the amount of energy and hence on the de Broglie wavelength of the electron. From Equation 1.61, it can be seen that the larger the energy, the smaller the de Broglie wavelength. From Equation 1.62, it is clear that if λ_n is small compared to L, more half de Broglie wavelengths will fit into L and the spacing between adjacent levels, $E_{n+1} - E_n$, will become small compared to L. This means that as for $\lambda_n/2 \ll L$, the electron has a nearly continuous range of energies that approaches the continuous variation of a free electron as E_n increases and λ_n decreases.

The converse is that for energies such that $\lambda/2 \sim L$, the energy is discrete and the electron is strongly influenced by the confinement. The importance here is that for half de Broglie wavelengths $\lambda/2 \sim L$, the influence of quantum confinement is significant.

1.4 ELECTRONS IN PERIODIC STRUCTURES

1.4.1 SOLID STATE PHYSICS

In nanoelectronics, it is important to understand the behavior of electrons* in the solids that make up typical devices. Often, the materials are crystals with repeating structures, or they can be treated as such. This leads to the well-known

---

* The behavior of holes if present as charge carriers is modeled in a similar manner.

ideas in conventional solid state physics, which include band theory, density of states, effective mass, and various statistical distributions (such as Fermi–Dirac and Bose–Einstein distributions). Many textbooks cover these ideas (Ziman, 1972; Ashcroft and Mermin, 1976; Kittel, 2005; Hofmann, 2008; Holgate, 2009; Ibach and Lüth, 2009) and some familiarity with them will be assumed in this chapter. However, some of the concepts of solid state physics required to understand nanoelectronic structures and devices will be discussed, and in the course of the development of the models of nanostructures various formulae from this area will be used.

1.4.1.1 Band Theory

The regularly repeating structure of crystals or other repeating structures (such as carbon nanotubes (CNTs)) gives rise to a band structure in the allowed energy states of the electrons. The Kronig–Penney model (de L. Kronig and Penney, 1931; Ashcroft and Mermin, 1976) employs a particularly simple form of the periodic potential consisting of repeating rectangular wells constant in time. In one dimension, for a lattice of period $l = l_1 + l_2$ this takes the form

$$V(x) = \begin{cases} 0 & \text{if } 0 \leq x \leq l_1, \\ V_0 & \text{if } -l_2 \leq x \leq 0, \end{cases} \qquad V(x) = V(x+l). \qquad (1.63)$$

With the help of Bloch's theorem, the Schrödinger's equation can be solved exactly and an energy band structure results. Figure 1.6 shows band energy E versus Bloch wave number k for the Kronig and Penney model. This arises from a transcendental equation that is easily solved numerically (de L. Kronig and Penney, 1931).

The Kronig–Penney model is straightforward to work with and produces a band structure but quantitatively it is not accurate; more realistic models can be employed at the expense of an increase in mathematical complexity. However, it is difficult to

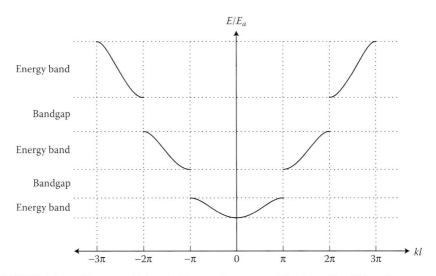

FIGURE 1.6 E/E_0 versus kl for the Kronig–Penney model, where $E_l = \hbar^2/(2ml^2)$ serves to make the energy scale dimensionless.

model all of the features of the band structure of typical electronic materials such as Ge and Si (see Sze (1969) for some realistic bandstructure plots of Ge and Si). Nonetheless, because the charge carriers that are most important in interactions with the material and external influences are usually in restricted regions of the band structure, relatively simple models can be used together with the effective mass concept mentioned next. The shape of the band structure in the region of interest is often close to parabolic, allowing the use of a parabolic $E–k$ relationship in many cases.

1.4.1.2 Effective Mass

Often, the regions of the bands that are of most interest in nanoelectronics are at the band edges comprising the top of the conduction band and the bottom of the valence band, since the charge carriers here are the most important for conduction. Therefore, the charge carries in these regions are usually the most influential. As mentioned earlier, the $E–k$ relation for the edges can be modeled as parabolic.

The band structure and parabolic shape are a result of the periodic potential of the structure. However, there is a parallel with the quantum treatment of an electron in free space; it can be shown that the $E–k$ relationship for the electron is parabolic (Ashcroft and Mermin, 1976). The exact shape of the parabola is fixed for the electron with mass m_e. The shape of the parabolas for the bandstructures of periodic materials is different from this, but this can be accounted for by introducing an effective mass for the carrier m^* and the carrier behaves as though it were a free electron but with mass m^* rather than m_e. It can be shown (Ferry and Bird, 2001) that a relationship between effective mass and $E(k)$ is given by

$$m^* = \hbar^2 k \left(\frac{\partial E}{\partial k} \right)^{-1} \tag{1.64}$$

Note that holes in the valence band have a negative effective mass.

1.4.1.3 Excitons

Electron–hole pairs are created when an electron moves from the valence band to the conduction band; each electron and hole can contribute to conduction but the electron–hole pairs are still bound together by Coulomb attraction and can behave like a hydrogen atom with the nucleus being replaced by the hole. These hydrogen-like electron–hole pairs are called excitons. Usually in conventional semiconductor devices in microelectronics, the excitons do not play a large role because the Coulomb attraction is not strong and they are easily broken. However, they play a greater role in nanoscale devices due to the confinement.

A hydrogen atom consists of a single electron of charge $-e$ and a proton of charge e. Schrödinger's equation can be solved exactly for the hydrogen atom (Bohm, 1951) resulting in discrete energy levels

$$E_n = \frac{m_r e^4}{8\varepsilon_0^2 h^2 n^2}, \quad n = 1, 2, 3, \ldots \tag{1.65}$$

where $m_r = m_e m_p/(m_e + m_p)$ ($\approx m_e$, since $m_p \gg m_e$). The electron–hole pair of the exciton, however, moves through a semiconductor with relative permittivity ε_r.

Also, the masses of each are effective masses. For the exciton, we therefore modify Equation 1.65 to give

$$E_n = \frac{m_r^* e^4}{8\varepsilon_0^2\varepsilon_0^2 h^2 n^2},$$

(1.66)

where m_r^* is the reduced effective mass

$$m_r^* = \frac{m_e^* m_h^*}{m_e^* + m_h^*}.$$

(1.67)

The effective masses of the electron and hole, m_e^* and m_h^*, do not differ greatly and hence Equation 1.67 cannot be approximated as done for the hydrogen atom.

1.4.1.4   Density of States

The density of states expressions introduced here are valid when the density is fairly high, that is, when the effects of quantum confinement are not too large. For structures where the confinement size is of the order of the de Broglie wavelength, the density of states is modified. This will be examined in Section 1.5.

1.4.1.4.1   Three Dimensional

The electron is confined in a three-dimensional infinite potential well and the potential inside the well is zero. An extension of the one-dimensional calculation done in Section 1.3.4 for a confined electron to three dimensions—as will be seen when we consider a quantum dot in Section 1.5.4—gives, for confinement L in the x, y, and z directions, the energy levels

$$E_n = \frac{\hbar^2 \pi^2}{(2m_e^* L^2)} n^3,$$

(1.68)

where
$n^2 = n_x^2 + n_y^2 + n_z^2$
$n_\alpha = 1, 2, 3, \ldots, \alpha = x, y, z$

The number of states below a given energy E_n is equal to the number of states in an n-space octant of radius $n = \sqrt{E_n/E_1}$ multiplied by 2—we consider an octant rather than a sphere because the n_α do not take negative values, and the factor of 2 accounts for electron spin. Therefore, for energies $E_n \gg E_1$, the total number of states N_{E_n} having energy less than E_n is approximately 2 times the volume of the octant, or the volume of the quarter-sphere,

$$\frac{1}{4}\left(\frac{4}{3}\pi n^3\right) = \frac{\pi}{6}\left(\frac{E_n}{E_1}\right)^{3/2} = N_{E_n}.$$

(1.69)

Thus, the total number of states in the range E_n to E_{n-1} $(=E_n - \Delta E_n)$ is $\Delta N_{E_n} = N_{E_n} - N_{E_n - \Delta E_n}$. Using Equation 1.68 and the fact that $\Delta E_n/E_n \gg 1$, it can be shown that the number of states per unit volume is

$$\frac{\Delta N_{E_n}}{V_{cube}} \approx \frac{2^{1/2}(m_e^*)^{3/2} E^{1/2}}{\hbar^3 \pi^2} \Delta E_n,$$

(1.70)

where $V_{cube} = L^3$ is the confined volume. The density of states $N(E)$ is the number of states per unit volume per unit energy around an energy E.

Assuming that we have many energy levels closely spaced (so that the quantum confinement is not significant in that it does not make the energy levels very discrete: they are closely spaced), we can consider E_n to be a continuous variable so that $\Delta E_n \to dE$ and $\Delta N_{E_n} \to dN_E$. Now, since the density of states $N(E)$ is the number of states per unit volume* per unit energy around an energy E, we have

$$\frac{dN_E}{V_{cube}} = \frac{2^{1/2}(m_e^*)^{3/2} E^{1/2}}{\hbar^3 \pi^2} dE = N(E)dE, \tag{1.71}$$

which gives,

$$N(E) = \frac{2^{1/2}(m_e^*)^{3/2}}{\hbar^3 \pi^2} E^{1/2} \quad \text{(3D density of states).} \tag{1.72}$$

This is for an electron in a region with zero potential energy. If the potential energy has a constant value V_0, the density of states is obtained from Equation 1.72 by making the replacement $E \to E - V_0$. Although a semiconductor crystal does not have such a potential, the effective mass m_e^* takes this into account.

1.4.1.4.2 Two Dimensional

The energy expression for E_n in Equation 1.68 is still valid, but now the spatial region is an area L^2 and the number of states N_{E_n} lies in twice the area of a quarter circle, that is, the area of a semicircle of radius n: $\pi n^2/2$. The density of states is the number of states per unit area around an energy E. Following a similar argument to the one for the three-dimensional case, it can be shown that

$$N(E) = \frac{m_e^*}{\pi \hbar^2} \quad \text{(2D density of states),} \tag{1.73}$$

which is valid for $E > V_0$.

1.4.1.4.3 One Dimensional

Again, Equation 1.68 for E_n can be used. The spatial region is a line of length L, and the number of states lies in n-space along a line of length n. The density of states is now the number of available states per unit length per unit energy around energy E, and a similar calculation to the three-dimensional one yields

$$N(E) = \frac{\sqrt{2m_e^*}}{\pi \hbar} E^{-1/2} \quad \text{(1D density of states).} \tag{1.74}$$

(Use $E \to E - V_0$ if the potential is V_0.)

---

* This derivation applies to bulk materials; the volumes considered would be far from any surface. So the starting volume in our calculation is a cube within the material and it is not necessary to consider the more general energy expression in which instead of L in Equation 1.68 there would be lengths L_x, L_y, and L_z forming a volume $V_{rect} = L_x L_y L_z$.

1.4.1.5 Fermi–Dirac Distribution

Statistical distributions in solid state physics give the probability that a state at an energy E will be occupied. The Fermi–Dirac distribution is of interest concerning the electronic properties of materials because it applies to indistinguishable particles when only one particle can occupy a particular state, such as electrons. It is given by

$$f(E) = \frac{1}{e^{\frac{E-E_F}{k_B T}} + 1}.$$ (1.75)

Figure 1.7 shows that at $T = 0$ the Fermi–Dirac distribution is a step function such that all states below the Fermi energy E_F are occupied ($f(E < E_F) = 1$) and those above the Fermi energy are empty ($f(E > E_F) = 1$). So E_F is the energy to which the states would be filled at $T = 0$ (absolute zero). For temperatures $T \gg 0$, as can be seen also from Figure 1.7, although the step becomes less sharp, only states within a few $k_B T$ of the Fermi energy are likely to be excited. A key point here is that in many materials, even at rather high temperatures, $k_B T \ll E_F$, implying that the electrons with energies at, or close to, E_F are of principal importance.

The distribution for holes is given by $1 - f(E)$, since the probability of finding a hole is the probability that an electron will not occupy a level.

1.4.2 GRAPHENE AND CARBON NANORIBBONS

The material in Section 1.4.1 is standard solid state theory. Graphene and CNTs are relatively new and are important for nanoelectronics because they can be fabricated at nanoscale sizes and have interesting transport properties.

1.4.2.1 Graphene

Graphene is actually a single layer of graphite with a repeating structure, as shown in Figure 1.8. The lattice basis vectors are $\alpha_1 = (\sqrt{3}\hat{a}_x + \hat{a}_y)(a/2)$ and $\alpha_2 = (\sqrt{3}\hat{a}_x - \hat{a}_y)(a/2)$, where $a = \sqrt{3}a_{C-C}$, with a_{C-C} being the interatomic distance between nearest neighboring carbon atoms in graphene.

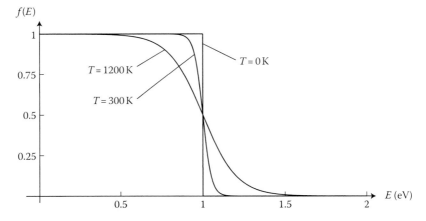

FIGURE 1.7 The Fermi–Dirac distribution plotted at temperatures $T = 0$, 300, and 1200 K.

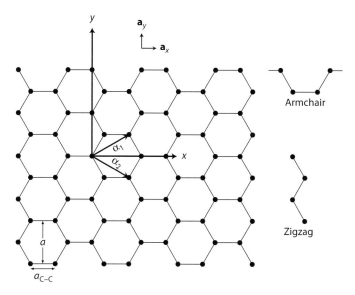

FIGURE 1.8 Structure of graphene.

Because the structure of graphene is periodic, it has a similar energy band structure typical of semiconductor structures such as Si and GaAs. The most important bands have an E–k relationship well approximated by (Wallace, 1947; Castro Neto et al., 2009)

$$E(k_x, k_y) = \frac{-s\kappa_0 f(k_x, k_y)}{1 + swf(k_x, k_y)}, \tag{1.76}$$

where

$w = 0.129$
$\kappa_0 = 3$ eV
$s = 1$ for bonding bands, -1 for antibonding bands

The function f is given by

$$f(k_x, k_y) = \sqrt{1 + 4\cos\left(\frac{\sqrt{3}k_x a}{2}\right)\cos\left(\frac{k_y a}{2}\right) + 4\cos^2\left(\frac{k_y a}{2}\right)}. \tag{1.77}$$

The Fermi surface is a collection of six points on the first Brillouin zone, which is hexagonal in shape (Castro Neto et al., 2009). At each of these Fermi points, $E = E_F = 0$; they are given by the (k_x, k_y) values $\left(\pm 2\pi/(\sqrt{3}a), \pm 2\pi/(3a)\right)$ and $(0, \pm 4\pi/(3a))$. The electronic properties of graphene are strongly influenced by what happens in the vicinity of these points, and there the E–k relation can be simplified further, being given approximately by

$$E(k_x, k_y) = -s\kappa_0 f(k_x, k_y). \tag{1.78}$$

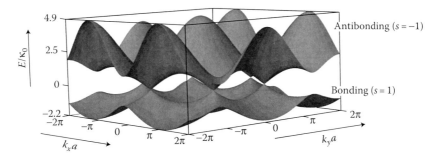

FIGURE 1.9 E versus k for graphene, given by Equation 1.78. The six K points occur in the plane $E = 0$ where the bonding and antibonding surfaces meet.

An E–k plot using Equation 1.78 is shown in Figure 1.9. The six points, in the plane $E = 0$, where the bonding and antibonding surfaces meet are called K points.

The Fermi points are important because in their vicinity electrons—referred to as π electrons in π orbitals—are weakly bound to the carbon atoms. These electrons are the most influential in electronic applications; only a small amount of energy is required to move them into the conduction band.

In contrast to semiconductors that have a parabolic band structure near the Fermi surface, the band structure near the Fermi points in graphene is linear. The electron dispersion near these points is

$$E = v_F \hbar k, \tag{1.79}$$

where $v_F = 9.7 \times 10^5$ m/s is the Fermi velocity. Comparing this with the $E = c\hbar k$ dispersion for photons (which are massless) shows that near the π band crossing points, the electrons behave like photons and are called Dirac fermions.

1.4.2.2 CNTs

1.4.2.2.1 Structure

One way to understand the CNT structure is by considering CNTs as being made by rolling up a graphene sheet.* This can be done in different ways. Referring to Figure 1.10a, one possibility is to roll the sheet up so that points A_i and A_i' merge. This would result in zigzag-shaped ends (Figure 1.10c). Another possibility is to roll the sheet up so that points B and B' in Figure 1.10b merge, resulting in ends that look like a top view of side-by-side armchairs (when viewing the tube from the side), as shown in Figure 1.10d. Other possible roll-up directions would produce a spiral pattern on the surface of the tube, like a screw (Figure 1.10e) which, unlike the armchair or zigzag type, would look different if reflected in a mirror. For this reason, the spiral-like CNTs are described as being chiral.

In general, the roll-up configuration is defined by the vector

$$\mathbf{C} = n\mathbf{\alpha}_1 + m\mathbf{\alpha}_2, \tag{1.80}$$

---

* This is a conceptual aid. CNTs are not actually made like this. One method of producing CNTs is by arc discharge of carbon electrodes (Iijima, 1991).

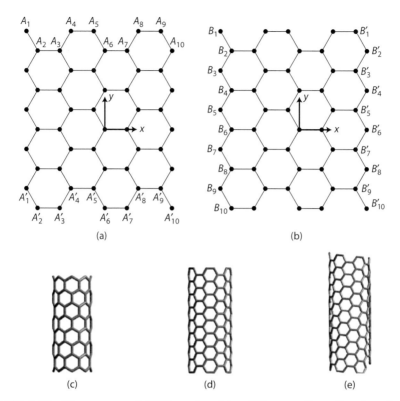

FIGURE 1.10 Three types of CNT produced by different roll-up directions for the graphene sheet. (a) Depiction of a graphene sheet that is rolled up to form a zigzag CNT. (b) Graphene sheet that rolls up to form an armchair CNT. (c) Zigzag CNT. (d) Armchair CNT. (e) Chiral CNT. Note that (a) and (b) are for illustration only. The tubes depicted in (c) and (d) are constructed from sheets with a greater number of cells. Also, for clarity, the tubes are orientated such that the cells at the back are hidden by those at the front.

called a chiral vector, indexed by the integers n and m. Zigzag CNTs have indexes (n, n); armchair CNTs are $(n, 0)$; chiral CNTs are indexed by (m, n), with $m \neq n$ and $m > 0$.

1.4.2.2.2 E–k Relation

If a CNT is considered to be a tube of infinite length (in practice this would apply to a tube whose length is large compared to its diameter), then it is periodic in the axial direction and the edge of the Brillouin zone can be shown (Tersoff and Ruoff, 1994) to be given by $k_x = \pi/(3a_{C-C})$ for armchair CNTs, and $k_y = \pi/(\sqrt{3}a_{C-C})$ for the zigzag type. The transverse direction is finite, and so the wave number is quantized according to

$$k_{x,l} = \frac{2\pi l}{3na_{C-C}} \quad \text{(armchair CNTs)}, \tag{1.81}$$

and

$$k_{y,l} = \frac{2\pi l}{\sqrt{3}na_{C-C}} \quad \text{(zigzag CNTs).} \tag{1.82}$$

The E–k relations for CNT π electrons can be found by substituting for k_x and k_y in Equation 1.78 using Equations 1.81 and 1.82.

1.4.2.2.3 Properties

One of the very interesting properties of CNTs is that their electronic behavior can be either metallic or semiconducting, depending on the roll-up geometry as defined by the chiral vector $\mathbf{C}$ in Equation 1.80. The general rule for determining this (Dresselhaus et al., 1995) is as follows:

$$\text{if } \frac{2n+m}{3} \in \mathbb{I}^+ \text{ CNT is metallic,}$$

$$\text{if } \frac{2n+m}{3} \notin \mathbb{I}^+ \text{ CNT is semiconducting,}$$

where $\mathbb{I}^+$ denotes the set of positive integers.

Note that these equations for CNTs are useful for understanding the basic behavior described, but do not take into account the curvature of the tube. A more complete analysis would need to take this into account because the curvature affects the carbon–carbon bonds.

Some other attractive properties of CNTs are as follows:

- Strength, due to the bonds between the atoms being very strong
- High stiffness
- High thermal conductivity
- The ability to carry very high current densities

More details of CNT properties and their many applications can readily be found in the literature, including Tersoff and Ruoff (1994), Dresselhaus et al. (1995), and Saito (1998).

1.5 SEMICONDUCTOR QUANTUM WELLS, WIRES, AND DOTS

In this section, the effect of quantum confinement in semiconductor nanostructures is considered in one, two, and three dimensions, for quantum wells, wires, and dots, respectively.

1.5.1 Quantum Confinement

In Section 1.3, it was shown that for confined electrons with energies that produce de Broglie wavelengths of the order of the confinement size, the quantum discreteness of the energy levels is significant. In a semiconductor, the most important electrons, as brought out in Section 1.4.1, are those with energies close to the Fermi energy and so the de Broglie wavelength at the Fermi energy, called the Fermi wavelength, λ_F, sets the confinement size required for quantum discretization effects to become significant.

We can solve Schrödinger's equation as in Section 1.3, but now the electron mass m_e is replaced with an effective mass m^* which would be given by m_e^* for electrons and m_h^* for holes. This accounts for the presence of the material instead of empty space (as explained in Section 1.4.1).

Replacing m by m^* in Equation 1.53 gives the $E–k$ relation

$$E = \frac{\hbar^2 k^2}{2m^*}. \tag{1.83}$$

This can be used since, typically, it holds near the edges of the conduction and valence bands in semiconductors where the influential carriers lie. Hence, with $k = 2\pi/\lambda$, the following expression for the Fermi wavelength can be obtained from Equation 1.83:

$$\lambda_F = \frac{h}{\sqrt{2m^* E_F}}. \tag{1.84}$$

If the semiconductor confinement of the carrier is large compared to λ_F, it will behave approximately like a free particle of mass m^*. If λ_F is of the order of or larger than the confinement size, the quantum confinement will be apparent.

A numerical calculation shows that typical values of λ_F for metals are in the range $\lambda_F \sim 0.5–1$ nm (Mizutani, 2001), but are considerably larger for semiconductors at $\lambda_F \sim 10–100$ nm (Chazalviel, 1999). Therefore, semiconductor structures that have at least one dimension at the nanoscale are expected to be strongly influenced by quantum discreteness; we refer to this as quantum confinement.

We next look at confinement ion in one dimension: quantum wells; two dimensions: quantum wires; and three dimensions: quantum dots.

1.5.2 QUANTUM WELLS IN SEMICONDUCTOR HETEROSTRUCTURES

A quantum well refers to a structure in which quantum confinement is apparent in only one dimension. If the well width is L_x, then we will have $L_x \lesssim \lambda_F$, with L_y and L_z for the other dimensions being $L_y, L_z \gg \lambda_F$.

Such a well can be formed from a semiconductor heterostructure consisting of a thin layer of a small band-gap material like GaAs sandwiched between two thick layers of wide band-gap material such as AlGaAs. This is depicted in Figure 1.11. Assuming that the heterostructure is composed of crystalline semiconductor material, and the Bloch functions representing the crystal structure of the layers are similar in each layer, to a good approximation it can be modeled by replacing the crystal structure with a potential energy profile and using an effective mass (Bastard, 1991). Quantum wells will form in both the conduction and valence bands of the GaAs layer, that is, discrete energy levels will be apparent in these bands.

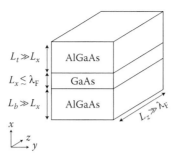

FIGURE 1.11   Semiconductor heterostructure forming a quantum well.

1.5.2.1   Infinite Well Mathematical Model

For the potential energy $V(x, y, z) = V(x)$, the simplest approximation is an infinitely deep well. Such a well was considered in Section 1.3.4; so here we can use Equation 1.51 for $V(x)$ (with $L = L_x$).

The wavefunction, using the separation of variables technique, can be written as

$$\psi(x, y, z) = \psi_x(x)\psi_y(y)\psi_z(z),   \tag{1.85}$$

which on substituting into Equation 1.21, Schrödinger's time-independent equation, gives

$$\frac{1}{\psi_x}\frac{d^2\psi_x}{dx^2} + \frac{1}{\psi_y}\frac{d^2\psi_y}{dy^2} + \frac{1}{\psi_z}\frac{d^2\psi_z}{dz^2} + \frac{2m^*E}{\hbar^2} = 0, \quad \text{in the well.}   \tag{1.86}$$

In the unconstrained directions, we can follow the same procedure as for the unbounded electron case in Section 1.3.2. Hence, we write

$$\frac{1}{\psi_y}\frac{d^2\psi_y}{dy^2} = -k_y^2 \quad \text{and} \quad \frac{1}{\psi_z}\frac{d^2\psi_z}{dz^2} = -k_z^2,   \tag{1.87}$$

with solutions such that

$$\psi_y(y)\psi_z(z) = Ae^{ik_y y}e^{ik_z z},   \tag{1.88}$$

in which the wave vector components k_y and k_z can be positive or negative, and vary continuously.

It now remains to solve

$$\left[-\frac{\hbar^2}{2m_e}\frac{d^2}{dx^2} + V_{eff}(x) \right]\psi_x(x) = E\psi_x(x),   \tag{1.89}$$

where

$$V_{eff} = \frac{\hbar^2}{2m_e}(k_y^2 + k_z^2)   \tag{1.90}$$

is defined as an effective potential. The problem of solving Equation 1.89 to find $\psi_x(x)$ is of the same form as the one-dimensional infinite well problem in Section 1.3.4 except that the potential in the well is now $V_{\text{eff}}(x)$ instead of zero; this can be accounted for by replacing E in that problem by $E - V_{\text{eff}}(x)$. Thus, the solution is

$$\psi_x(x) = \left(\frac{2}{L_x}\right)^{1/2} \sin\frac{n\pi}{L_x}x, \tag{1.91}$$

and

$$k_x = k_{x,n} = \frac{n\pi}{L_x}, \quad n = 1, 2, \ldots . \tag{1.92}$$

Finally, it follows that the normalized wavefunction (for which $A = 1$) is

$$\psi(x, y, z) = \left(\frac{2}{L_x}\right)^{1/2} \sin\frac{n\pi}{L_x}x e^{ik_y y} e^{ik_z z}, \tag{1.93}$$

and the energy is

$$E = E_n + \frac{\hbar^2}{2m^*}k_{yz}^2, \tag{1.94}$$

in which

$$E_n = \frac{\hbar^2}{2m^*}\left(\frac{n\pi}{L_x}\right)^2 \quad \text{and} \quad k_{yz}^2 = k_y^2 + k_z^2. \tag{1.95}$$

The energy is quantized according to Equation 1.94, in what are known as subbands with E_n being the lowest energy for the nth subband. Figure 1.12 shows the first few subbands.

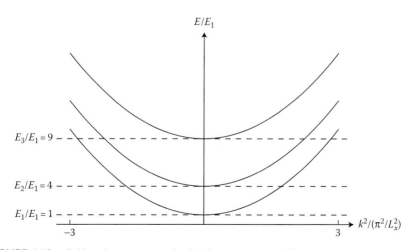

FIGURE 1.12 Subband energy quantization in a quantum well.

1.5.2.1.1 Scope of the Model

The infinite-well model used here is not a very accurate approximation of a real heterostructure; but it does represent the basic behavior fairly well, and it illustrates how quantum effects are more pronounced than in structures that in all dimensions are large compared to the de Broglie wavelength. A fuller treatment taking into account the finiteness of the wells and other factors is given in Bastard (1991).

1.5.2.2 Density of States in a Quantum Well

It is also useful to look at the influence of the well confinement on the density of states. For the quantum well case, a two-dimensional density of states is appropriate. The expression in Equation 1.73 is valid, but such a density of states is associated with each subband leading to the steps plotted in Figure 1.13. Formally, this can be represented, with the help of the Heaviside step function H, as

$$N_{2D}(E) = \frac{m^*}{\pi\hbar^2} \sum_n H(E - E_n). \tag{1.96}$$

1.5.2.3 Influence of Thermal Energy

The influence of thermal energy has not been considered so far. It is important to note that the model introduced here is only applicable if the thermal energy k_BT is small compared to the difference between energy subbands, that is, if $[\hbar^2/(2m^*)](\pi/L_x)^2 \gg k_BT$. But if this condition is not met, the thermal energy will be enough for electrons to move up to the higher bands and the discrete nature of the subbands will become blurred.

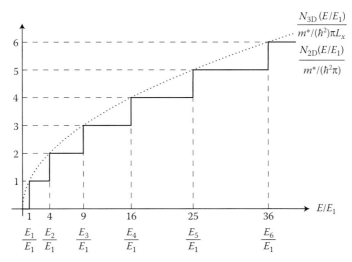

FIGURE 1.13 Two-dimensional step density of states for a quantum well. For comparison, the dotted curve is the density of states—from Equation 1.72—for a three-dimensional electron gas whose confinement is large in every direction compared to the de Broglie wavelength.

1.5.2.4 Quantum Well Energy Transitions

The quantum confinement creates quantum wells or discrete energy levels in both the conduction and valence bands in the confined layer (the GaAs in the aforementioned example). In this sense, the bands are not conventional bands in which the energy in them varies almost continuously. Now the bands are regions in which discrete energy levels are grouped, but there is still an energy band gap E_g between bands that is recognizable because it separates the groupings.

1.5.2.4.1 Interband Transitions

Transitions can occur from energy levels in one band to those in another; they are called interband transitions. For direct semiconductors,* incident photons can cause such transitions. The photon energies for which interband transitions can occur are given by

$$\hbar\omega = E_g + E_n^{m_h^*} + E_m^{m_e^*} = E_g + \frac{\hbar^2\pi^2}{2m_h^*L_x^2}n^2 + \frac{\hbar^2\pi^2}{2m_e^*L_x^2}m^2. \qquad (1.97)$$

Here, $E_m^{m_e^*}$ is measured from the bottom of the conduction band and $E_m^{m_h^*}$ from the top of the valence band. The situation is illustrated in Figure 1.14, which shows the allowed transitions that are set by Fermi's golden rule (Fox, 2001).

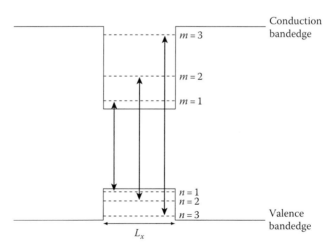

FIGURE 1.14   Interband transitions in a quantum well.

---

* In direct semiconductors, transition from the valence to the conduction band occurs with little change in k and an incident photon can cause the transition since it has small k. In contrast, indirect semiconductors require a larger k change for a transition that cannot be achieved with a photon alone; in this case a two-step photon–phonon interaction would be able to cause the transition (Ashcroft and Mermin, 1976).

1.5.2.4.2 Intraband Transitions

Transitions within each band, which may be called intraband transitions, can also occur; these involve lower energies than interband transitions, as is clear from Figure 1.14. Intraband energy transitions occur at following energies:

$$\Delta E = E_{n_1}^{m_h^*} - E_{n_2}^{m_h^*} \quad \text{(conduction band transitions),} \tag{1.98}$$

$$\Delta E = E_{m_1}^{m_e^*} - E_{m_2}^{m_e^*} \quad \text{(valence band transitions).} \tag{1.99}$$

Selection rules (Fox, 2001) only allow transitions for which $m_1 - m_2$ and $n_1 - n_2$ are odd.

1.5.2.5 Excitonic Effects

The bound electron–hole pair excitons mentioned in Section 1.4.1 play a more prominent role in quantum well structures than in bulk semiconductors because the confinement enhances the binding energy. In bulk semiconductors, this is of the order of a few millielectron volt, which, except at low temperatures, is easily overcome by thermal energy so that the excitons break up. The increased binding energy in quantum wells means that the excitons are more influential as they do not break up as easily at room temperature. The excitonic effect is evident in absorption measurements (Fox, 1996), where it produces sharp absorption peaks.

1.5.3 Quantum Wires

1.5.3.1 Mathematical Model

For a quantum wire, the confinement of the order of λ_F is in two dimensions, with only one dimension—along the length of the wire—being much larger than this so that the charge carriers behave as though they are unbounded only along this length. A schematic of a quantum wire of rectangular cross section is given in Figure 1.15.

The mathematical analysis is easiest for rectangular geometry and we employ a similar model to that used for the quantum well. But now, since the confinement is in two dimensions, the Schrödinger's equation to solve is

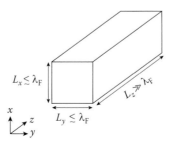

FIGURE 1.15 A rectangular cross-section quantum wire.

$$\left[-\frac{\hbar^2}{2m^*}\nabla^2 + V(x,y)\right]\psi(x)(\mathbf{r}) = E\psi(\mathbf{r}), \quad \text{in the wire,} \tag{1.100}$$

and the wavefunction will take the form

$$\psi(\mathbf{r}) = Ae^{ik_z z}\psi(x,y), \tag{1.101}$$

following the geometry of Figure 1.15.

An infinite confining potential is again assumed, this time given by

$$V(x,y) = \begin{cases} 0 & \text{if } 0 \le x \le L_x, 0 \le y \le L_y, \\ \infty & \text{if } x < 0, x > L_x, y < 0, y > L_y. \end{cases} \tag{1.102}$$

It is easy to show, using a similar analysis to that in Section 1.5.2, that this potential gives

$$\psi(x,y) = \left(\frac{4}{L_x L_y}\right)^{1/2} \sin\frac{n_x\pi}{L_x}\sin\frac{n_y\pi}{L_y}, \tag{1.103}$$

and the normalized wavefunction is

$$\psi(x,y,z) = \left(\frac{4}{L_x L_y}\right)^{1/2} \sin\frac{n_x\pi}{L_x}\sin\frac{n_y\pi}{L_y}e^{ik_z z}, \tag{1.104}$$

$$n_x = 1, 2, \ldots, \quad n_y = 1, 2, \ldots \ . \tag{1.105}$$

Also, by a similar analysis, the energy is

$$E = E_{n_x,n_y} + \frac{\hbar^2}{2m^*}k_z^2, \tag{1.106}$$

where the k_z wavefunction component is continuous and the subband energy is

$$E_{n_x,n_y} = \frac{\hbar^2}{2m^*}(k_{x,n_x}^2 + k_{y,n_y}^2), \tag{1.107}$$

in which

$$k_{\alpha,n_\alpha} = \frac{n_\alpha\pi}{L_\alpha}, \quad \alpha = x, y \tag{1.108}$$

are discrete due to the confinement along x and y.

1.5.3.2 Some Quantum Wire Characteristics

1.5.3.2.1 Density of States

In the quantum wire, the confinement creates discrete energy levels so that we again have subbands. This will make the density of states different from that for

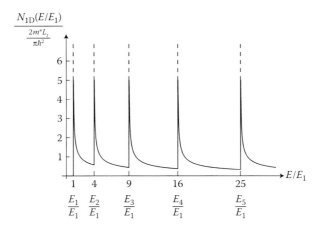

FIGURE 1.16 Density of states for a quantum wire for the case $L_x = L_y = L_t$, so that the discrete part of the energy reduces to $E_{n_x,n_y} = E_n = [(\hbar^2 \pi^2)/(2m^* L_t^2)]n^2$, where $n^2 = n_x^2 + n_y^2$. Asymptotes are marked by the dashed lines and illustrate the van Hove singularities.

the bulk in Equation 1.74. Now, each subband will have a density of states given by Equation 1.74, but for each band there is an effective potential $V_{eff} = E_{n_x,n_y}$. With these ideas, the quantum wire density of states can be expressed as a sum over the subbands given by

$$N_{1D}(E) = \frac{1}{\pi\hbar} \sum_{n_x,n_y} \left(\frac{2m^*}{E - E_{n_x,n_y}} \right)^{1/2} H(E - E_{n_x,n_y}), \quad E > E_{n_x,n_y}. \tag{1.109}$$

The higher the energy E, the greater the number of subbands filled. Figure 1.16 shows a plot of $N(E)$ versus E. The discontinuous jumps are called van Hove singularities. Density of states plots for semiconducting and metallic CNTs* can be found in Saito et al. (1992).

1.5.3.2.2 Energy Transitions

Various optical transitions occur in quantum wires governed by the energy levels, the location of van Hove singularities, and selection rules (Bockelmann and Bastard, 1991; Mariani et al., 2000).

1.5.3.2.3 Excitons

As with quantum wells, the quantum confinement in quantum wires makes excitonic effects more important at room temperature than they are in bulk materials (Goldoni et al., 1997).

---

* CNTs can behave like quantum wires if the axial direction is large compared to λ_F and the radius is $\lesssim \lambda_F$.

1.5.3.2.4 Carrier Transport

The carrier transport in quantum wires can be quite different from bulk materials. The behavior tends toward ballistic transport. This means that the transport can be much faster than in conventional materials. One advantage of this is in high-frequency applications. Hanson (2008) gives an introductory account of ballistic transport which require high speed carrier transport.

1.5.4 QUANTUM DOTS

1.5.4.1 Mathematical Model

Again we use an infinite well model similar to that in Section 1.5. But now the quantum confinement is in all three dimensions, as illustrated in Figure 1.17. For the geometry in the figure, the following Schrödinger's equation must be solved:

$$\left[-\frac{\hbar^2}{2m^*} \nabla^2 + V(x,y,z) \right] \psi(x)(\mathbf{r}) = E\psi(\mathbf{r}s), \quad \text{in the dot.} \tag{1.110}$$

The three-dimensional infinite well is described by

$$V(x,y,z) = \begin{cases} 0 & \text{if } 0 \le x \le L_x, 0 \le y \le L_y, 0 \le z \le L_z, \\ \infty & \text{if } x < 0, x > L_x, y < 0, y > L_y, z < 0, z > L_z. \end{cases} \tag{1.111}$$

Using a similar method given in Section 1.5 yields the normalized wavefunction

$$\psi(x,y,z) = \left(\frac{8}{L_x L_y L_z} \right)^{1/2} \sin\frac{n_x \pi}{L_x} \sin\frac{n_y \pi}{L_y} \sin\frac{n_z \pi}{L_z}, \tag{1.112}$$

$$n_x = 1, 2, \ldots, \ n_y = 1, 2, \ldots, \ n_z = 1, 2, \ldots, \tag{1.113}$$

and the corresponding eigenenergies are

$$E = E_{n_x, n_y, n_z} = \frac{\hbar^2 \pi^2}{2m^*} \left[\left(\frac{n_x}{L_x} \right)^2 + \left(\frac{n_y}{L_y} \right)^2 + \left(\frac{n_z}{L_z} \right)^2 \right]. \tag{1.114}$$

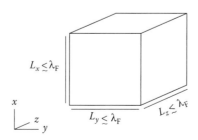

FIGURE 1.17   A quantum dot structure fitting rectangular Cartesian coordinates.

1.5.4.2 Some Quantum Dot Characteristics

1.5.4.2.1 Density of States

With quantum confinement in all three dimensions, no dimension is large compared to λ_F and so none of the density of states expressions in Section 1.4.1 is valid for any quantum dot dimension, and the dot can be thought of as zero dimensional. The energy is discrete in all directions and so can be expressed with the help of a delta function as

$$N_{0D}(E) = 2 \sum_{n_x,n_y,n_z} \delta(E - E_{n_x,n_y,n_z}), \qquad (1.115)$$

where the factor of 2 accounts for electron spin. This density of states is represented in Figure 1.18. However, in real quantum dots, the delta functions are broadened by electron collisions in the dot.

As with the other quantum structures, the discreteness of the density of states leads to different behavior compared to bulk semiconductors; the delta-function nature also distinguishes it from quantum wells and wires which at least are piecewise continuous. So the quantum discreteness is most evident in the quantum dot, as would be expected.

1.5.4.2.2 Other Properties

Some other properties are mentioned in the following:

Transport: In a quantum dot, electron transport is quite different as the lack of any continuity in the density of states means that there is no current flow in the conventional sense.

Thermal effects: These are markedly different as there are only a limited number of states that are widely separated, which limits the possible thermal excitations.

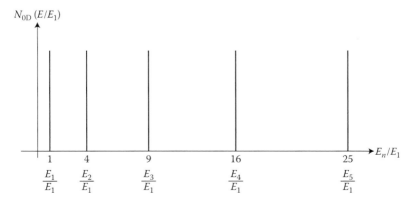

FIGURE 1.18 Density of states for a quantum dot for the case $L_x = L_y = L_z = L_d$, so that the energy reduces to $E_{n_x,n_y,n_z} = E_n = [(\hbar^2 \pi^2)/(2m^* L_d^2)]n^2$, where $n^2 = n_x^2 + n_y^2 + n_z^2$.

Optical spectrum: The limited number of discrete states also has a strong influence on the spectrum of optical radiation that is emitted; the spectrum is very narrow, even at quite high temperatures. Therefore, quantum dots are attractive in laser applications.

Excitons: As with the other quantum structures considered, excitons play a greater role at room temperature than in bulk materials. In quantum dots, excitons often play a dominant role in determining the optical properties at room temperature (Takagahara and Takeda, 1992; Verzelen et al., 2002; Karabulut et al., 2008).

1.6 QUANTUM TUNNELING

Quantum tunneling has no counterpart in classical physics. In essence, it allows particles with less energy than a finite potential barrier to have some probability of going through the barrier—this is absolutely not allowed classically. However, in an attempt to recover some kind of classical picture, the term tunneling is used to invoke a picture of a particle tunneling through the barrier, but literal tunneling, as could be done classically, is not implied. However, in some way, described but not explained by standard quantum theory, quantum tunneling is a real effect responsible, for example, for the emission of radioactive particles from nuclei, which, without tunneling, would not be able to escape the potential binding them.

1.6.1 A SIMPLE EXAMPLE ILLUSTRATING TUNNELING

We give a straightforward example to show how tunneling occurs in quantum theory by considering the one-dimensional rectangular potential barrier of Figure 1.19. The equation describing it is

$$V(x) = \begin{cases} V_0 & \text{if } 0 \le w, \\ 0 & \text{if } x < 0, x > w. \end{cases} \tag{1.116}$$

Although this is of simple form it can model, at least qualitatively, several real situations such as an electron bound to a quantum dot.

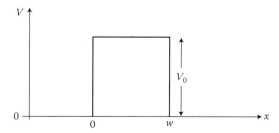

FIGURE 1.19 A rectangular potential barrier of height V_0. There is a finite probability that a quantum particle with energy $E < V_0$ will pass through the barrier.

Classically, for $E > V_0$, the particle would definitely pass the barrier, and for $E < V_0$, it would definitely be reflected. To find out what would happen quantum mechanically, Schrödinger's equation

$$\left[-\frac{\hbar^2}{2m} \frac{d^2}{dx^2} + V(x) \right] \psi(x) = E\psi(x) \tag{1.117}$$

is solved in the three regions: $x < 0$, with wave number k_1; $0 \leq x \leq w$, with wave number k_2; $x > 0$, with wave number $k_3 = k_1$ (assuming either side of the barrier is the same medium). It is not difficult to show that the solutions for a particle incident at the $x = 0$ boundary are given by

$$\psi(x) = \begin{cases} \psi_1(x) = A_1 e^{ik_1 x} + B_1 e^{-ik_1 x} & \text{if } x < 0, \\ \psi_2(x) = A_2 e^{ik_2 x} + B_2 e^{-ik_2 x} & \text{if } 0 \leq x \leq w, \\ \psi_3(x) = A_3 e^{ik_1 x} & \text{if } x > 0. \end{cases} \tag{1.118}$$

There is no backward traveling wave for ψ_3 because the particle, if it passes the barrier, will meet nothing else to reflect from. However, reflections at $x = 0$ and $x = w$ are possible, which gives rise to the B_1 and B_2 terms.

The wave numbers are given by

$$k_1^2 = \frac{2mE}{\hbar^2} \quad \text{and} \quad k_2^2 = \frac{2m(E - V_0)}{\hbar^2}. \tag{1.119}$$

From this, it is clear that k_1 and k_2 are real for $E > V_0$ and imaginary for $E < V_0$.

Applying the boundary conditions for continuity of the wavefunction and its derivative (as discussed in Section 1.2.4) gives, after some algebra,

$$\frac{B_1}{A_1} = \frac{(k_1^2 - k_2^2)(1 - e^{2iwk_2})}{(k_1 + k_2)^2 - (k_1 - k_2)^2 e^{2iwk_2}}, \tag{1.120}$$

$$\frac{A_3}{A_1} = \frac{4k_1 k_2 e^{i(k_2 - k_1)w}}{(k_1 + k_2)^2 - (k_1 - k_2)^2 e^{2iwk_2}}. \tag{1.121}$$

From the probabilistic interpretation of the wavefunction (see Equation 1.39), it then follows that a tunneling probability t and reflection probability r can be defined as

$$T(E) = \left| \frac{A_3}{A_1} \right|^2 = \left[1 + \frac{(V_0^2)}{4E(E - V_0)} \sin^2(k_2 w) \right]^{-1} \tag{1.122}$$

and

$$R(E) = \left| \frac{B_1}{A_1} \right|^2 = \left[1 + \frac{4E(E - V_0)}{V_0^2 \sin^2(k_2 w)} \right]^{-1}. \tag{1.123}$$

For $E < V_0$ there can be a nonzero probability of tunneling $0 < t < 1$, as illustrated in Figure 1.20, meaning that there is a chance that the particle will tunnel through the barrier, even though $E < V_0$. Figure 1.20 also shows that the probability of tunneling

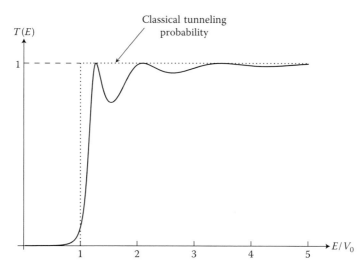

FIGURE 1.20 The probability T that a particle will tunnel through the rectangular potential barrier given in Equation 1.122, plotted against E/V_0 for $w\sqrt{2m^* V_0 / \hbar^2} = 6$. The dotted line shows the familiar classical behavior for comparison.

is increased for smaller barrier widths. This is why tunneling becomes more important at the nanoscale.

Note that the classical values of $t = 0$ for $E < V_0$ and $t = 1$ for $E > V_0$ are recovered in the limits $E \ll V_0$ and $E \gg V_0$, respectively. This is illustrated in Figure 1.20.

1.6.2 TUNNELING AND NANOELECTRONIC DEVICES

In nanoelectronic devices, tunneling is useful for various applications such as connecting quantum dots to leads. This turns out to be a practical way of connecting the dots to their surroundings and allows for some interesting applications and effects to be produced (Kouwenhoven et al., 1997; De Franceschi et al., 2001).

Sometimes, though, tunneling can cause unwanted current flows. One example of this is tunneling across the gate oxide of MOSFETs when the oxide thickness is of the order of a few nanometers (Muller et al., 1999; Hirose et al., 2000).

REFERENCES

N. W. Ashcroft and N. D. Mermin. *Solid State Physics*. Brooks Cole, Belmont, CA, 1976.

G. Bastard. *Wave Mechanics Applied to Semiconductor Heterostructures*. Wiley-Interscience, New York, 1991.

U. Bockelmann and G. Bastard. Interband optical transitions in semiconductor quantum wires: Selection rules and absorption spectra. *Europhysics Letters*. 15:215–220, 1991.

D. Bohm. *Quantum Theory*. Prentice Hall, New York, 1951. Reprint: Dover Publications, New York, 1989.

D. Bohm and B. Hiley. *The Undivided Universe*. Routledge, Oxon, 1993.

M. Born and P. Jordan. Zur quantenmechanik (English title: On quantum mechanics). *Zeitschrift für Physik.* 34:858–888, 1925. [English translation in: B. L. van der Waerden, editor, *Sources of Quantum Mechanics.* Dover Publications, Mineola, NY, 1968.]

M. Born, W. Heisenberg, and P. Jordan. Zur quantenmechanik II (English title: On quantum mechanics II). *Zeitschrift für Physik.* 35:557–615, 1926. [English translation in: B. L. van der Waerden, editor, *Sources of Quantum Mechanics.* Dover Publications, Mineola, NY, 1968.]

A. H. Castro Neto, F. Guinea, N. M. R. Peres, K. S. Novoselov, and A. K. Geim. The electronic properties of graphene. *Reviews of Modern Physics.* 81:109–162, 2009.

J.-N. Chazalviel. *Coulomb Screening by Mobile Charges: Applications to Material Science, Chemistry and Biology*, Section I-C-2. Birkhäuser, Boston, MA, 1999.

P. C. W. Davies and J. R. Brown, editors. *The Ghost in the Atom: A Discussion of the Mysteries of Quantum Physics.* Cambridge University Press, Cambridge, 1993.

C. Davisson and L. H. Germer. Diffraction of electrons by a crystal of nickel. *Physical Review.* 30:705–740, 1927.

L. de Broglie. Recherches sur la théorie des quanta (English title: Researches on the quantum theory). *Annales de Physique.* 10:22–128, 1925. [English translation reprinted in: *Annales De la Fondation Louis de Broglie*, 17:92, 1992.]

S. De Franceschi, S. Sasaki, J. M. Elzerman, W. G. van der Wiel, S. Tarucha, and L. P. Kouwenhoven. Electron cotunneling in a semiconductor quantum dot. *Physical Review Letter.* 86:878–881, 2001.

R. de L. Kronig and W. G. Penney. Quantum mechanics of electrons in crystal lattices. *Proceedings of the Royal Society of London. Series A, Containing Papers of a Mathematical and Physical Character.* 130:499–513, 1931.

B. S. DeWitt and R. N. Graham, editors. *The Many-Worlds Interpretation of Quantum Mechanics, Princeton Series in Physics.* Princeton University Press, Upper Saddle River, NJ, 1973.

M. S. Dresselhaus, G. Dresselhaus, and R. Saito. Physics of carbon nanotubes. *Carbon.* 33:883–891, 1995.

D. K. Ferry and J. P. Bird. *Electronic Materials and Devices.* Academic Press, San Diego, 2001.

A. M. Fox. Optoelectronics in quantum well structures. *Contemporary Physics.* 37:111–125, 1996.

A. M. Fox. *Optical Properties of Solids.* Oxford University Press, Oxford, 2001.

G. Goldoni, F. Rossi, and E. Molinari. Excitonic effects in quantum wires. *Physics Status Solidi A.* 164:265–271, 1997.

D. Halliday, R. Resnick, and J. Walker. *Fundamentals of Physics Extended*, 8th edition, Chapter 35. Wiley, New York, 2008.

G. W. Hanson. *Fundamentals of Nanoelectronics.* Pearson Prentice Hall, Upper Saddle River, NJ, 2008.

W. Heisenberg. Über quantentheoretische umdeutung kinematischer und mechanischer beziehungen (English title: Quantum-theoretical re-interpretation of kinematic and mechanical relations). *Zeitschrift für Physik.* 33:879–893, 1925. [English translation in: B. L. van der Waerden, editor, *Sources of Quantum Mechanics.* Dover Publications, Mineola, NY, 1968.]

W. Heisenberg. über den anschaulichen inhalt der quantentheoretischen kinematik und mechanik. *Zeitschrift für Physik.* 43:172–198, 1927. [English translation in: J. A. Wheeler and H. Zurek, The Physical Content of Quantum Kinematics and Mechanics. *Quantum Theory and Measurement.* Princeton University Press, Princeton, NJ, 1983, pp. 62–84.]

M. Hirose, M. Koh, W. Mizubayashi, H. Murakami, K. Shibahara, and S. Miyazaki. Fundamental limit of gate oxide thickness scaling in advanced MOSFETs. *Semiconductor Science Technology.* 15:485–490, 2000.

P. Hofmann. *Solid State Physics: An Introduction.* Wiley-VCH, Weinheim, 2008.

S. A. Holgate. *Understanding Solid State Physics.* Taylor & Francis, London, 2009.

F. Hund. *The History of Quantum Theory*. Harper & Row, New York, 1974.

H. Ibach and H. Lüth. *Solid-State Physics: An Introduction to Principles of Materials Science*, 4th edition. Springer, Heidelberg, 2009.

S. Iijima. Helical microtubules of graphitic carbon. *Nature*. 354:56–58, 1991.

J. Kalckar, editor. *Foundations of Quantum Physics I (1926–1932)*, volume 6 of *Niels Bohr Collected Works*. North Holland, Amsterdam, 1985.

J. Kalckar, editor. *Foundations of Quantum Physics II*, volume 7 of *Niels Bohr Collected Works*. North Holland, Amsterdam, 1996.

I. Karabulut, H. Safak, and M. Tomak. Excitonic effects on the nonlinear optical properties of small quantum dots. *Journal of Physics D: Applied Physics*. 41:155104, 2008.

C. Kittel. *Introduction to Solid State Physics*, 8th edition. Wiley, New York, 2005.

L. P. Kouwenhoven, C. M. Marcus, P. L. McEuen, S. Tarucha, R. M. Westervelt, and N. S. Wingreen. Electron transport in quantum dots. In L. L. Sohn, L. P. Kouwenhoven, and G. Schön, editors, *Mesoscopic Electron Transport*, vol. 345 of *NATO ASI Series E*. Kluwer, Dordrecht, 1997, pp. 105–214.

E. Mariani, M. Sassetti, and B. Kramer. New selection rules for resonant Raman scattering on quantum wires. *Europhysics Letters*. 49:224–230, 2000.

M. Massimi. Exclusion principle (or Pauli exclusion principle). In D. Greeberger, K. Hentschel, and F. Weinert, editors, *Compendium of Quantum Physics: Concepts, Experiments, History and Philosophy*. Springer, Heidelberg, 2009, pp. 220–222.

U. Mizutani. *Introduction to the Electron Theory of Metals*, Chapter 2. Cambridge University Press, Cambridge, 2001.

G. E. Moore. Cramming more components onto integrated circuits. *Electronics*. 38:114–117, 1965.

D. A. Muller, T. Sorsch, S. Moccio, F. H. Baumann, K. Evans-Lutterodt, and G. Timp. The electronic structure at the atomic scale of ultrathin gate oxides. *Nature*. 399:758–761, 1999.

A. I. M. Rae. *Quantum Physics: Illusion or Reality?* 2nd edition. Cambridge University Press, Cambridge, 2004.

M. Riordan and L. Hoddeson. *Crystal Fire: The Invention of the Transistor and the Birth of the Information Age*. W. W. Norton and Company, New York, 1998.

R. Saito. *Physical Properties of Carbon Nanotubes*. Imperial College Press, London, 1998.

R. Saito, M. Fujita, G. Dresselhaus, and M. S. Dresselhaus. Electronic structure of chiral graphene tubules. *Applied Physics Letters*. 60:2204–2206, 1992.

E. Schrödinger. Über das verhältnis der heisenberg-born-jordanschen quantenmechanik zu der meinen (English title: On the relation between the quantum mechanics of Heisenberg, Born, and Jordan, and that of Schrödinger). *Annalen der Physik*. 79, 1926. [English translation in: J. F. Shearer, translater, *Collected Papers on Wave Mechanics*. Chelsea Publishing Company, New York, 1927, pp. 45–61.]

S. M. Sze. *Physics of Semiconductor Devices*. Wiley, New York, 1969.

T. Takagahara and K. Takeda. Theory of quantum confinement effect on excitons in quantum dots of indirect band-gap materials. *Physics Review B*. 46:15578–15581, 1992.

J. Tersoff and R. S. Ruoff. Structural properties of a carbon-nanotube crystal. *Physics Review Letters*. 73:676–679, 1994.

A. Tonomura, J. Endo, T. Matsuda, T. Kawasaki, and H. Ezawa. Demonstration of single-electron buildup of an interference pattern. *American Journal of Physics*. 57:117–120, 1989.

R. Tumulka. Pauli spin matrices. In D. Greeberger, K. Hentschel, and F. Weinert, editors, *Compendium of Quantum Physics: Concepts, Experiments, History and Philosophy*. Springer, Heidelberg, 2009, pp. 470–472.

O. Verzelen, R. Ferreira, and G. Bastard. Excitonic polarons in semiconductor quantum dots. *Physics Review Letters*. 88:146803, 2002.

J. von Neuman. *Mathematical Foundations of Quantum Mechanics*. Princeton University Press, Princeton, NJ, 1955. [Translated by R. T. Beyer, original edition 1932.]

P. R. Wallace. The band theory of graphite. *Physics Review.* 71:622–634, 1947.

F. Weinert. Wrong theory—right experiment: The significance of the Stern–Gerlach experiments. *Studies in History and Philosophy of Modern Physics.* 26B:75–86, 1995.

T. Young. Experimental demonstration of the general law of the interference of light. *Philosophical Transactions of the Royal Society of London.* 94, 1804.

N. Zettili. *Quantum Mechanics: Concepts and Applications*, Chapter 3. Wiley, Chichester, 2009.

J. M. Ziman. *Principles of the Theory of Solids*, 2nd edition. Cambridge University Press, Cambridge, 1972.

2 Carbon-Based Materials Concepts and Basic Physics

Mohammad Taghi Ahmadi, Jeffrey Frank Webb,
Razali Ismail, and Moones Rahmandoust

CONTENTS

2.1   INTRODUCTION

This chapter presents a description of key solid-state concepts required to understand nanotransistors. A methodical introduction to these perceptions can be found in Datta (2009).

2.2   QUANTUM THEORY OF SOLIDS

Analytical solution of Schrödinger equation is available for simple atoms such as hydrogen; however, a numerical solution is required for most of the materials. The finite difference method has been used for obtaining numerical solutions. It is based on two common modifications of both wave function and differential operator. In this method, the wave function and the differential operator will be converted into a column vector and a matrix, respectively. In other words, a partial differential equation form of Schrödinger equation will be converted into a matrix equation based on a discrete lattice assumption (Datta, 2009). As shown in Figure 2.1, in simple form of the material derived from the one-dimensional (1D) view, the position variable x can be separated by a lattice as $x_n = na$, where a is the distance between two atoms.

The wave function $\psi(x, t)$ can be formed by a column vector $\{\psi_1(t)\ \psi_2(t)\ \ldots\}^T$, where "$T$" stands for transpose and indicates its magnitude around each lattice point at time t. Besides, the time variable is suppressed to make it clear, resulting in

$$\{\psi_1\psi_2\ldots\} = \{\psi(x_1)\psi(x_2)\ldots\}$$

which will be exact when distance between two atoms (a) goes closer to zero as Hamiltonian matrix can then be obtained and energy can be calculated accordingly.

FIGURE 2.1   One-dimensional lattice whose unit cell consists of one atom.

2.3 FERMI–DIRAC DISTRIBUTION FUNCTION

Quantum state at energy E can be filled by electrons that may have upward or downward spin characteristics. Moreover, the probability of occupation $f(E)$ can be discussed in the form of Fermi–Dirac distribution function as (Karamdel et al., 2008; Polash and Huq, 2008)

$$f(E) = \frac{1}{e^{\frac{E-E_F}{k_B T}} + 1} \tag{2.1}$$

where
E_F is the Fermi energy
k_B is the Boltzmann constant
T is the temperature

At energy E equal to the Fermi energy ($E = E_F$), the probability of occupation is half. In the Maxwell–Boltzmann (nondegenerate) approximation, as the Fermi energy E_F is below the conduction band edge (E_{CND}), the factor 1 in the denominator is neglected. In other words, the distribution function can be approximated as

$$f(E) \approx \frac{1}{e^{\frac{E-E_F}{k_B T}}} = e^{-\frac{E-E_F}{k_B T}} \tag{2.2}$$

On the other hand, if the Fermi energy lies in regions above the conduction band edge, the degenerate approximation is appreciated and brought up to the Fermi level where the probability of occupation is equal to one, while it is zero when the approximation is above the Fermi level.

2.4 THREE-DIMENSIONAL MATERIALS

Three-dimensional (3D) visions of the materials illustrate the large-scale size in all the directions. In other words, based on the Cartesian coordinates x, y, and z, in directions placed at right angles to each other, the size in each direction is much longer than the de Broglie wavelength. It implies that in all three directions, carrier movement is continuous and capable of being modeled by traveling waves (Figure 2.2):

$$L_{x,y,z} \gg \lambda_D \approx 10 \text{ nm (bulk 3D)}$$

The conventional 3D materials illustrate continues energy spectrum in all three directions as

$$E_k = E_0 \pm \left(\frac{\hbar^2 k_x^2}{2m^*} + \frac{\hbar^2 k_y^2}{2m^*} + \frac{\hbar^2 k_z^2}{2m^*} \right) \tag{2.3}$$

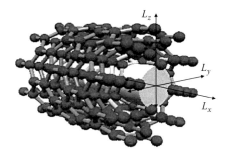

FIGURE 2.2 **(See color insert.)** In a bulk semiconductor all three directions are more than the de Broglie λ_D wavelength.

where

$k_{x,y,z}$ are the continuous wave vectors

$\hbar$ is the Planck constant

m^* carrier effective mass

E_0 indicates the equilibrium energy band

(+) and (−) are used for electrons and holes, respectively

The quantized momentum is assumed to be $k_x = n_x 2\pi/L_x$, $k_y = n_y 2\pi/L_y$, and $k_z = n_z 2\pi/L_z$ in x, y, and z directions, respectively; therefore, the gradient of momentum can be written as $\Delta k = 2\Delta n\pi/L$ resulting in the number of quantum states (Bera et al., 2006; Lansbergen et al., 2007) to be given as

$$\mathrm{d}n_x \mathrm{d}n_y \mathrm{d}n_z = \frac{L_x L_y L_z}{(2\pi)^3} \, \mathrm{d}k_x \mathrm{d}k_y \mathrm{d}k_z \tag{2.4}$$

In the k space, the differential volume is given as $4\pi k^2 \mathrm{d}k$ by employing $k = (2m^*/\hbar^2)^{1/2}(E - E_{co})^{1/2}$. Combining the Fermi–Dirac distribution function $f(E)$ with the density of states (DOS), the carrier concentration $\mathrm{d}n$ between E and $E + \mathrm{d}E$ can be calculated (Appendix A). Note that the Fermi–Dirac integral (FDI) of order i is defined as

$$\Im_i(\eta) = \frac{1}{\Gamma(i+1)} \int\limits_0^\infty \frac{x^i}{e^{x-\eta}+1} \, \mathrm{d}x \tag{2.5}$$

In general, $\Gamma(i+1) = \int_0^\infty e^{-x} x^i \mathrm{d}x = i!$ if i is an integer applicable to Gamma function $\Gamma(1/2) = \sqrt{\pi}$ and $\Gamma(3/2) = (1/2)\Gamma(1/2) = \sqrt{\pi}/2$. It is significant to observe the general properties of the FDI in the nondegenerate and the strongly degenerate limits. The nondegenerate limit occurs when "1" is neglected in the denominator. In this limit, in spite of the value of normalized Fermi energy η, the FDI is always e^η:

$$\Im_j(\eta) \approx e^\eta \quad \text{(nondegenerate)} \tag{2.6}$$

For the reason of simplicity of nondegenerate limit, this is the most used limit in the analysis of semiconductor devices. Nevertheless, this limit is not suitable to the heavily doped semiconductors and nanoscale regimes. In the other extreme, we come up with

$$\frac{1}{e^{x-\eta}+1} \approx \begin{cases} 1 & x < \eta \\ 0 & x > \eta \end{cases} \quad \text{(strongly degenerate)} \tag{2.7}$$

In this limit, the FDI of Equation 2.5 is simplified as

$$\Im_i(\eta) = \frac{1}{\Gamma(i+1)} \int_0^\eta x^i dx = \frac{1}{\Gamma(i+1)} \frac{\eta^{i+1}}{i+1} \tag{2.8}$$

The FDI is accessible in the simple form only for $i = 0$ as

$$\Im_0(\eta) = \ln(e^\eta + 1) \tag{2.9}$$

The appearance for holes can be found similarly. In the valence band, we move downward from the valence band edge E_{vo} toward bottom of the valence band $E_{bottom} = -\infty$.

2.4.1 NONDEGENERATE APPROXIMATION

In the nondegenerate limit, when Equation 2.6 is applicable for approximating the FDI, the carrier concentration for both electrons and holes for the 3D materials can be expressed (Ahmadi, Saad et al., 2008; Fallahpour et al., 2009; Sundaram and Mizel, 2004) as

$$n = N_c e^{\frac{E_F - E_{co}}{k_B T}} = N_c e^{-\frac{E_{co} - E_F}{k_B T}} \tag{2.10}$$

$$p = N_v e^{\frac{E_{vo} - E_F}{k_B T}} = N_{v3} e^{-\frac{E_F - E_{vo}}{k_B T}} \tag{2.11}$$

2.4.2 STRONG DEGENERATE LIMIT

In order to improve the electrical property of a conducting channel, semiconductors need to be heavily doped. As shown in Figure 2.3, degenerate and nondegenerate regions can be defined by $3K_B T$ distance from conduction or valence band edge. Normally, the initiation of strong degeneracy occurs when the Fermi level passes through the conduction band from the forbidden band gap, and so the carrier concentration is given by (Ahmadi et al., 2009a; Petrov and Rotkin, 2004)

$$n_{deg} = N_c \Im_{1/2}(0) \tag{2.12}$$

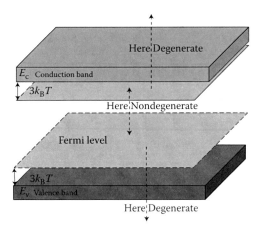

FIGURE 2.3 Schematic illustration of degenerate and nondegenerate regions in a semiconductor.

The strong degeneracy occurs when $n \gg n_{deg}$, which leads to the narrowing of band gaps. In spite of being small, this narrowing does influence the intrinsic carrier concentration through the exponential function. The enhancement of the carrier concentration due to the band gap narrowing $\Delta E_g = E_{g\,deg} - E_{go}$ may control the noise behavior of the device. The other noticeable effect emerges in the carrier statistics. The Fermi level in this regime is weakly dependent on temperature and strongly dependent on its carrier concentration.

The simplified form of occupancy factors known as the Maxwell–Boltzmann type function also accounts for the energy distribution of molecules at a high temperature. The Maxwell–Boltzmann type function leads directly to nondegenerate relationships. In closed form relationships, we find limited usage in device analyses, since the nondegenerate relationships are obviously valid for an intrinsic semiconductor (Pierret, 2003).

2.5   QUASI-TWO-DIMENSIONAL

As one direction is shrunk to be less than the de Broglie wavelength, $L_z < \lambda_D$, and the other two maintain their bulky character, $L_{x,y} \gg \lambda_D$, a quasi-one-dimensional (Q1D) quantum well emerges. In quasi-two-dimensional (Q2D) quantum well, continued type levels come into view only in two dimensions, though the third one goes quantum. In 2D semiconductors, two Cartesian directions with the length less than the de Broglie wavelength can be assumed (Figure 2.4). Therefore, the energy spectrum is

$$E_n = E_{co} + \frac{\hbar^2 \left(k_x^2 + k_y^2 \right)}{2m^*} + n^2 E_{ez} = E_c + \frac{\hbar^2 \left(k_x^2 + k_y^2 \right)}{2m^*} \tag{2.13}$$

with

$$E_c = E_{co} + n^2 E_{ez} \tag{2.14}$$

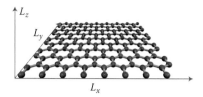

FIGURE 2.4 Schematics of 2D quantum limits.

In the quantum limit ($n = 1$), where the spacing between the levels in the z-direction is very large, it is impractical for large states to be occupied as $E_c = E_{co} + E_{ez}$. In this limit, Q2D layers turn truly 2D. The conduction band edge will be elevated by what is known as zero-point energy of electrons E_{ez}. Similarly, the valence band $E_v = E_{vo} - E_{hz}$ jumps down by zero-point energy of the holes E_{hz}, and thus the effective band gap will be increased by $E_g = E_{go} + E_{ez} + E_{hz}$.

It is possible that mobility in a quantum channel be barrier-limited and controlled by processes different from a micro-channel. This possibility is a reflection of quantum waves from contacts likely to yield an exponential behavior (See appendix B).

2.6 ONE-DIMENSIONAL DEVICES

A Q1D quantum well or nanowire emerges as the y- and z-directions are squeezed with $L_{yz} < \lambda_D$, and the remaining direction maintains its bulky character with $L_x \gg \lambda_D$, as shown in Figure 2.5. In such a Q1D quantum well, the analog type levels appear only in one dimension, while the other two go quantum (Johari et al., 2011):

$$E_{nk} = E_{co} + \frac{\hbar^2 k_x^2}{2m_e^*} + m^2 E_{ey} + n^2 E_{ez} = E_c + \frac{\hbar^2 k_x^2}{2m_e^*} \tag{2.15}$$

with

$$E_c = E_{co} + m^2 E_y + n^2 E_z \tag{2.16}$$

In the extreme case of ($m, n = 1$), in which the spacing between the levels in the y- and z-directions is very large, it is impossible for large states to be populated as $E_c = E_{co} + E_y + E_z$. In this extreme, Q1D layer becomes truly 1D (a nanowire). The conduction band is lifted by what is known as zero-point energy of electrons $E_{ey} + E_{ez}$. Similarly, the valence band $E_v = E_{vo} - E_y - E_z$ drops by zero-point energy of the holes

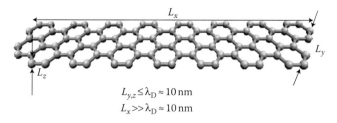

$$L_{y,z} \leq \lambda_D \approx 10\,\text{nm}$$
$$L_x \gg \lambda_D \approx 10\,\text{nm}$$

FIGURE 2.5 Schematics of 1D quantum limits.

$E_{hy} + E_{hz}$. The effective band gap now increases as $E_g = E_{go} + E_{ey} + E_{hy} + E_{ez} + E_{hz}$. Following the same pattern as before (with spin 2 factor), the DOS $D_{e(h)}$ for electrons and for holes are given by

$$D_{1e}(E) \equiv \frac{\Delta n_x}{L_x dE} = \frac{2}{(2\pi)} \left(\frac{2m_e^*}{\hbar^2} \right)^{1/2} \frac{1}{2} (E - E_c)^{-1/2} = \frac{1}{2\pi} \left(\frac{2m_e^*}{\hbar^2} \right)^{1/2} (E - E_c)^{-1/2} \quad (2.17)$$

$$D_{1h}(E) \equiv \frac{\Delta n_x}{L_x dE} = \frac{2}{(2\pi)} \left(\frac{2m_h^*}{\hbar^2} \right)^{1/2} \frac{1}{2} (E_v - E)^{-1/2} = \frac{1}{2\pi} \left(\frac{2m_h^*}{\hbar^2} \right)^{1/2} (E_v - E)^{-1/2} \quad (2.18)$$

In a Q1D device, two directions are less than the de Broglie wavelength as shown in Figure 2.5 (Arora, 2006). Using Equation 2.10 in this case, we found the nondegenerate condition:

$$n = N_c e^{-\left(\frac{E_{c1} - E_F}{K_B T} \right)}, \quad E_c = E_{co} + \varepsilon_{oy} + \varepsilon_{ox} \quad (2.19)$$

A nondegenerate simplified function has been widely employed in determining the transport parameters, which is true for nondegenerately doped semiconductors. On the other hand, most nanoelectronic devices today are degenerately doped. Hence, any design based on the Maxwellian distribution function cannot be accurate. However, for the degenerate ones, we have

$$n_1 = \left[\frac{8 m^*(E_F - E_c)}{\pi^2 \hbar^2} \right]^{\frac{1}{2}} \quad (2.20)$$

In Q1D semiconductor such as nanowire transistor and carbon nanotubes (CNTs), the FDI is proportional to the exponential of η in nondegenerate approximation and also proportional to $2/\sqrt{\pi} e^{1/2}$ in degenerate approximation, which shows that the Fermi order $-1/2$ is closely approximated by exponential of η when $\eta \leq -3$ in nondegenerate regime and approximated by $2/\sqrt{\pi} e^{1/2}$ in degenerate regime for $\eta \geq 6$.

2.6.1 FERMI ENERGY OF NONDEGENERATE AND DEGENERATE REGIMES

To distinguish between the nondegenerately and degenerately doped devices, the approximation of solving the FDI given in Equation 2.8 is essential. In any respect, based on Boltzmann approximation, the solution of the FDI can be estimated by

$$\mathfrak{I}_i(\eta) = e^{\eta} \quad (2.21)$$

This is only applicable for nondegenerate semiconductor devices for all dimensions. For the degenerately doped sample, the approximation can be given in the following general equation, where the order i is used to distinguish between three, two, and one dimensions of the device structure. The FDIs of order one-half $\mathfrak{I}_{1/2}(\eta)$ and of order zero $\mathfrak{I}_0(\eta)$ are employed for the bulk or 3D semiconductor devices and the 2D devices, respectively. In the case of 1D, the FDI of order minus one-half $\Im_{-1/2}(\eta)$ is utilized

$$\Im_i(\eta) = \frac{1}{\Gamma(i+1)} \frac{\eta^{i+1}}{i+1} \tag{2.22}$$

Based on Equation 2.22, the results of FDI approximation in degenerately doped devices for Q3D, Q2D, and Q1D systems are, respectively, given as (Ahmadi et al., 2009b)

$$\Im_{1/2}(\eta) = \frac{4}{3\sqrt{\pi}} \eta^{3/2} \tag{2.23}$$

$$\Im_0(\eta_2) = \eta \tag{2.24}$$

$$\Im_{-1/2}(\eta_1) = \frac{2}{\sqrt{\pi}} \eta^{1/2} \tag{2.25}$$

In the nondegenerate region, in which η_{Fd} is small, the FDI is well approximated by Equation 2.21 for all dimensions.

2.7 CARBON NANOTUBE

Although the electrical structure of CNT was the first to attract the attention of scientists, several other outstanding properties of the structure, such as its unique mechanical properties, made these nanomaterials very remarkable candidates for several interesting applications that are presented and discussed in this chapter. Furthermore, this chapter presents the discovery of CNT and its methods of fabrication as well as depicts the properties of these unique nanomaterials. The imperfections associated with the structure and the methods of simulating its behavior in different conditions are also explained briefly. Carbon is a very unique element. It is very flexible in terms of bonding with various atoms and hence presents a variety of exceptional physical, chemical, and biological properties. Among the types of bonding that carbon exhibits, the sp, sp^3, and sp^2 bonding that takes place in diamond and graphite and the existence of sp^2 bonding called "π-electron bonding" lead to the use of carbon-based materials in a wide range of applications. The materials with extended π-electron clouds are called "π-electron materials," which include graphite, CNTs, fullerenes, and other various carbonaceous materials.

As unique valuable nanostructures with many outstanding electronic and mechanical properties, CNTs have attracted significant attention since their discovery, leading to their use in a wide variety of novel and amazing applications. The extraordinary properties of CNT stem mostly from its perfect hexagonal structure, as well as its high length to diameter aspect ratio, which is specific to most of the nanostructured materials. In other terms, nanomaterials are all uniquely beneficial in many of today's emerging applications, only as a result of their high aspect ratio, which make them interact with their surrounding environment more efficiently, especially when it comes to adsorption properties and interaction in a gaseous environment. CNT was first reported by Iijima (1991) in the NEC Laboratory in Tsukuba, Japan. The structure observed by using a high-resolution transmission electron microscopy (HRTEM) was the multiwalled CNT (MWCNT). The single-walled CNT (SWCNT) was discovered after about 2 years at the same laboratory (Iijima and Ichihashi, 1993). At about the same time, Russian scientists also reported observing CNT and nanotube bundles (Kosakovskaya et al., 1992). Their discovered structures had a much smaller length to diameter ratio compared to what was found in Iijima's laboratory. Bethune also reported the experimental discovery of SWCNT in 1993 (Dresselhaus et al., 1993). These experimental observations and the subsequent theoretical studies over the discovered unique nanostructures led to the finding of many amazing properties and applications for the structure, which will be explained later in this chapter.

2.7.1 HISTORICAL OVERVIEW OF CNT

Nanoscale (<10 nm) carbon fibers in the 1970s and 1980s in the synthesis of vapor grown carbon filaments through the decomposition of hydrocarbons at high temperatures with transition metal with diameters of <10 nm as a catalyst particles were produced (Endo, 1975; Oberlin et al., 1976). However, no academic studies of CNT were reported until 1991. The first observation was reported by Iijima about MWCNT (see Figure 2.6). In 1980 and 1991, he introduced CNT to the world, and by then, many scientists started their work in the field of CNT. First, using the simulation study concerning the remarkable 1D quantum effects, its electronic properties were predicted, and accordingly such remarkable structures and properties of CNT have led to some unique applications. Preliminary experimental observations of MWCNT and SWCNT were the basis for a huge body of theoretical studies and predictions. The most unusual of these theoretical studies was the prediction that CNT could be considered as either semiconductor or metal depending on its chirality and characteristics. Although predicted in 1992, the important parameters are their diameters and the direction of their hexagons with respect to the CNT axis (chiral angle) (Dresselhaus et al., 1992; Hamada et al., 1992). It was not until 1998 that these predicted amazing electronic and mechanical properties were supported experimentally (Hamada et al., 1992; Wildoer et al., 1998).

2.7.2 TYPES OF CNTS

CNTs can be considered as a 1D quantum nanowire. An ideal CNT is fundamentally formed as either a single atomic layer of hexagonal network of carbon atoms (i.e., a graphene layer) rolled into a hollow cylinder provoking the SWCNT with the diameter as small as 0.7 nm, or as a coaxial set of these hollow cylinders producing the MWCNT.

As explained earlier, CNTs are usually found as either SWCNT or MWCNT. The constituent layer of MWCNT is actually 2–50 coaxial SWCNTs separated from each other by an approximate distance of 0.34 nm, which is as much as the interlayer

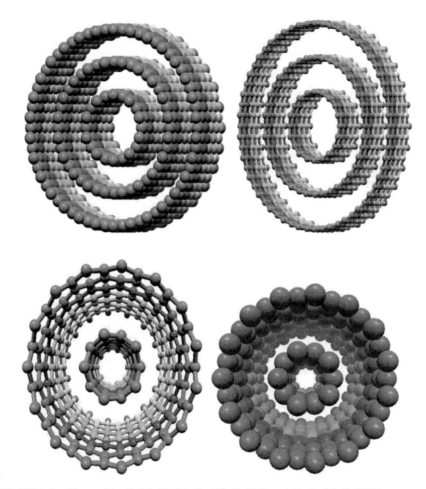

FIGURE 2.6 The multiwall (10, 0), (20, 0), (30, 0) CNT and (5, 0), (10, 0) CNT.

lattice constant of graphite with the diameter generally ranging from 4 to 30 nm (Dresselhaus et al., 1996).

An SWCNT, on the other hand, can be imagined as a graphene sheet having been rolled into a tube or cylinder with a radius of about 0.4–3 nm. The thickness of the tube's wall, in most of the models presented so far in literature, is considered equal to that of a graphene sheet, which is about 0.34 nm. As carbon atoms in this structure are arranged in a hexagonal array (Figure 2.7), each of them has three nearest neighbors.

The atomic structure of nanotubes can be described in terms of the tube chirality, which is defined by the chiral vector $\overrightarrow{C_h}$ and the chiral angle θ, as shown in Figure 2.7.

The chiral vector known as the roll-up vector is defined by the following equation:

$$\overrightarrow{C_h} = n\overrightarrow{a_1} + m\overrightarrow{a_2} \qquad (2.26)$$

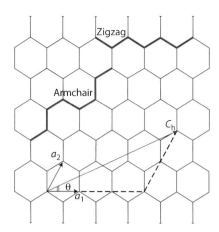

FIGURE 2.7 Schematic diagram of a hexagonal graphene sheet. (From Rahmandoust, M. and Ochsner, A., *J. Nano Res.*, 6, 185, 2009.)

where the integers n and m are the number of steps along the unit vectors $\vec{a_1}$ and $\vec{a_2}$. The chiral angle θ determines the amount of "twist" of the CNT.

Conceptually, the graphene sheet can be rolled up in different ways, including zigzag, armchair, and generally in other chiral forms. This structure is defined in the way that for any specific chiral vector $\vec{C_h}$, the tip of the vector touches its tail visualizing an (m, n) CNT. For a chiral (m, n) tube in general, the chiral angle θ is between the two limits ($0° < \theta < 30°$), or in terms of chiral vector, the integers m and n are nonzero and nonequal, that is, $m \neq n \neq 0$. Zigzag and armchair tubes are the two limiting cases of twisting, where in terms of chiral angle, by ($\theta = 0°$), or in terms of chiral vector, by ($n = 0$), the zigzag structure is formed, and by having $\theta = 30°$, or in other words, by $m = n$, the armchair structures are obtained. These two cases are the symmetric CNTs.

The radius of the CNT can be calculated using the chiral vector integers, m and n, as follows:

$$R_{CNT} = \frac{\text{Length of } \vec{C_h}}{2\pi} = \frac{a_0 \sqrt{m^2 + mn + n^2}}{2\pi} \qquad (2.27)$$

where a_0 is the length of each unit vector. For SWCNTs, the carbon–carbon bond length is $b = 0.142$ nm and therefore $a_0 = \sqrt{3}b$ (Dresselhaus et al., 1996).

The structure of CNT has been discovered by high-resolution transmission electron microscopy (TEM) techniques. These techniques show direct evidence that CNTs are seamless cylinders resulted from the honeycomb lattice from a single atomic layer of a crystalline graphite called a graphene sheet. The structure of an SWCNT is easily explained in terms of its 1D unit cell defined by the vectors $\vec{C}$ (chiral vector) and $\vec{T}$ (translation vector) as shown in Figure 2.8. The circumference of any CNT is articulated in terms of the chiral vector $\vec{C} = n\vec{a_1} + m\vec{a_2}$, which connects the two crystalographically equivalent sites on a 2D graphene sheet (Dresselhaus

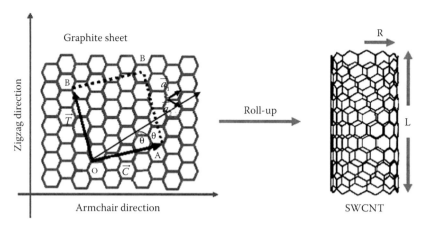

FIGURE 2.8 The chiral vector $\vec{C} = n\vec{a}_1 + m\vec{a}_2$ is defined on the honeycomb lattice of carbon atoms by unit vectors $\vec{a}_1$ and $\vec{a}_2$ and the chiral angle θ with respect to the zigzag axis ($\theta = 0$). The diagram is constructed for $(n, m) = (4, 2)$.

et al., 1992). The structure of Figure 2.8 depends uniquely on the pair of integers (n, m) that specify the chiral vector. The angle between the chiral vector $\vec{C}$ and the zigzag direction ($\theta = 0$) is called the chiral angle θ.

Three different types of CNT structures can be generated by rolling up the graphene sheet into a cylinder as shown in Figure 2.9. The zigzag and armchair CNTs matched with chiral angles of $\theta = 0°$ and $\theta = 30°$, respectively, and chiral CNTs related to $0° < \theta < 30°$. The connection of the vector [T] (which is normal to $\vec{C}$) with the first lattice point concludes the fundamental 1D translation vector [T]. The unit

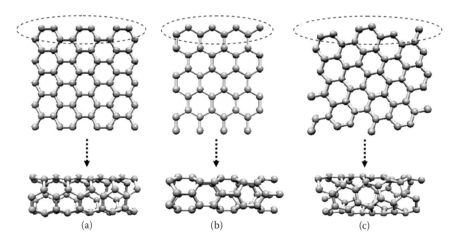

FIGURE 2.9 Schematic models of SWCNTs; the CNT axis is normal to the chiral vector. The latter is along (a) the $\theta = 30°$ direction for an (n, n) armchair CNT, (b) the $\theta = 0$ direction for a $(n, 0)$ zigzag CNT, and (c) a general θ direction with $0 < \theta < 30°$ for an (n, m) chiral CNT.

cell of the 1D CNT lattice is the rectangle that can be shown by the vectors $\vec{C}$ and $\vec{T}$. The cylinder connecting the two hemispherical lids of the CNT (see Figure 2.9) is formed by applying the ends of the vector $\vec{C}$ and the cylinder made along the two lines OB and AB' in Figure 2.8. The lines OB and AB' are both perpendicular to vector $\vec{C}$ at each end of $\vec{C}$ (Dresselhaus et al., 1992; Odom et al., 1998). In the (n, m) notation for $\vec{C} = n\vec{a}_1 + m\vec{a}_2$, the vectors $(n, 0)$ or $(0, m)$ denote zigzag CNTs, whereas the vectors (n, m) matched with chiral CNTs (Saito et al., 1992).

The CNT diameter d_{CNT} is given by

$$d_{CNT} = \frac{|\vec{C}|}{\pi} = \frac{a_{C-C}\sqrt{3(m^2 + nm + n^2)}}{\pi} \qquad (2.28)$$

where

$|\vec{C}|$ is the length of $\vec{C}$

a_{C-C} is the C–C bond length (1.42 Å)

The differences in the CNT diameter d_{CNT} and chiral angle θ give different electrical properties of the CNTs. The number of hexagons in each unit cell of a CNT (N) defunded by integers (n, m) is given by

$$N = \frac{2(m^2 + nm + n^2)}{d_R} \qquad (2.29)$$

where

$d_R = d$, if $(n - m)$ is not a multiple of $3d$

$d_R = 3d$, if $(n - m)$ is a multiple of $3d$, and d is defined as the greatest common divisor (gcd) of (n, m)

Each honeycomb lattice of a hexagon contains two carbon atoms. The unit cell of a graphene layer is N times smaller than that of a CNT. Table 2.1 provides a summary of relations useful for describing the structure of SWCNTs (Reich et al., 2004).

2.7.3 OTHER TYPES OF RELATED CARBON MATERIALS

The carbon element possessing six electrons is considered as the first element of group IV in the periodic table. These electrons will occupy the 1s, 2s, and part of 2p atomic orbitals. They relate strongly to other forms of carbon, particularly to the 2D graphene layers. Carbon-based materials are exceptional in many ways. They can form, for instance, many different possible configurations of bonding with their neighbor atoms.

The 1s orbital contains two strongly bound electrons known as core electrons. The energy difference between the 2s and 2p valence orbital bands that are occupied by the other four electrons is small compared with the binding energy of the chemical bonds. Therefore, the electronic wave functions for these four electrons can be readily mixed with each other leading to a rise in 2s, $2p_x$, $2p_y$, and $2p_z$ orbitals. This phenomenon known as the hybridization enhances the binding energy of the carbon atom with its neighboring atoms (Dresselhaus et al., 1996; Saito et al., 1998).

Mixing of a single 2s electron with one, two, or three 2p electrons is called, in general, sp^n hybridization with $n = 1$, 2, and 3. Thus, there are three possible hybridizations for carbon, namely, sp, sp^2, and sp^3, while the rest of group IV elements such as silicon (Si) and germanium (Ge) exhibit mostly sp^3 hybridization.

Carbon behaves differently from Si and Ge because it does not have inner atomic orbitals, except for the 1s core orbital. The absence of nearby inner orbitals makes the hybridizations process of carbon element very easy. Various bonding states result in certain structural arrangements in the way that sp, sp^2, and sp^3 bonding can give rise to chain structures, planar structures, and tetrahedral structures, respectively (Dresselhaus et al., 2001).

2.7.4 FULLERENES

An SWCNT has many edge atoms on its ends with dangling bonds corresponding to high energy states. Therefore, in order to minimize the energy level, these dangling bonds will be eliminated even at the cost of an increase in the system's strain energy. The strain energy of the system increases as the structure constructs a cap on the tube end by allowing some bond-switching from hexagonal to pentagonal defects, and thereby promotes the formation of closed cage clusters such as fullerenes and CNTs (Endo et al., 1992).

The Kroto research team at Rice University speculated that an SWCNT might be a limiting case of a fullerene molecule (Kroto et al., 1985).

The relation between SWCNT and fullerene became more interesting when it was found that the smallest reported diameter for a CNT resulting from the isolated pentagon rule is the same as the diameter of the C_{60} molecule, which is the smallest fullerene. The isolated pentagon rule states that no two pentagons can be adjacent to one another in these nanostructures, thereby achieving the least strain energy in the structure as well (Dresselhaus et al., 2001).

2.8 PROPERTIES AND APPLICATIONS

There are many emerging functional applications for CNTs found either in individual forms or ensemble modes as in composite materials. Some of these new applications are referred to as field emission–based flat-panel display, electrochemical capacitors, high-capacity hydrogen storage media, sensors, conductivity and photoconductivity, photovoltaic cells and photodiodes, optical limiting devices, Schottky diodes, high-resolution printable conductors, electromagnetic absorbers, electromechanical actuators, and different types of transistors. Individually, CNTs have been used as field emission sources, tips for scanning probe microscopy, and nano-tweezers. They have also been applied as field effect transistors, single-electron transistors, and rectifying diodes.

Understanding the CNT's properties and applying them in industry is a very important motivation for scientists worldwide to make them try to comprehend all aspects of these nanostructures more perfectly.

CNTs have exceptional electrical and mechanical properties such as high Young's modulus and tensile strength. Yet, an SWNT can behave as a well-defined metallic, semiconducting, or semi-metallic wire depending on two key structural parameters, including chirality and diameter.

Thermally, CNTs are more conductive than other crystals (Hone et al., 2000). After the discovery of carbon fullerenes (Kroto et al., 1985) followed by the discovery of CNTs (Iijima, 1991), it was found out that polyhedral structures are the thermodynamically stable form of carbon located under a constraint, in which the number of atoms is not allowed to grow beyond a certain limit.

By charge injection of such an extraordinary nanomaterial, the change in its dimension is so remarkable that it is not proper to call the effect piezoelectricity; and as a fact, it is not. Chemically, CNTs are inert everywhere along their length, but not in their ends or at the site of a bend, kink, or defect. Finally, they respond when exposed to light by changing both their conductivity and shape (Zhang et al., 1999).

2.8.1 ATOMIC STRUCTURE OF CNTS

Hybridization of carbon atomic orbital in the form of sp, sp^2, and sp^3 produces different structural forms with different physical and chemical properties. As mentioned earlier, the covalent bonding in a graphene sheet or in CNTs is sp^2, in which each atom is joined to three other neighbors in a planar trigonal arrangement forming a hexagonal sheet.

In case of an MWCNT, individual sheets or cylinders interact with each other by noncovalent interactions that can be adequately described by weak van der Waals force using the Lennard-Jones potential (Kalamkarov et al., 2006).

$$V_{LJ} = 4\varepsilon \left[\left(\frac{\sigma}{r} \right)^{12} - \left(\frac{\sigma}{r} \right)^{6} \right] \tag{2.30}$$

where the terms σ (in nm) and ε (in kJ mol^{-1}) are defined as the Lennard-Jones parameters. The Lennard-Jones parameters are material-specific and determine the nature and strength of the interaction. For a carbon–carbon noncovalent interaction, the values of Lennard-Jones parameters are $\sigma = 0.3851$ nm and $\varepsilon = 0.4396$ kJ $\cdot$ mol^{-1}. The diagram of the Lennard-Jones potential interacting between two separate carbon atoms is presented in Figure 2.10.

As shown in Figure 2.10, σ is defined as the distance at which the potential between the interacting particles becomes zero, and ε is the minimum potential acting on them. As a result of the Lennard-Jones potential, the Lennard-Jones force acts on the particles. Lennard-Jones force is strongly repulsive when the particles approach close to each other, and it is mildly attractive when they are far away. The Lennard-Jones force, given in the following equation and presented in Figure 2.11, is generally considered as a weak force. The critical distance at which the force becomes zero is r_0 nd, and $r_0 = \sqrt[6]{2}\sigma$

$$F_{LJ} = \frac{dV_{LJ}}{dr} = \frac{4\varepsilon}{r} \left[-12 \left(\frac{\sigma}{r} \right)^{12} + 6 \left(\frac{\sigma}{r} \right)^{6} \right] \tag{2.31}$$

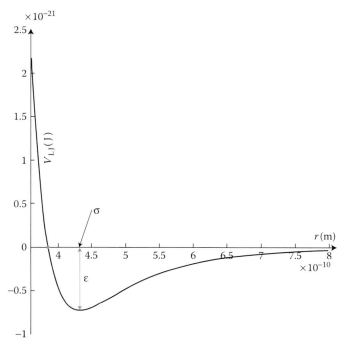

FIGURE 2.10 Carbon–carbon Lennard-Jones interaction. (From Rahmandoust, M. and Ochsner, A., *J. Nanosci. Nanotechnol.*, 2012.)

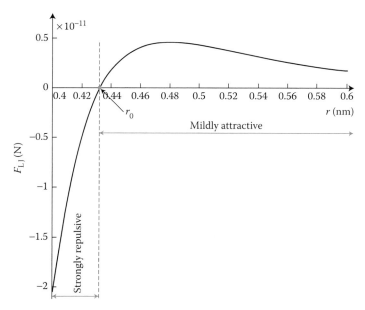

FIGURE 2.11 Lennard-Jones force. (From Rahmandoust, M. and Ochsner, A., *J. Nanosci. Nanotechnol.*, 2012.)

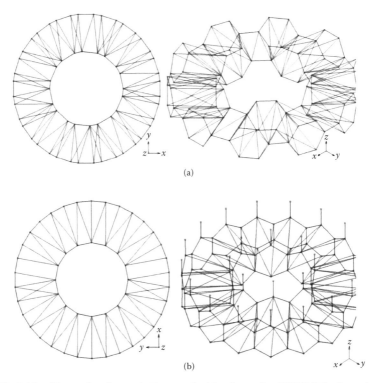

FIGURE 2.12   (See color insert.) Top- and side-view of a DWCNT's first ring for a (a) (10, 10), (5, 5) DWCT and (b) (17, 0), (8, 0) DWCNT.

As described, the weak Lennard-Jones force is a noncovalent force acting between the layers in an MWCNT. If a double-walled CNT (DWCNT) is considered as an example, a schematic model of the top view and side view of the two samples of the DWCNT representing the ring are shown in Figure 2.12 for better understanding. In this figure, the radial red lines represent the weak van der Waals force acting between separate layers of carbons, and the surrounding black lines indicate the strong covalent C–C bonding, contributing in making very strong mechanical structures by connecting them together.

2.8.2 ELECTRICAL PROPERTIES AND DEVICES

SWCNTs, as mentioned earlier, can be either a 1D metal or a semiconductor with a band gap ranging from approximately 20 meV to 2 eV. Although the structure of SWCNT is supposed to be very much similar to that of a graphene sheet, its electrical properties are in contrast with graphite, which is found either acting as a semimetal or behaving as a metal; furthermore, it is capable of sustaining current densities that are hundreds of times greater than those of a bulk metal (Dresselhaus et al., 1996). It can also exhibit quantum effects under certain circumstances.

Direct measurements of electrical conductivity have shown that metallic nanotubes exhibit discrete electron states due to quantum effect confinements and also that coherent electron transport can be maintained through them as nanowires.

2.8.3 CNT Conductivity

As mentioned earlier, depending on its configuration, an (m, n) CNT will become either metallic or semiconductive. Many experimental and theoretical studies have been devoted to this topic to find out the relationship between the CNT's atomic structures and its electronic structure, as well as the electron–electron and electron–phonon interaction effects (Dekker, 1999).

For an (m, n) CNT, CNT is metallic, i.e., the band gap length is zero, if $2m + n = 3N$, and it is a semiconductor if $2m + n \neq 3N$, where N is a positive integer and assuming that there is no bond alternation of C–C bond distances. This condition can be explained by the analytical expression of band structure for any (m, n) CNT,

$$E(k)_N^{\mp} = \alpha \mp \beta \left[1 + 4\cos\left(\frac{2\pi N}{m} - \frac{m+2n}{2m} kl \right) \cos\left(\frac{kl}{2} \right) + 4\cos^2\left(\frac{kl}{2} \right) \right]^{1/2} \quad (2.32)$$

where
  α and β represent the Coulomb and transfer integrals, respectively
  l is the translation length of the unit cell
  $N = 1, 2, \ldots, m$ (Tanaka et al., 1999)

Regardless of the fact that it is necessary to refine this condition by consideration of bond alternation, experiments on the electrical-conductivity measurement and studies on the other related solid-state properties of CNTs confirm that this rather simple but impressive prediction is capable of giving realistic analysis of the structure's electrical behavior.

As a result of the technological difficulties in purification of the CNT samples and direct measurement of the electrical conductivity, using such predictions becomes vital. After the consideration of bond alternation, the aforementioned condition can be simplified as shown in Table 2.1.

In MWCNTs, modification of the electronic structure, especially in the metallic state, shows that due to symmetry reasons, the electronic structure of the inner tube does not seriously affect the metallic property of the outermost tube within

TABLE 2.1
Electrical Conductivity of CNTs

Type	Definition	Electrical Conductivity
Armchair	(m, m)	Metallic
Zigzag	$(3m, 0)$	Narrow-gap semiconductor
		Narrow-gap semiconductor; if $2m + n = 3N$
Chiral	(m, n)	Wide-gap semiconductor; all other tubes

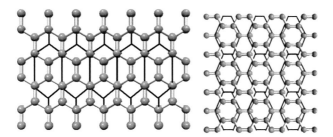

FIGURE 2.13 Stacking patterns in two adjacent layers of zigzag and armchair DWCNTs. (From Tanaka K., et al., *Carbon*, 35, 121, 1997.)

the stacking modes (Satio et al., 1998). The stacked layer patterns considered are illustrated in Figure 2.13. This property is almost similar to the case in the interlayer interaction of two graphene sheets.

2.8.4 CNT STRUCTURE

Carbon materials can be seen in many structural forms such as fullerenes, graphite, carbon fibers, CNTs, and diamond. Because of the different types of orbital hybridization, a carbon atom shows a variety of forms. Dimensionality of carbon-based molecules and solids is dependent on the sp^n hybridization. In the periodic table, only carbon shows zero- to three-dimensional properties. Number of σ bonds in sp^n hybridization forms a skeleton for the n-dimensional carbon structure. As an example in carbine, two σ bonds form a 1D chain structure from sp hybridization. Also in the 2D graphite, sp^2 hybridization makes a 2D structure delineating interestingly a planar structure in the fullerene family (zero-dimensional) and 1D CNTs. The amorphous graphite with sp^2 hybridization is made of the randomly piled graphite layers. Weak interaction between two layers in graphite allows them to slide easily across one another. Four σ bonds of diamond structure make them 3D components.

CNTs as the interesting nanostructures are a good example of 1D quantum wire. The building block of CNTs is one atom in thickness with tens of them in boundary (with typical diameter ~1.4 nm) and many atoms in cylindrical direction. CNTs have attracted the researchers' interest because of their excellent electronic properties, and this attention continues as other remarkable properties are discovered and practical applications are developed. In this chapter, a historical review of CNT and SNW research is presented, and basic definitions related to their structural properties are presented. Finally, the applications of CNT and SNW in electronics are discussed. (This is a repetition of what was written in the earlier sections of this chapter!)

2.8.5 ELECTRONIC STRUCTURE

Applying the zone folding (Hamada et al., 1992), the tight-binding method (Popov, 2004), and the density functional theory (Reich et al., 2002), the energy distribution relations of SWCNTs can be explained. The energy distribution relations of CNTs can be obtained by folding those of graphene.

2.8.5.1 Electronic Band Structure of Graphene

Applying the tight-binding technique, the eigenvalue problem for a Hamiltonian H linked by two carbon atoms in the graphene unit cell can be calculated. In the Slater–Koster system one gets

$$H = \begin{bmatrix} 0 & f(k) \\ -f(k) & 0 \end{bmatrix} \tag{2.33}$$

where

$$f(k) = -t(1 + e^{i\vec{k}\cdot\vec{a}_1} + e^{i\vec{k}\cdot\vec{a}_2}) = -t(1 + 2e^{\sqrt{3}k_x a/2}\cos(k_y a/2))$$

the nearest neighbor C–C tight binding overlap energy is $t = 2.7$ eV (Wildöer et al., 1998)

Solution of the scalar equation $\det(H - EI) = 0$ shows

$$E^{\pm}(k) = \pm t\sqrt{1 + 4\cos\left(\frac{\sqrt{3}k_x a}{2}\right)\cos\left(\frac{k_y a}{2}\right) + \cos^2\left(\frac{k_y a}{2}\right)} \tag{2.34}$$

where the E^+ and E^- correspond to the π^* and the π energy bands, respectively.

2.8.5.2 Electronic Band Structure of SWCNTs

The electronic structure of an SWCNT as a rolled-up layer of graphene can be calculated from graphene. Applying periodic boundary conditions in the circumferential direction matched with the chiral vector $\vec{C}$, the wave vector associated with the $\vec{C}$ direction shows quantized performance, though the wave vector associated with the translational vector $\vec{T}$ (along the CNT axis) remains continuous for a CNT of infinite length. Terminology for the reciprocal lattice vectors $\vec{K}_2$ along the CNT axis ($\vec{C}\cdot\vec{K}_2 = 0$, $\vec{T}\cdot\vec{K}_2 = 2\pi$) and $\vec{K}_1$ in the circumferential direction ($\vec{C}\cdot\vec{K}_1 = 0$, $\vec{T}\cdot\vec{K}_1 = 2\pi$) is given by

$$\vec{K}_1 = \frac{1}{N}\left(-t_2\vec{b}_1 + t_1\vec{b}_2\right), \quad \vec{K}_2 = \frac{1}{N}\left(m\vec{b}_1 - n\vec{b}_2\right) \tag{2.35}$$

The 1D energy distribution relation of an SWCNT is given by

$$E_{CNT}(k) = E\left(k\frac{\vec{K}_2}{|\vec{K}_1|} + v\vec{K}_1\right) \tag{2.36}$$

where $(-\pi/T) < k < (\pi/T)$ is a 1D wave-vector along the CNT axis and quantized $v = 1, \ldots, N$. N quantized k values in the circumferential direction come from the periodic boundary condition. The 2D energy distribution of graphene gives N pair of energy distribution curves corresponding to the cross sections of graphene given by Equation 2.6. In Figure 2.15, a number of cutting lines closer to one of the K points are shown. The distinction between two nearby lines and the length of the cutting lines is known by $|\vec{K}_1| = 2/d_{CNT}$ and $|\vec{K}_2| = 2\pi/T$, respectively. The situation of cutting line is very important. In other words, if the cutting line gets ahead of a K-point of the 2D Brillouin zone (Figure 2.14a), where the π and π^* energy bands of graphene

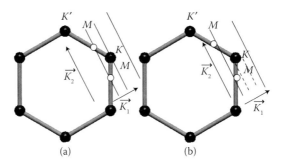

FIGURE 2.14 One-dimensional wave-vectors K is shown in the Brillouin zone of graphene as bold lines for (a) metallic and (b) semiconducting CNTs.

are degenerate by symmetry, then the 1D energy bands show a zero energy band gap. In the other case when the K-point is passed through between two cutting lines, the 1D energy band shows a nonzero energy band gap. Clearly, K is always located in the position at one-third of the distance between two nearby $\vec{K}_1$ lines (Figure 2.14b), and therefore a semiconducting CNT with a limited energy gap can be formed. This conjures up the (n, m) CNT showing metallic performance if $n - m$ is divisible by 3, otherwise CNTs are considered semiconducting.

Generally, (n, n) armchair CNTs give up $4n$ energy subbands with $2n$ conduction and $2n$ valence bands. Two of them are nondegenerate, and $(n - 1)$ is doubly degenerate. The degeneracy arises from this fact that there are two subbands with different v-values but same energy distribution. In the armchair CNTs, band degeneracy can be seen between the highest valence and the lowest conduction bands (Figure 2.15a). On the other hand, in zigzag CNTs, the lowest conduction and the highest valence bands show double degeneracy (Figure 2.15b and c). Because of the symmetry with respect to $k = 0$, the band of an armchair CNT shows a minimum rate at point $k = 2\pi/3a$, a mirror minimum at point $k = -2\pi/3a$, and consequently two equivalent valleys can be seen around the point $\pm 2\pi/3a$. The bands of zigzag and chiral CNTs can show at least one valley (Figure 2.15b and c). In armchair CNTs, the bands pass through the Fermi level at $k = \pm 2\pi/3a$. Thus, all of them are predicted to exhibit metallic conduction (Saito et al., 1998). The energy gap is zero for the $(9, 0)$ CNT at $k = 0$, while the $(10, 0)$ CNT shows nonzero energy gap. Electrical conduction of CNT depends on states around the Fermi energy. Therefore, it is constructive to expand an approximate relation that illustrates the distribution relations around the Fermi energy E_F $= 0$. This is possible by replacing the expression for $f(k) = -t(1 + 2e^{\sqrt{3}k_x a/2} \cos k_x a/2)$ with a Taylor expansion around the point with zero band gap $(0, \pm 4\pi/3a)$ and $f(k) = 0$. It is simple to demonstrate that $f(k) = (i\sqrt{3}at/2)(k_x mi\beta_y)$ with $\beta_y = k_y \mp 4\pi/3a$. The related energy distribution relation is given by

$$E(k) = \pm |f(k)| = \pm \frac{\sqrt{3}at}{2} \sqrt{k_x^2 + \beta_y^2} \qquad (2.37)$$

$(n, 0)$ zigzag CNTs energy bands can be found by applying the periodic boundary condition, which describes the number of allowed wave-vectors k_y in the

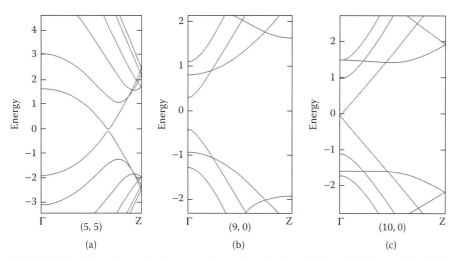

FIGURE 2.15 One-dimensional energy dispersion relations of (a) the (5, 5) armchair CNT, (b) the (9, 0) zigzag CNT, and (c) the (10, 0) zigzag CNT. k_{max} for armchair and zigzag CNTs correspond to $k_{max} = \pi/a$ and $k_{max} = \pi/\sqrt{3}a$, respectively.

circumferential direction as $nk_y a = 2\pi v$ ($v = 1, ..., 2n$). By this method, 1D distribution relation of states for the $(n, 0)$ zigzag CNT can be obtained as

$$E_{CNT}(k) = \pm \frac{\sqrt{3}at}{2} \sqrt{k_x^2 + \left[\frac{4\pi}{3a}\left(\frac{3v}{2n} - 1\right)\right]^2} - \pi/\sqrt{3}a \langle k_x \langle \pi/\sqrt{3}a \qquad (2.38)$$

Consequently, the energy gap is related to the subband v as the variation between the energies of the "+" and "−" parts at $k_x = 0$ can be achieved. The minimum value of zero for band gap is related to $v = 2n/3$. When n is not a multiple of three, the minimum value of $v - 2n/3$ has to be 1/3. In other words, the minimum energy gap is given by

$$E_g = \frac{\sqrt{3}at}{3}\frac{2\pi}{na} = \frac{2a_{C-C}t}{d_{CNT}} \approx \frac{0.8}{d_{CNT}} \text{ eV nm} \qquad (2.39)$$

where $d_{CNT} = {}^{na}\!/_\pi$ is the diameter of the CNT in nanometers. The DOS for semiconducting zigzag CNTs is given by

$$DOS = \sum_v \frac{8}{3a_{C-C}t\pi}\frac{E}{\sqrt{E - E_g/2}} \qquad (2.40)$$

This approximation is valid when $(E - E_F) \ll t$ (Datta, 2005). At minimum and maximum energy, Van Hove singularities in the DOS become visible, which is essential for determining the properties of CNTs (Kasuya et al., 1997; Kazaoui et al., 1999). The (9, 0) zigzag CNT shows metallic behavior, and the (10, 0) zigzag CNT has semiconducting behavior (Saito et al., 1998). Since the dispersion relations around the Fermi energy are nearly linear, the DOS per unit length along the CNT axis can be a constant, which is equal to $8/3\pi a_{C-C}t$ (Saito et al., 1998).

2.9 CNT DOPING

Principally, nanotubes and pure carbon nanostructures depending on the chirality and diameter include interesting mechanical and electronic properties (Dresselhaus et al., 1996; Sloan et al., 2002). However, in sp^2 hybridization of carbon nanosystems, it is achievable to modify the electronic, vibration, chemical, and mechanical properties by applying impurities, such as non-carbon atoms or molecules with less concentrations (from parts per million to small weight percentages). This fact has been defined as doping. The three main doping categories are (see Figure 2.16)

1. Endohedral doping or encapsulation
2. Exohedral doping or intercalation
3. In-plane or substitutional doping

2.9.1 INTERCALATION OR EXOHEDRAL DOPING

Similar to graphite and bulk C_{60}, the "bundled" form of SWNT electronic properties can be modified by doping with donors, acceptors, small molecules, and non-carbon atoms existing in the interstitial channels. Chemical doping increases the density of carriers and thereby improves the thermal and electrical conductivity in the SWNT. In Raman scattering studies, it is necessary to explore the amphoteric behavior of the SWNT bundles. A material is "amphoteric" when it can donate (or accept) carriers from added impurity atoms or molecules. The work of Lee et al. (1997) shows that at room temperature, SWNT resistance decreases by several orders of magnitude in potassium-doped SWNT bundles compared to undoped SWNTs. The inserted alkali-metal atoms operate as donors, which make the C–C bonds in the SWNTs weak (Dresselhaus and Dresselhaus, 1981). Optical absorption spectrum of doped

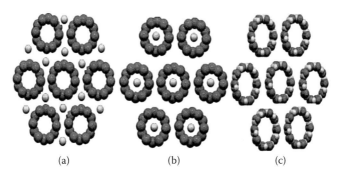

(a)   (b)   (c)

FIGURE 2.16   Schematic models of (a) endohedral doping, (b) exohedral doping or intercalation, and (c) in-plane doping (substitutional) in CNT.

SWNT thin films specifies which carrier states are controlled by charge-transfer interaction between CNT and dopant. DC-resistance measurement in combination with optical absorption, to observe the doping behavior of metallic and semiconducting SWNTs, has been reported by Kazaoui et al. (1999). Both electron acceptors (Br_2, I_2) and donors (K, Cs) with controlled stoichiometry are used as impurities. Band absorption vanishes at 0.68 and 1.2 eV in semiconducting SWNTs and at 1.8 eV in metallic SWNTs.

2.9.2 ENDOHEDRAL DOPING

The understanding of capillarity phenomena in CNTs opened up an innovative research area, regarding the issue that molecules or atoms can be encapsulated inside nanotubes. Pederson and Broughton's (1992) theoretical work was the groundbreaking research for claiming the nanocapillarity of CNTs. This field started with the theoretical and experimental work of Ajayan and Iijima (1993), which involved filling of MWNT cores with Pb or Pb oxide by heating the metallic Pb in air together with tubes. The filling of CNTs has been reviewed in some papers (Bethune et al., 1993; Corio et al., 2004; Iijima and Ichihashi, 1993; Kiang et al., 1999; Luzzi and Smith 2000; Meyer et al., 2000; Monthioux, 2002; Sloan et al., 1999; Terrones et al., 1999; Wang et al., 2006). Another very crucial improvement in CNT science was the discovery of C_{60} and SWNTs. DWCNTs can be obtained from the C_{60} after heat treatment. Furthermore, by using an electronic microscope, this system permits the observation of nanoscale phenomena in real time, such as diffusion (Hirahara et al., 2000). 1D crystal has been produced (Khlobystov et al., 2004) by using the encapsulation of metallofullerenes. DWNTs as effective holders of C_{60} molecules have been reported (Khlobystov et al., 2004). Detection of water inside nanotubes by neutron-diffraction, neutron-scattering experiments, and molecular dynamics simulations have been reported (Li et al., 2005). Encapsulation metallocenes as an organic molecules inside SWNTs, have been investigated (Li et al., 2005). There are some problems in assembling very homogeneous samples of packed nanotubes. On the other hand, in this case TEM probing is difficult to perform (Kataura et al., 2001).

2.9.3 SUBSTITUTIONAL DOPING

Boron and nitrogen substitutional doping within graphene nanocylinders is notable. This method will introduce localized electronic properties in valence or conduction bands. According to the location and concentration of dopants, it will enhance the number of electronic states at the Fermi level (E_F). Substituting boron for carbon within an SWNT, regarding the fact that B has one electron less than C, under the Fermi level (valence band), the sharply localized states (three-coordinated B) appear. Considering the existence of holes in the structure, sharp localized states appear in the valence band, thus leading to the tube being assumed as a p-type nanoconductor. A p-type nanoconductor can act in response with donor-type molecules. There are two kinds of C–N bonds in N-doped SWNTs. The first is a three-coordinated N atom inside the sp^2 hybridized network (Terrones et al., 2008).

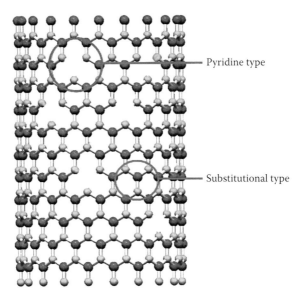

FIGURE 2.17 Schematic model of an N-doped CNT, presenting two types of nitrogen: (1) pyridine type and (2) substitutional N atoms, which are bonded to three carbon atoms.

The next type of substitutional N doping SWNT is the pyridine type (2D-N) that can be a part of the SWNT lattice (Figure 2.17). This type of doping makes localized states below and above the Fermi level. Therefore, substitutional N doping in SWNTs leads to n-type conducting behavior. On the other hand, pyridine-type N creates a p-type or n-type conductor related to the number of N atoms, the level of doping, and the number of removed C atoms within the hexagonal sheet. Nevertheless, when the N concentration is <0.5%, the mechanical properties are roughly constant. It is also important to make it clear that other atoms such as Si and P can be used for SWNT doping (Baierle et al., 2001). Based on the theoretical work (Baierle et al., 2001; Fagan et al., 2003), substituting silicon persuades deformation of the (outward) cylindrical surface of the tubes and leads to a more reactive surface compared to the undoped tubes (Terrones et al., 2008).

APPENDIX A: Q3D MODEL

A.1   DOS FOR 3D

$$g_{3se} = \frac{1}{V} \frac{dN}{dk} \frac{dk}{dE} \qquad (2.A.1)$$

Number of electronics stated dN between k and $k + dk$

$$\frac{dN}{dk} = 2 \frac{4\pi k^2}{(2\pi/L)^3} = \frac{Vk^2}{\pi^2}, \quad \text{where } L^3 = V \qquad (2.A.2)$$

Energy momentum dispersion

$$E = E_{co} + \frac{\hbar^2 \left(k_x^2 + k_y^2 + k_z^2 \right)^2}{2m^*} \Rightarrow k = \sqrt{\frac{2m^*(E - E_{co})}{\hbar^2}} \qquad (2.A.3)$$

$$\frac{dk}{dE} = \frac{1}{2} \left[\frac{2m^*(E - E_{co})}{\hbar^2} \right]^{-1/2} (2m^*) = m^* \left[\frac{2m^*(E - E_{co})}{\hbar^2} \right]^{-1/2} = \frac{m^*}{\hbar^2 k} \qquad (2.A.4)$$

DOS is given as

$$g_{3se} = \frac{1}{L^3} \frac{dN}{dk} \frac{dk}{dE} = \frac{1}{V} \frac{Vk^2}{\pi^2} \frac{m^*}{\hbar^2 k} \qquad (2.A.5)$$

$$= \frac{k}{\pi^2} \frac{m^*}{\hbar^2} = \sqrt{\frac{2m^*(E - E_{co})}{\hbar^2}} \frac{1}{\pi^2} \frac{m^*}{\hbar^2} \qquad (2.A.6)$$

$$= \frac{2}{2} \frac{1}{\pi^2} \frac{m^*}{\hbar^2} (2m^*)^{1/2} \frac{(E - E_{co})^{1/2}}{\hbar} \qquad (2.A.7)$$

$$= \frac{1}{2\pi^2} \frac{2m^*}{\hbar^3} (2m^*)^{1/2} (E - E_{co})^{1/2} \qquad (2.A.8)$$

$$= \frac{1}{2\pi^2} \left(\frac{2m^*}{\hbar^2} \right)^{3/2} (E - E_{co})^{1/2} \qquad (2.A.9)$$

A.2   ELECTRON CONCENTRATION FOR 3D

$$n_2 = \int_{E_C}^{top \approx \infty} g_{2se}(E) f(E) dE \qquad (2.A.10)$$

$$= g_{2se} \int_{E_C}^{top \approx \infty} \frac{1}{e^{\frac{E - E_F}{k_B T}} + 1} dE \qquad (2.A.11)$$

$$= g_{2se} k_B T \int_{E_C}^{top \approx \infty} (k_B T x)^0 \frac{1}{e^{x - \eta} + 1} dx \qquad (2.A.12)$$

$$= \frac{m^* k_B T}{\pi \hbar^2} \mathfrak{I}_0(\eta_c) = N_{2D} \mathfrak{I}_0(\eta_c) \qquad (2.A.13)$$

A.3   FERMI VELOCITY FOR 3D

$$n_3 = \int_{E_C}^{top \approx \infty} g_{3se}(E) f(E) dE = N_C \mathfrak{I}_{1/2}(\eta_c) \qquad (2.A.14)$$

$$\text{where } \mathfrak{I}_{1/2}(\eta_c) \approx \frac{2}{\sqrt{\pi}} \frac{\eta_c^{3/2}}{3/2} \Rightarrow n_3 \approx N_c \frac{4}{3\sqrt{\pi}} \eta_c^{3/2} \tag{2.A.15}$$

$$E_F - E_c \approx \left[\frac{3\sqrt{\pi}}{4} \frac{n_3}{N_c} \right]^{2/3} k_B T = \left[\frac{3\sqrt{\pi}}{4} \frac{n_3}{1} 2 \left(\frac{m^* k_B T}{2\pi \hbar^2} \right)^{-3/2} \right]^{2/3} k_B T \tag{2.A.16}$$

$$= \left[\frac{3\sqrt{\pi}}{2} \frac{n_3}{1} \left(\frac{2\pi \hbar^2}{m^* k_B T} \right)^{3/2} \right]^{2/3} k_B T \tag{2.A.17}$$

$$= \left(3\sqrt{\pi} n_3 \right)^{2/3} \left(\frac{\pi \hbar^2}{m^*} \right) = \frac{\hbar^2 (3\pi^2 n_3)^{2/3}}{2 m_n^*} = \frac{\hbar^2 k_F^2}{2 m_n^*} \tag{2.A.18}$$

$$k_F \approx (3\pi^2 n_3)^{1/3} \qquad v_F \approx \frac{\hbar (3\pi^2 n_3)^{1/3}}{m^*} \tag{2.A.19}$$

A.4   INTRINSIC VELOCITY FOR 3D

$$v_{i3} = \frac{1}{n_3} \int_{E_C}^{top \approx \infty} |v| g_{3se}(E) f(E) dE = \frac{1}{n_3} \int_{E_C}^{\infty} \sqrt{\frac{2(E - E_c)}{m}} g_{3se}(E) f(E) dE \tag{2.A.20}$$

$$= \frac{1}{n_3} \frac{1}{2\pi^2} \left(\frac{2m^*}{\hbar^2} \right)^{3/2} \int_{E_C}^{\infty} \sqrt{\frac{2}{m}} (E - E_{co}) \frac{1}{e^{\frac{E_k - E_F}{k_B T}} + 1} dE \tag{2.A.21}$$

$$= \frac{1}{n_3} \frac{1}{2\pi^2} \left(\frac{2m^*}{\hbar^2} \right)^{3/2} \sqrt{\frac{2}{m}} \int_{E_C}^{\infty} (k_B Tx) \frac{1}{e^{\frac{E_k - E_F}{k_B T}} + 1} k_B T dx \tag{2.A.22}$$

$$= \frac{1}{n_3} k_B T \frac{1}{2\pi^2} \left(\frac{2m^*}{\hbar^2} \right)^{3/2} \int_{E_C}^{\infty} \sqrt{\frac{2}{m}} \frac{(k_B T)}{1} \frac{x}{e^{\frac{E_k - E_F}{k_B T}} + 1} dx \tag{2.A.23}$$

$$= \frac{1}{n_3} (k_B T) \left(\sqrt{k_B T} \right) \frac{1}{2\pi^2} \left(\frac{2m^*}{\hbar^2} \right)^{3/2} \sqrt{\frac{2k_B T}{m}} \int_{E_C}^{\infty} \frac{x}{e^{x - \eta} + 1} k_B T dx \tag{2.A.24}$$

$$= \frac{2}{\sqrt{\pi}} \frac{1}{n_3} \frac{\sqrt{\pi}}{2} \frac{1}{2\pi^2} \left(\frac{2m^* k_B T}{\hbar^2} \right)^{3/2} \sqrt{\frac{2k_B T}{m}} \int_{E_C}^{\infty} \frac{x}{e^{x - \eta} + 1} k_B T dx \tag{2.A.25}$$

$$= \frac{1}{\Gamma(3/2) \mathfrak{I}_{1/2}(\eta_c)} v_{th} \Gamma(2) \mathfrak{I}_1(\eta_c) \tag{2.A.26}$$

$$= \frac{\Gamma(2)\mathfrak{I}_1(\eta_c)}{\Gamma(3/2)\mathfrak{I}_{1/2}(\eta_c)} v_{th} \tag{2.A.27}$$

$$= \frac{2}{\sqrt{\pi}} \frac{\mathfrak{I}_1(\eta_c)}{\mathfrak{I}_{1/2}(\eta_c)} v_{th} \tag{2.A.28}$$

APPENDIX B: Q2D MODEL

B.1 DOS FOR 2D

$$g_{2se} = \frac{1}{A} \frac{dN}{dk} \frac{dk}{dE} \tag{2.B.1}$$

Number of electronics stated dN between k and $k + dk$

$$\frac{dN}{dk} = 2 \frac{2\pi k}{(2\pi/L)^2} = \frac{Ak}{\pi}, \quad \text{where } L^2 = A \tag{2.B.2}$$

Energy momentum dispersion

$$E = E_{co} + \frac{\hbar^2 k_x^2}{2m^*} + \frac{\hbar^2 k_y^2}{2m^*} + n^2 \varepsilon_{oz} \Rightarrow k_{x,y} = \sqrt{\frac{2m^*(E - E_{cn})}{\hbar^2}} \tag{2.B.3}$$

$$\frac{dk}{dE} = \frac{1}{2}\left[\frac{2m^*(E - E_{cn})}{\hbar^2}\right]^{-1/2} \quad (2m^*) = m^* \cdot \left[\frac{2m^*(E - E_{cn})}{\hbar^2}\right]^{-1/2} = \frac{m^*}{\hbar^2 k} \tag{2.B.4}$$

DOS is given as

$$g_{2se} = \frac{1}{L^2} \frac{dN}{dk} \frac{dk}{dE} = \frac{1}{A} \frac{Ak}{\pi} \frac{m^*}{\hbar^2 k} = \frac{m^*}{\pi \hbar^2}, \quad \text{where } m^* \text{ is } N_{vi}\sqrt{m_1 m_2} \tag{2.B.5}$$

B.2 ELECTRON CONCENTRATION FOR 2D

$$n_2 = \int_{E_C}^{top \approx \infty} g_{2se}(E) f(E) dE \tag{2.B.6}$$

$$= g_{2se} \int_{E_C}^{top \approx \infty} \frac{1}{e^{\frac{E - E_F}{k_B T}} + 1} dE \tag{2.B.7}$$

$$= g_{2se} k_B T \int_{E_C}^{top \approx \infty} (k_B T x)^0 \frac{1}{e^{x - \eta} + 1} dx \tag{2.B.8}$$

$$= \frac{m^* k_B T}{\pi \hbar^2} \mathfrak{I}_0(\eta_c) = N_{2D} \mathfrak{I}_0(\eta_c) \tag{2.B.9}$$

B.3   FERMI VELOCITY FOR 2D

$$n_2 = \int_{E_C}^{top \approx \infty} g_{2se}(E)f(E)dE = N_C \mathfrak{I}_0(\eta_c)$$ (2.B.10)

where $\mathfrak{I}_0(\eta_c) \approx \eta_c \Rightarrow n_2 \approx N_c \eta_c$

$$E_F - E_c \approx \left[\frac{n_2}{N_c}\right] k_B T = \left[\frac{\pi \hbar^2}{m^* k_B T} \frac{n_2}{1}\right] k_B T = \left[n_2 \frac{\pi \hbar^2}{m^*}\right] = \frac{\hbar^2 k_F^2}{2m_n}$$ (2.B.11)

$$k_F \approx (2n_2\pi)^{1/2} \quad v_F \approx \frac{\hbar(2n_2\pi)^{1/2}}{m^*}$$ (2.B.12)

B.4   INTRINSIC VELOCITY FOR 2D

$$v_{i2} = \frac{1}{n_2} \int_{E_C}^{top \approx \infty} |v| g_{2se}(E)f(E)dE = \frac{1}{n_3} \int_{E_C}^{\infty} \sqrt{\frac{2(E-E_c)}{m}} g_{3se}(E)f(E)dE$$ (2.B.13)

$$= \frac{1}{n_3} \frac{m^*}{\pi \hbar^2}(E - E_{co})^0 \int_{E_C}^{\infty} \sqrt{\frac{2}{m}}(E - E_{co})^{1/2} \frac{1}{e^{\frac{E_k - E_F}{k_B T}} + 1} dE$$ (2.B.14)

$$= \frac{1}{n_2} \frac{m^*}{\pi \hbar^2} \sqrt{\frac{2}{m}} \int_{E_C}^{\infty} (k_B T x)^{1/2} \frac{1}{e^{\frac{E_k - E_F}{k_B T}} + 1} k_B T dx$$ (2.B.15)

$$= \frac{1}{n_2} \frac{m^* k_B T}{\pi \hbar^2} \sqrt{\frac{2k_B T}{m}} \int_{E_C}^{\infty} (x)^{1/2} \frac{1}{e^{\frac{E_k - E_F}{k_B T}} + 1} k_B T dx$$ (2.B.16)

$$= \frac{1}{n_2} \frac{m^* k_B T}{\pi \hbar^2} \sqrt{\frac{2k_B T}{m}} \int_{E_C}^{\infty} \frac{x^{1/2}}{e^{\frac{E_k - E_F}{k_B T}} + 1} dx$$ (2.B.17)

$$= \frac{1}{n_2} \frac{m^* k_B T}{\pi \hbar^2} \sqrt{\frac{2k_B T}{m}} \int_{E_C}^{\infty} \frac{x^{1/2}}{e^{x - \eta} + 1} dx$$ (2.B.18)

$$= \frac{\Gamma(3/2)\mathfrak{I}_{1/2}(\eta_c)}{\Gamma(1)\mathfrak{I}_0(\eta_c)} v_{th}$$ (2.B.19)

$$= \frac{\sqrt{\pi}}{2} \frac{\mathfrak{I}_{1/2}(\eta_c)}{\mathfrak{I}_0(\eta_c)} v_{th}$$ (2.B.20)

REFERENCES

Ahmadi M. T., H. H. Lau, R. Ismail, and V. K. Arora. 2009a. Current-voltage characteristics of a SNW transistor. *Microelectronics J.* 40, 547–549.

Ahmadi M. T., M. L. P. Tan, R. Ismail, and V. K. Arora. 2009b. The high-field drift velocity in degenerately-doped silicon nanowires. *Int. J. Nanotechnol.* 6(7/8), 601–617.

Ajayan P. M., and S. Iijima. 1993. Capillarity-induced filling of carbon nanotubes. *Nature*, 361, 333.

Arora V. K. 2006. Failure of Ohm's law: Its implications on the design of nanoelectronic devices and circuits. *Proceedings of the IEEE International Conference on Microelectronics*, May 14–17, Belgrade, Serbia and Montenegro, pp. 17–24.

Baierle R. J., S. B. Fagan, R. Mota, A. J. R. da Silva, and A. Fazzio. 2001. Electronic and structural properties of silicon-doped CNTs. *Phys. Rev.* 64, 085413.

Bera L. K., H. S. Nguyen, et al. 2006. Three dimensionally stacked SiGe nanowire array and gate-all-around p-MOSFETs. *2006 International Electron Devices Meeting*, vols. 1 and 2, pp. 298–301.

Bethune D. S., C. H. Kiang, M. S. D. Vries, G. Gorman, R. Savoy, J. Vazquez, and R. Beyers. 1993. Cobalt-catalysed growth of CNTs with single-atomic layer walls. *Nature*. 363, 605–607.

Corio P., A. P. Santos, M. L. A. Temperini, V. W. Brar, M. A. Pimenta, and M. S. Dresselhaus. 2004. Characterization of single wall carbon nanotubes filled with silver and with chromium compounds. *Chem. Phys. Lett.* 383, 475–480.

Datta S. 2005. *Quantum Transport: From Atoms to Transistors*. Cambridge, New York, Cambridge University Press.

Datta S. 2009. *ECE 659 Quantum Transport: Atoms to Transistors*. Cambridge, New York, Cambridge University Press.

Dekker, C. 1999. Carbon nanotubes as molecular wires. *Phys. Today*. 52, 22–28.

Dresselhaus M. S. and G. Dresselhaus. 1981. Intercalation compounds of graphite. *Adv. Phys.* 30, 139.

Dresselhaus M. S., G. Dresselhaus, and R. Saito. 1992. Carbon fibers based on C60 and their symmetry. *Phys. Rev. B*. 45(11), 6234–6242.

Dresselhaus, et al. 1993. Graphite fibers and filaments. *Springer Series in materials Sciences*, Berlin, Springer-Verlag, Vol. 5.

Dresselhaus M. S., G. Dresselhaus, and P. C. Eklund. 1996. *Science of Fullerenes and CNTs*. New York, Academic Press.

Dresselhaus M. S., G. Dresselhaus, Ph. Avouris. (Eds) 2001. *Carbon Nanotubes*. Springer, Berlin.

Endo M. 1975. Mecanisme de Croissance en Phase Vapeur de Fibres de Carbone [The growth mechanism of vapor-grown carbon fibers]. PhD Dissertation, University of Orleans, Orleans, France.

Endo, et al. 1993. The production and structure of pyrolytic carbon nanotubes (PCNTs). *J. Phys. Chem. Solids*. 54(12), 1841–1848.

Fagan S. B., et al. 2003. Ab initio study of an organic molecule interacting with a silicon-doped carbon nanotube. *Diam. Relat. Mater*. 12, 861–863.

Fallahpour A. H., M. T. Ahmadi, et al. 2009. Analytical study of drift velocity in N-type silicon nanowires. *2009 1st Asia Symposium on Quality Electronic Design*. pp. 252–254.

Hamada N., S. Sawada, and A. Oshiyama. 1992. New one-dimensional conductors: Graphitic microtubules. *Phys. Rev. Lett*. 68(10), 1579–1581.

Hirahara K., K. Suenaga, S. Bandow, H. Kato, T. Okazaki, H. Shinohara, and S. Iijima. 2000. One-dimensional metallofulerene crystal generated inside single-walled CNTs. *Phys. Rev. Lett*. 85, 5384.

Hone, et al. 2000. Electrical and thermal transport of magnetically aligned single wall carbon nanotube films. *Appl. Phys. Lett*. 77, 666–668.

Iijima S. 1991. Helical microtubules of graphitic carbon. *Nature* (London). 354(6348), 56–58.

Iijima S. and T. Ichihashi. 1993. Single-shell CNTs of 1 nm diameter. *Nature*. 363, 603–605.

Johari Z., N. A. Amin, et al. 2011. Modeling of quantum capacitance in graphene nanoribbon. *2010 International Conference on Enabling Science and Nanotechnology*, AIP Conference Proceedings, vol. 1341, pp. 384–387.

Kalamkarov A. L., et al. 2006. Analytical and numerical techniques to predict carbon nanotubes properties. *Int. J. Solids Struct.* 43, 6832–6854.

Karamdel J., M. T. Ahmadi, et al. 2008. Formulation and simulation for electrical properties of a (5,3) single wall carbon nanotube. *ICSE: 2008 IEEE International Conference on Semiconductor Electronics, Proceedings*, Nov 25–27, Persada Johor Intl. Convention Ctr./The Puteri Pac, Johor Bahru, Malaysia, pp. 545–548.

Kasuya A., Y. Sasaki, Y. Saito, K. Tohji, and Y. Nishina. 1997. Evidence for size-dependent discrete dispersions in single-wall nanotubes. *Phys. Rev. Lett.* 78(23), 4434–4437.

Kataura H., Y. Maniwa, T. Kodama, K. Kikuchi, K. Hirahara, S. Iijima, S. Suzuki, W. Krätschmer, and Y. Achiba. 2001. Fullerene-peapods: Synthesis, structure, and Raman spectroscopy. *AIP Proc.* 591(1), 251–255.

Kazaoui S., N. Minami, R. Jacquemin, H. Kataura, and Y. Achiba. 1999. Amphoteric doping of single-wall carbon-nanotube thin films as probed by optical absorption spectroscopy. *Phys. Rev. B.* 60, 13339.

Khlobystov A., D. A. Britz, A. Ardavan, and G. A. D. Briggs. 2004. Observation of ordered phases of fullerenes in CNTs. *Phys. Rev. Lett.* 92, 245507.

Kiang C. H., J. S. Choi, T. T. Tran, and A. D. Bacher. 1999. Molecular nanowires of 1 nm diameter from capillary filling of single-walled CNTs. *J. Phys. Chem. B.* 103, 7449.

Kosakovskaya, et al. 1992. Nanofilament carbon structure. *JETP Lett.* 56, 26–30.

Kroto H. W., J. R. Heath, S. C. O'Brien, R. F. Curl, R. E. Smalley. 1985. C_{60}: Buckminsterfullerene. *Nature.* 318, 162.

Lansbergen G. P., H. Sellier, et al. 2007. One-dimensional sub-threshold channels in nanoscale triple-gate silicon transistors. *Phys. Semiconductors, Pt. A B.* 893, 1397–1398.

Lee R. S., H. J. Kim, J. E. Fischer, A. Thess, and R. E. Smalley. 1997. Conductivity enhancement in single-walled CNT bundles doped with K and Br. *Nature.* 388, 255–257.

Li L. J., A. N. Khlobystov, J. G. Wiltshire, G. A. Briggs, and R. J. Nicholas. 2005. Diameter-selective encapsulation of metallocenes in single-walled carbon nanotubes. *Nature Mater.* 4, 481–485.

Luzzi D. E. and B. W. Smith. 2000. Carbon cage structures in single wall CNTs: A new class of materials. *Carbon.* 38, 1751–1756.

Meyer R. R., J. Sloan, R. E. Dunin-Borkowski, A. Kirkland, M. C. Novotny, S. R. Bailey, J. L. Hutchison, and M. L. H. Green. 2000. Discrete atom imaging of one-dimensional crystals formed within single-walled CNTs. *Science.* 289, 1324–1326.

Monthioux M. 2002. Filling single-wall CNTs. *Carbon.* 40, 1809–1823.

Oberlin A., M. Endo, and T. Koyama. 1976. Filamentous growth of carbon through benzene decomposition. *J. Cryst. Growth.* 32(3), 335–349.

Odom T. W., J. L. Huang, P. Kim, and C. M. Lieber. 1998. Atomic structure and electronic properties of single-walled CNTs. *Nature* (London). 391(6), 62–64.

Pederson M. R. and J. Q. Broughton. 1992. Nanocapillarity in fullerene tubules. *Phys. Rev. Lett.* 69, 2689.

Petrov A. G. and S. V. Rotkin. 2004. Transport in nanotubes: Effect of remote impurity scattering. *Phys. Rev. B.* 70(3), 035408.

Pierret R. F. 2003. *Advanced Semiconductor Fundamentals*, Upper Saddle River, N.J.: Prentice Hall.

Polash B. and H. F. Huq. 2008. Analytical model of carbon nanotube field effect transistors for NEMS applications. *2008 IEEE International 51st Midwest Symposium on Circuits and Systems*, vols 1 and 2, pp. 61–64.

Popov V. 2004. Curvature effects on the structural, electronic and optical properties of isolated single-walled CNTs within a symmetry-adapted non-orthogonal tight-binding model. *New J. Phys.* 6(17), 1–17.

Rahmandoust, M. and Ochsner, A. 2009. The influence of structural imperfections and doping on the mechanical properties of single-walled carbon nanotubes. *J. Nano Res.* 6, 185–196.

Rahmandoust, M. and Ochsner, A. 2012 (in press). On finite element modeling of single- and multiwalled carbon nanotubes. *J. Nanosci. Nanotechnol.*

Reich S., C. Thomsen, and P. Ordejón. 2002. Electronic band structure of isolated and bundled CNTs. *Phys. Rev. B.* 65, 155411.

Reich S., C. Thomsen, and J. Maultzsch. 2004. *CNTs: Basic Concepts and Physical Properties.* Weinheim, Cambridge, Wiley.

Saito R., M. Fujita, G. Dresselhaus, and M. S. Dresselhaus. 1992. Electronic structure of chiral graphene tubules. *Appl. Phys. Lett.* 60(18), 2240–2206.

Saito R., G. Dresselhaus, and M. Dresselhaus. 1998. *Physical Properties of CNTs.* London, Imperial College Press.

Sloan J., M. Terrones, S. Nufer, S. Friedrichs, S. R. Bailey, H. G. Woo, M. Ruhle, J. L. Hutchison, and M. L. H. Green. 2002. Metastable one-dimensional AgCl1-xIx solid-solution wurzite "tunnel" crystals formed within singlewalled CNTs. *J. Am. Chem. Soc.* 124, 2116.

Sundaram V. S. and A. Mizel. 2004. Surface effects on nanowire transport: A numerical investigation using the Boltzmann equation. *J. Phys. Conden. Matter.* 16(26), 4697–4709.

Tan M. L. P., V. K. Arora, I. Saad, M. T. Ahmadi, and R. Ismail. 2009. The drain velocity overshoot in an 80 nm metal-oxide-semiconductor field-effect transistor. *J. Appl. Phys.* 105, 074503.

Tanaka K., H. Aoki, H. Ago, T. Yamabe, and K. Okahara. 1997. Interlayer interaction of two graphene sheets as a model of double-layer carbon nanotube. *Carbon.* 35, 121–125.

Tanaka, K., T. Yamabe, and K. Fukui. (1999). *The Science and Technology of Carbon Nanotubes.* Elsevier, Amsterdam, The Netherlands.

Tasis D., N. Tagmatarchis, A. Bianco, and M. Prato. 2006. Chemistry of CNTs. *Chem. Rev.* 106, 1105.

Terrones M., N. Grobert, W. K. Hsu, Y. Q. Zhu, W. B. Hu, H. Terrones, J. P. Hare, H. W. Kroto, and D. R. M. Walton. 1999. Advances in the creation of filled nanotubes and novel nanowires. *Mater. Res. Soc. Bull.* 24, 43.

Terrones M., A. G. Souza Filho, and A. M. Rao. 2008. Doped carbon nanotubes: Synthesis, characterization and applications. *Top. Appl. Phys.* 111/2008, 531–566.

Wang Z. Y., Z. B. Zhao, and J. S. Qiu. 2006. Development of filling CNTs. *Prog. Chem.* 18, 563.

Wildoer J. W. G., L. C. Venema, A. G. Rinzler, R. E. Smalley, and C. Dekker. 1998. Electronic structure of atomically resolved CNTs. *Nature* (London). 391(6662), 59–62.

Zhang Y., et al. 1999. Heterostructures of single-walled carbon nanotubes and carbide nanorods. *Science.* 285(5434), 1719–1722.

3 Carbon Nanotube Field Effect Transistor Model

Mohammad Taghi Ahmadi and Razali Ismail

CONTENTS

3.1 INTRODUCTION

This chapter provides a perspective of the carbon nanotube (CNT)–based field effect transistor (FET) to help us understand nanoscale device physics. Primarily, a review of the key concepts on quantum electronics and semiconductor physics, which are presented in the previous two chapters, is recommended as a starting point for this chapter. Besides, more details about the fundamental concepts can be found in the presented references, for example, Datta (2005). It is assumed that the reader is familiar with the basics of semiconductor physics, for example, Pierret (2003). For the quantum mechanical underpinning, a review of Datta (2005) is recommended. In this chapter, based on the quantum confinement effect, the carrier transport phenomenon in carbon nanotube field effect transistors (CNTFETs) is discussed.

3.2 TYPES OF CNTFET

The structure of a CNT transistor is quite similar to that of a silicon-based transistor, with both being fabricated by using available silicon-based technology (Li et al., 2003). Due to the amazing characteristics of CNTs consisting of graphite cylinders, smaller transistors and on-chip interconnections can be produced with atoms of slight thickness (McEuen et al., 2002; Saito et al., 1992). In this section, we summarize the three possible types of CNTFETs fabricated so far (Heinze et al., 2002):

1. Schottky-barrier (SB) CNTFET (Figure 3.1)
2. Partially gated (PG) CNTFET (Figure 3.2)
3. Doped-S/D CNTFET (Figure 3.3)

3.2.1 SCHOTTKY-BARRIER CNTFET

CNT-based nanoscale transistors are considered as promising alternatives for the future of nanoelectronics (Auth and Plummer, 1998; Guo et al., 2002). Since the small diameter makes transport of carriers one dimensional (1D), very little carrier–phonon interaction is notable (Winstead and Ravaioli, 2000). The structure of Schottky-barrier (SB) CNTFETs is shown in Figure 3.1. In this type, the drain current is controlled by the SB potential at the source and drain ends (Appenzeller et al., 2005; Javey et al., 2005; Lin et al., 2005). SB CNTFET structure is planar in nature

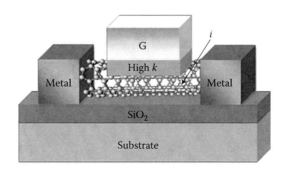

FIGURE 3.1 Schottky-barrier CNT field effect transistors.

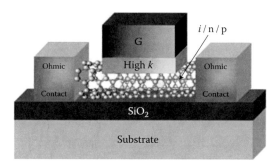

FIGURE 3.2 Partially gated CNT field effect transistor (PG-CNTFET).

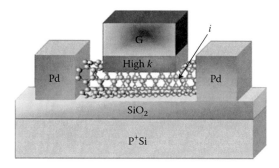

FIGURE 3.3 Doped-sources/drain CNT field effect transistor.

(Lin et al., 2005). However, to modulate the SB, the gate-all-around structures produce better operating conditions (Appenzeller et al., 2005; Tans et al., 1998). Also in the nanoscale regime because of very thin oxide layers, material with very large dielectric constant (high k) has been used extensively.

Nowadays, CNTFETs are generally known as p-type devices, and the holes are considered as current carriers; hence, if there is a negative gate bias, the devices will be ON. The n-type CNTFETs can be achieved by many methods such as direct doping of the tube with an electropositive component or a simple annealing process of p-type CNTFETs. The manufacturing challenge seems to be in the alignment of the SB and gate electrode. In this type, tunneling transmission through an SB between the source-metal and the nanotube channel is controlled by the gate. The SB-CNTFET shows bipolar characteristics that make it difficult to use in the conventional complementary metal–oxide–semiconductor (CMOS)-like logic families (Martel et al., 1998a,b), because based on the gate bias, their functionality can be changed from n-type to p-type. More significantly, the I_{on}/I_{off} ratio is quite small. Currently, the am-bipolar characteristic has been explored to assemble some novel logic architectures, which are different from conventional CMOS logic.

3.2.2 PARTIALLY GATED CNTFET

The PG-CNTFETs are presented in Figure 3.2. They are homogeneously doped or homogeneously intrinsic, with the ohmic contacts at both ends of the CNTs being notable. Depending on the type of doping, the PG-CNTFETs can be n-type or p-type when n-doped or p-doped, respectively.

These devices operate in depletion mode (uniformly n–p doped). The gate locally reduces the carriers in the nanotube and turns *off* the n-type device with an efficiently negative threshold voltage that comes closer to the theoretical limit under room-temperature operation. The "source exhaustion" phenomenon shows a limit on the *on*-current for these devices, $I_{D(on)} = qQ_L v_T$, where Q_L is the carrier density per unit length and v_T is the unidirectional thermal velocity. In case of the intrinsic CNT, CNTFETs exhibit n- or p-type unipolar behavior tunable by electrostatic, which illustrates the best operation in the improvement mode (Appenzeller et al., 2002).

3.2.3 DOPED-SOURCES/DRAIN CNTFET

Doped-S/D CNTFETs given in Figure 3.3 are combinations of heavily or lightly n/p-doped ungated parts. The amount of charge controls the *on* current; the charges can be induced in the channel by the gate, which is independent of doping at the source.

Based on the principle of barrier height accounting for the modulation through applying a gate potential, the doped-S/D CNTFETs work in a pure p-type or n-type enhancement mode or in a depletion mode (Burke, 2004).

3.3 CNTFET DESIGN

There are two different CNTFET design methods (Appenzeller et al., 2002). First, the conventional metal–oxide–semiconductor field-effect transistors (MOSFETs) are like CNTFETs with a p-i-p or n-i-n doping scheme; these are similar to the conventional p- or n-MOSFETs in principle, the so-called CNTFETs. However, when these devices are intensively scaled, the "charge pileup" phenomenon appears in the channel, which significantly brings down the *off* state performance and limits the I_{on}/I_{off} ratio. Second are the tunneling (T) CNTFETs, which are CNTFETs with n-i-p doping scheme. These kinds of CNTFETs operate based on a band-to-band tunneling effect. The location of valence and the conduction band situation controls the tunneling current in the devices. These tunneling devices reduce the charge pileup effect that is one of the drawbacks of the CNTFETs. Also, this device shows very good switching speed and acceptable power expenditure (Svizhenko and Anantram, 2005).

3.4 CNTFET MODELS

In 1998, the first CNTFET was fabricated, after which CNTFET modeling and simulations have been done by many researchers (Appenzeller et al., 2002; Burke, 2004; Svizhenko and Anantram, 2005). MOSFET-like fabrication procedure has been reported by many (Javey et al., 2004; Lin et al., 2004; Wind et al., 2003) and CNTFETs with doped source/drain regions have also been studied (Chen et al., 2004; Javey et al., 2005; Radosavlijevic et al., 2004). The core in the earlier models was based on charge or nonequilibrium Green's function calculations (see Table 3.1). For nanoscale transistor modeling, a velocity approach has been employed on the Maxwell–Boltzmann

TABLE 3.1
Various Models That Used for CNTFET Modeling

References	Authors	Method
Hashempour and Lombardi (2008)	H. Hashempour, F. Lombardi	Charge base (Fermi level, and bias variations, while significantly improving simulation) performance
Ossaimee et al. (2008)	M. I. Ossaimee, S. H. Gamal, K. A. Kirah, and O. A. Omar	Charge base (based on a self-consistent solution of Poisson's equation and the carrier transport equation)
Najari et al. (2008); Sinha et al. (2008)	M. Najari, S. Frégonèse, C. Maneux, T. Zimmer, H. Mnif, and N. Masmoudi; S. Sinha, A. Balijepalli, and Y. Cao	Charge base (tunnel current through a SB based on the Wentzel–Kramers–Brillouin, "WKB" approximation of the transmission coefficient)
Martin et al. (2006); Johnson et al. (2007)	I. Martin-Bragado, S. Tian, M. Johnson, P. Castrillo, and R. Pinacho	TCAD simulations using kinetic Monte Carlo
Xia et al. (2005)	T.-S. Xia, L. F. Register, and S. K. Banerjee	Charge base (discredited energy levels effecton the I–V curve)
Pourfath et al. (2005); Javey et al. (2005)	M. Pourfath, E. Ungersboeck, A. Gehring, B. H. Cheong, W. J. Park, H. Kosina, and S. Selberherr; A. Javey -A. Svizhenko -M. Pourfath	Charge base (nonequilibrium Green's function)
Javey et al. (2004); Raychowdhury et al. (2004)	A. Javey, J. Guo, D. B. Farmer, Q. Wang, D. Wang, R. G. Gordon, M. Lundstorm, and H. Dai; A. Raychowdhury, S. Mukhopadhyay, and K. Roy	Charge base (ballistic model approach)
Wind et al. (2003)	S. J. Wind, J. Appenzeller, and P. Avouris	Charge and velocity base (based on the modified mobility and carrier velocity saturation models including the high field effects along the entire channel)
Appenzeller, et al. (2002); Lundstrom and Rhew (2002)	J. Appenzeller, J. Knoch, V. Derycke, R. Martel, S. Wind, and P. Avouris; M. Lundstrom and J.-H. Rhew	Charge-base Boltzmann statistic (theory of the ballistic MOSFET is introduced by presenting a simple derivation based on Boltzmann statistics)
Tans and Martel (1998)	S. J. Tans and R. Martel	CNTFET first fabrication

approximation (nondegenerate regime). Since it is obvious that nanoscale devices operate in the degenerate regime, the velocity approach can be used for nanoscale transistor modeling in both degenerate and nondegenerate regimes. In a high electric field, velocity vectors are streamlined and therefore the ultimate drift velocity due to the high-electric-field streaming is based on the asymmetrical distribution function that converts randomness in zero field to a streamlined one in a very high electric field. The results obtained can be applied to the modeling of the current–voltage characteristic of a SNW transistor and CNTFET (three-dimensional (3D) and two–dimensional (2D) models are presented in Appendix B to comparison).

3.5   CARRIER TRANSPORT MODEL

In the nanoscale dimension, a variety of striking transport phenomena have been discovered. In nanodevices, and particularly metallic and semiconducting nanotubes known as 1D materials, the quantum effects are dominant (Lundstrom, 1997). These effects include Coulomb blockage and Luttinger-liquid behavior, which are not seen in classical physics. A semiclassical model that adopts quantum correction insertion is still being sought after due to its simplicity in calculation and numerical stability. This top-down approach is questionable in molecular scale devices. Therefore, the modeling for CNT should incorporate quantum effects, when the device size is of the order of the de Broglie electron wavelength. When carrier scattering can be neglected, the most efficient approach to model a current in a nanodevice is to utilize the quantum ballistic transport models by solving the Schrödinger equation with its corresponding boundary condition. The last transport model is a dissipative quantum transport modeling, which has an added advantage in that it combines both coherent carrier motion and carrier scattering. One option is the computationally complex nonequilibrium Green's function (NEGF) that is capable of providing accurate results on 1D CNTFET and SNW analysis and atomistic descriptions of devices (Pierret, 2003). Another alternative is applying the Wigner function given in density matrix representation.

3.6   CARBON NANOTUBE TRANSISTOR MODEL

New materials with innovative properties are needed for technology development. Material modification depends on the atomic composition, chemical bonding, and the dimensions of the material. Interesting properties take place when a material system moves toward the molecular scale. At such small nanometer-scale dimensions, materials receive some amazing properties, resulting in unique physical and chemical characteristics. Nobel Laureate Richard Feynman (1959) predicted nano-regime phenomena when he said, "there is plenty of room at the bottom." Since then, new artificial approaches have produced a spectrum of materials with very small dimensions. One main interesting example of nanostructured materials is the CNTs. Nanotubes are 1D structure nanomaterials with atomically smooth and well-defined surfaces. In recent years, major developments have been made in their synthesis, purification, and assembly, in understanding the fundamental properties, in developing novel electronic device designs, and in technological applications. By reporting the observation of CNTs in the *Journal of Nature* in (1991), Sumio Iijima created a worldwide interest in developing the field of nanotubes. The first-discovered nanotubes were synthesized by the arc-discharge method; since then, a number of other methods such as chemical vapor deposition and laser ablation have been used for enabling higher purity of nanotubes. Due to the small diameters, as small as 0.04 nm, and lengths as long as a few centimeters, CNTs are perhaps the closest analog to an ideal 1D system. Their unique C–C bonding and 1D structure results in interesting properties, including remarkable electron transport properties and band structures. The high electron and hole mobility (10,000 cm²/Vs) of semiconductor nanotubes, their compatibility with high-*k* gate dielectrics for coaxially gated devices, the enhanced electrostatics and reduced short channel effects due to their molecular-scale diameters, and ability to readily form metal ohmic contacts make these miniaturized structures an ideal material for high performance nanoscale transistors. Furthermore, the high conductance of metallic nanotubes makes them highly promising in nanoscale interconnects for future integrated circuits. Another unique property of CNTs is their large surface-area-to-volume ratio with every atom being exposed to the surface. As a result, CNTs are highly sensitive to the environment, and unpassivated nanotube devices are capable of highly sensitive detection of a wide range of analyses. Device simulation of compact current transport model of CNTFET and nanowire FETs, known as an alternative for transistors, is essential in aiding the development of new technology for continuous improvements in the density and performance of electronic systems. The carrier transport properties of long-channel MOSFET models are no longer capable of describing carrier transport accurately, even for sub-100 nm MOSFETs. For future downscaling of active devices, such as FETs, the emergence of new structures and materials is required. In addition, CNTs can make this miniaturization possible. They are composed of source, gate, and drain regions with extensions of only a few nanometers in the directions of confinement (Pourfath et al., 2008).

3.6.1 CNT BAND STRUCTURE

A single-walled CNT (SWCNT) is a sheet of graphite (called graphene) rolled up into a cylinder with a diameter of the order of a nanometer, see Figure 3.4.

Therefore, an understanding of CNT band structure starts from the graphite sheet (graphene) band structure. The band energy throughout the entire Brillouin zone of graphene is (Lundestrom and Guo, 2006)

$$E(\vec{k}) = \pm t \sqrt{1 + 4\cos\left(\frac{k_x 3 a_{C-C}}{2}\right)\cos\left(\frac{k_y \sqrt{3} a_{C-C}}{2}\right) + 4\cos^2\left(\frac{k_y \sqrt{3} a_{C-C}}{2}\right)} \quad (3.1)$$

where

$a_{C-C} = 1.42$ Å is the carbon–carbon (C–C) bond length

$t = 2.7$ (eV) is the nearest neighbor C–C tight-binding overlap energy

$k_{x,y,z}$ is the wave vector component (Jishi et al., 1994; Wildoer et al., 1998)

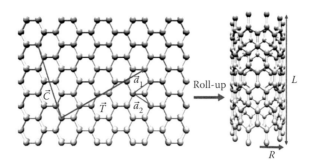

FIGURE 3.4 A prototype SW CNT with length much more than de Broglie wavelength $L \gg \lambda_D$ and diameter ($d = 2R$) less than de Broglie wavelength $R \gg \lambda_D$.

By using a Taylor series expansion for the cosine function near the Fermi point we find

$$E(\vec{k}) = \pm t \frac{3a_{C-C}}{2} \left| \vec{k} - \vec{k}_F \right| = \pm \frac{t\, 3a_{C-C}}{2} \sqrt{(k_x - k_{Fx})^2 + (k_y - k_{Fy})^2} \tag{3.2}$$

Due to the approximation for the graphene band structure near the Fermi point, the $E(k)$ relation of the CNT is

$$E(\vec{k}) = t \frac{3a_{C-C}}{2} \left| \vec{k}_{CN} \right| = \frac{t\,3a_{C-C}}{2} \sqrt{k_{cv}^2 + k_t^2} \tag{3.3}$$

where
  k_{cv} is the wave vector component along the circular direction, which is quantized
      by the periodic boundary condition
  k_t is the wave vector along the length of the nanotube

For the lowest band of CNT, k_{cv} is minimum, and for the metallic CNT, minimum value for k_{cv} is zero. Therefore, metallic CNT shows simple density of state (DOS) as

$$D(E) = D_0 = \frac{8}{3a_{C-C}t\pi} \tag{3.4}$$

For semiconducting CNT, the minimum magnitude of the circumferential wave vector is $k_{cv} = 2/3d$ (recall that d is the diameter of the CNT). By substituting this equation into $E(k)$, approximations for the semiconducting CNT are

$$E(k) = \pm \frac{t3a_{C-C}}{2} \sqrt{\left(\frac{2}{3d}\right)^2 + k_t^2} \tag{3.5}$$

According to Equation 3.5, the conduction and valence bands of a semi-conducting CNT are mirror images of each other. The first band gap is

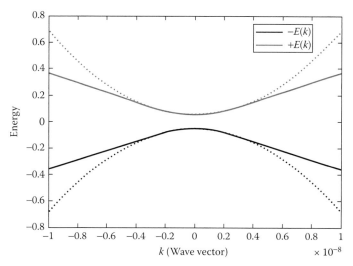

FIGURE 3.5 The band structure of CNT near the minimum energy are parabolic.

$E_G = 2a_{C-C}t/d = 0.8$ (eV)/d (nm). In other words, the energy $E(k)$ and band gap in semiconducting CNT are functions of the CNT diameter. For the CNT near $k = 0$, the band structure is parabolic as shown in Figure 3.5. The E–k graph shows this clearly; therefore, Equation 3.5 can be simplified as follows:

$$E(k) = \frac{E_G}{2}\sqrt{1+\left(\frac{3k_xT}{2}\right)^2} \qquad (3.6)$$

By expanding the square root part $\sqrt{1+a} = 1 + \frac{1}{2}a$, the parabolic energy spectrum appears in our interpretation and band energy graph shows 1D behavior similar to that of silicon nanowire (SNW).

$$E \approx \frac{E_G}{2} + \frac{\hbar^2 k_x^2}{2m^*} \qquad (3.7)$$

where m^* is effective mass for CNT which equals to $m^* = 4\hbar^2/9a_{C-C}td$ or $m^* = 0.08m_0/d$ (nm).

One dimensional CNT using Equation 3.7 for the gradient of k together with the definition of DOS and effect of electron spin (Arora, 2000) leads to the following equation for DOS in quasi 1D (Q1D) CNT as shown in Figure 3.6:

$$\text{DOS} = \frac{\Delta n_x}{\Delta EL_x} = \frac{1}{2\pi}\left(E - \frac{E_G}{2}\right)^{-\frac{1}{2}}\left(\frac{2m^*}{\hbar^2}\right)^{\frac{1}{2}} \qquad (3.8)$$

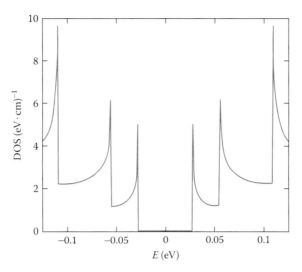

FIGURE 3.6 The density of states near the Fermi point.

When $E_G/2$ is replaced by E_{cd}, Equation 3.8 converts into DOS in the other 1D devices similar to SNW.

3.6.2 CARRIER STATISTICS IN PARABOLIC BAND

Substituting the DOS(E) and Fermi–Dirac distribution function $f(E)$ into the carrier concentration, the definitions for holes (p) and electrons are given as

$$p = N_v F_{\left(\frac{-1}{2}\right)}(\eta_h) \tag{3.9}$$

And for electrons:

$$n = N_c F_{\left(\frac{-1}{2}\right)}(\eta_e) \tag{3.10}$$

where
 $F_{-\frac{1}{2}}$ is the Fermi–Dirac integral of order $\frac{-1}{2}$
 $\eta_{dh} = E_{vd} - E_F/k_B T$ for holes and $\eta_{de} = E_F - E_{cd}/k_B T$ for electrons is the normalized Fermi energy

In a nondegenerate regime with Maxwellian approximation, the carrier concentration will be

$$n_1 = N_{c1} e^{-\left(\frac{E_F - E_{c1}}{k_B T}\right)}, \quad E_{c1} = E_{c0} - \varepsilon_{0y} - \varepsilon_{0x} \tag{3.11}$$

$$p_1 = N_{v1} e^{-\left(\frac{E_{v1} - E_F}{k_B T}\right)}, \quad E_{v1} = E_{v0} - \varepsilon_{0y} - \varepsilon_{0x} \tag{3.12}$$

This simplified distribution function is extensively used in determining the transport parameters (Arora, 2000, 2006). This simplification is true for nondegenerately doped semiconductors, but in a degenerate regime we have

$$n_1 = \left[\frac{8m^*(E_F - E_v)}{\pi^2 \hbar^2} \right]^{\frac{1}{2}} \tag{3.13}$$

$$p_1 = \left[\frac{8m^*(E_c - E_F)}{\pi^2 \hbar^2} \right]^{\frac{1}{2}} \tag{3.14}$$

In other words, for a Q1D CNT semiconductor, the Fermi–Dirac integral is dependent on the η exponentially in nondegenerate approximation, but shows $2/\sqrt{\pi}e^{1/2}$ relation in degenerate approximation. Figure 3.7 shows Fermi order −1/2 in semiconducting CNTs, which indicates that minimum band energy is closely approximated by exponential of η when η ≤ −3 in nondegenerate regime, and for η ≥ 6 approximated by $2/\sqrt{\pi}e^{1/2}$ in degenerate regime.

3.6.3 CARRIER STATISTICS IN NONPARABOLIC BAND STRUCTURE

The number of electrons/cm³ and holes/cm³ with energies between E and $E + dE$ have been recognized to be DOS(E)$f(E)dE$ and DOS(E)[1 − $f(E)$]dE, respectively. Therefore, the total carrier concentration in a band is obtained simply by integrating the Fermi–Dirac distribution function over the energy band, that is

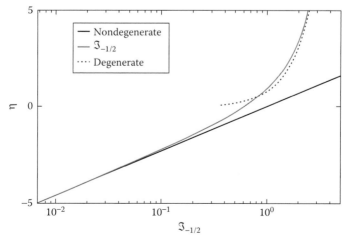

FIGURE 3.7 Comparison of the Fermi–Dirac integral, e^{η} and $2/\sqrt{\pi}e^{1/2}$.

$$n = \int_{E_C}^{E_{top}} \text{DOS}(E) f(E) \, dE \tag{3.15}$$

$$p = \int_{E_{bottom}}^{E_V} \text{DOS}(E) [1 - f(E)] \, dE \tag{3.16}$$

Substituting the DOS(E) and Fermi–Dirac distribution function $f(E)$ into Equation 3.15, the following equation is obtained:

$$n = \int_{E_C}^{E_{top}} D_0 \frac{E}{\sqrt{E^2 - (E_G/2)^2}} \frac{1}{e^{\frac{E_k - E_{F1}}{k_B T}} + 1} \, dE \tag{3.17}$$

In nonparabolic part, the square root approximation is not suitable. Unfortunately, the integral of Equation 3.17 cannot be done analytically; instead, a numerical method is required. Using some simplification for CNT, carrier concentration is given by

$$n_{CN} = N_{CN} \int_{E_C}^{E_{top}} \frac{x + E_G/2k_B T}{\sqrt{x^2 + x E_G/k_B T}} \left(\frac{1}{e^{x-\eta} + 1} \right) dx \tag{3.18}$$

where
$$N_{CN} = D_0 k_B T$$
$$\eta_{FCN} = \frac{E_F - E_c}{k_B T}$$
$$n_{CN} = N_{CN} F_{CN}(\eta_{FCN})$$

The inequalities adjacent to Equation 3.18 are simultaneously satisfied if the Fermi level lies in the band gap more than $3k_B T$ from either band edge (recall k_B is Boltzmann constant). For the cited positioning of the Fermi level, the semiconductor is said to be nondegenerate and Equation 3.18 is referred to as the nondegenerate relationship. Conversely, if the Fermi level is within $3k_B T$ of either band edge or lies inside a band, the semiconductor is said to be degenerate. The simplified form of the occupancy factors is a Maxwell–Boltzmann type function leading directly to the nondegenerate relationships. In Q1D devices, as shown in Figure 3.4, such as CNT and SNW, two directions should be less than the de Broglie wavelength. In nanotubes, the length is much more than the de Broglie wavelength $L \gg \lambda_D$ and the diameter is $d = 2R \ll \lambda_D$ less than the de Broglie wavelength. By using Equation 3.18 for carrier concentration in nondegenerate condition, we find

$$n = N_{CN} \int_{E_C}^{E_{top}} \frac{x + E_G/2k_B T}{\sqrt{x^2 + x E_G/k_B T}} \left(\frac{1}{e^{x-\eta} + 1} \right) dx \tag{3.19}$$

Similar to the 1D device in the nondegenerate regime, 1 can be neglected from the denominator of Equation 3.19, where the temperature effect is notable in Fermi energy

$$n = N_{CN} M e^{+\eta} \tag{3.20}$$

$$M = \int_{E_C}^{E_{top}} \frac{x + E_G/2k_BT}{\sqrt{x^2 + xE_G/k_BT}} e^{-x} dx \tag{3.21}$$

where N_{CN} is the carbon nanotube effective DOS. This simplified distribution function is used extensively in determining the transport parameters in a nondegenerate regime. This simplification is true only for nondegenerately doped CNTs near the minimum energy band. However, most nanoelectronic devices these days are degenerately doped. Hence, any design based on the Maxwellian distribution is not strictly correct and often leads to errors in our interpretation of the experimental results. For a degenerate regime, the exponential part of Equation 3.19 can be ignored as shown here:

$$n = N_{CN} \int_0^{\eta} \frac{x + E_G/2k_BT}{\sqrt{x^2 + xE_G/k_BT}} dx = N_{CN} F_{CN} \tag{3.22}$$

It is very clear that Equation 3.22 can be split into simple integral equations that can be solved by the numerical method:

$$n = D_0 k_B T \left(\int_0^{\eta} \frac{x}{\sqrt{x^2 + xE_G/k_BT}} dx + \int_0^{\eta} \frac{E_G/2k_BT}{\sqrt{x^2 + xE_G/k_BT}} dx \right) \tag{3.23}$$

Numerical solution of these integrals indicates that Fermi energy in a degenerate regime is independent of temperature but strongly depends on carrier concentration. Figure 3.8 shows this effect.

$$n = D_0 \left(\sqrt{(E_F - E_c)^2 + (E_F - E_c)E_G} \right) \tag{3.24}$$

The numerical solution of CNT Fermi–Dirac integrals in a nonparabolic band of Q1D CNTs shows a very interesting similarity with parabolic band structure. Furthermore, this solution indicates temperature dependence of Fermi energy for a nondegenerate regime. However, it shows carrier concentration dependence of Fermi energy in a degenerate regime.

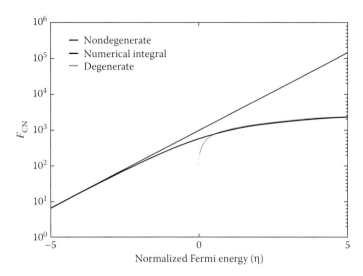

FIGURE 3.8 Comparison of the CNT Fermi–Dirac integral in degenerate and nondegenerate approximation.

3.6.4 Ballistic Drift Velocity in Parabolic Part

The cross-section geometry effect on the wave function especially in the quantum-confined directions can be seen in CNTs. This quantum-confinement influences the low field mobility of the channel, but does not alter the high-field drift velocity that depends on linear concentration (number of electrons per unit length) for any 1D system. In fact, the electrons subjected to very high magnetic field and those in CNT follow the same drift velocity pattern in high electric field. Only the scattering-limited mobility will be affected by the quantum confinement and cross-section of the confinement. Chai et al. (2007) reported on the transport of energetic electrons through single, well-aligned multiwall CNTs. In a parabolic band structure, by using the Fermi–Dirac distribution function where E_{F1} is the Fermi energy that is equivalent to chemical potential, the degenerate nature of the electron concentration is described. The Fermi energy level runs parallel to the conduction band edge. The Fermi level is in the band gap for the nondegenerate carrier concentration and within the conduction (or valence) band for the degenerate carrier (electron or hole) concentration. In the absence of an electric field, the bands are flat. The velocity vectors for the randomly moving stochastic electrons cancel each other, giving net drift equal to zero. In a homogenous CNT, equal numbers of electrons from left and right are entering the free path. The average of this intrinsic velocity $v_{i1,}$ as calculated from the average value of |v| with the Fermi–Dirac distribution function multiplied by the DOS is given by (see Appendix B)

$$v_{i1} = v_{th1} \frac{\Im_0(\eta_1)}{\Im_{-\frac{1}{2}}(\eta_1)} \tag{3.25}$$

for which

$$v_{th1} = \frac{1}{\sqrt{\pi}} v_{th} = \frac{1}{\sqrt{\pi}} \sqrt{\frac{2k_BT}{m^*}} \tag{3.26}$$

The normalized Fermi energy $\eta_1 = (E_{F1} - E_c)/k_BT$ is calculated from the carrier concentration n_1 per unit length in parabolic part of the CNT band structure as follows:

$$n_1 = N_{c1}\Im_{-\frac{1}{2}}(\eta_1) \tag{3.27}$$

with

$$N_{c1} = \left(\frac{2m^*k_BT}{\pi\hbar^2}\right)^{1/2} \tag{3.28}$$

The Fermi integral with Maxwellian approximation is always an exponential; therefore, velocity in this condition limits to thermal velocity.

$$v_{i1} = v_{th1}\frac{\Im_0(\eta_1)}{\Im_{-\frac{1}{2}}(\eta_1)} \tag{3.29}$$

Figure 3.9 indicates the ultimate velocity as a function of temperature. The graph for nondegenerate approximation is also shown (Arora, 2000, 2006). The velocity for low carrier concentration follows $T^{1/2}$ behavior independent of carrier concentration. The nondegenerate limit of intrinsic velocity v_{i1} is v_{th1}, from Equation 3.26

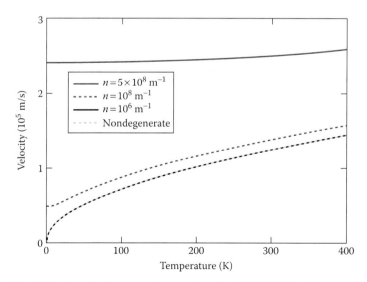

FIGURE 3.9 Intrinsic velocity versus temperature for (5,3) CNT for three concentration values also shown is the nondegenerate limit.

$$v_{i1} = v_{th1} \frac{e^{\eta}}{e^{\eta}} = v_{th1} \qquad (3.30)$$

In the strongly degenerated regime, the Fermi integral transforms into Equation 3.31, and the velocity limit goes to the Fermi velocity range. The degeneracy of the carriers sets in at $n_1 = N_{c1} = 2.0 \times 10^8\,\text{m}^{-1}$ for Q1D-CNT with chirality (5, 3) ($m^* = 0.189\,m_0$). The carriers are nondegenerate if the concentration n_1 is less than this value and degenerate if it is larger than this value. The threshold for the onset of degeneracy will change as chirality changes. For example, the effective mass is $m^* = 0.099\,m_0$ for (9, 2) chirality, and degeneracy sets in at $n_1 = 1.46 \times 10^8\,\text{m}^{-1}$. As concentration is increased to embrace the degenerate domain, the intrinsic velocity tends to be independent of temperature, but depends strongly on carrier concentration.

Figure 3.10 shows the graph of ultimate intrinsic velocity as a function of carrier concentration for three temperatures, $T = 4.2$ (liquid helium), 77 (liquid nitrogen), and 300 K (room temperature). As expected, at low temperature, carriers follow the degenerate statistics and hence their velocity is limited by an appropriate average of the Fermi velocity that is a function of carrier concentration (Arora, 2000, 2006). In the degenerate limit, the intrinsic velocity is only a function of carrier concentration independent of temperature:

$$v_{i1\text{Deg}} = \frac{\hbar}{4m^*}(n_1\pi) \qquad (3.31)$$

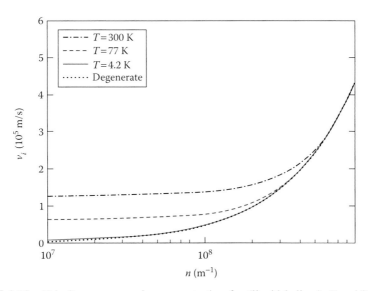

FIGURE 3.10 Velocity versus carrier concentration for (liquid helium), $T = 4.2$ (liquid nitrogen), $T = 77$, and $T = 300$ K (room temperature). The 4.2 K curve is closer to the degenerate limit.

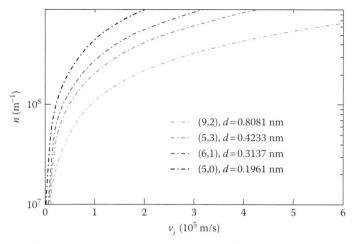

FIGURE 3.11 Velocity versus carrier concentration for $T = 300$ K (room temperature) the approaching to the degeneracy depends on the diameter.

It is clear that the property of CNT depends on the CNT diameter, and the diameter effect in the velocity has been presented in Figure 3.11. This figure shows semiconducting CNT in a degenerate regime with constant temperature, which indicates by increasing the diameter of the CNT, degeneracy occurs at lower values of carrier concentration.

The carrier effective mass in the CNT is proportional to the diameter of the CNT as well $m^*/m_0 = 0.08/d$ (nm). According to this relation, increasing the diameter decreases effective mass, as shown in the following. Whereas, in the nondegenerate, saturation velocity increases by increasing CNT diameter as shown in Figure 3.12.

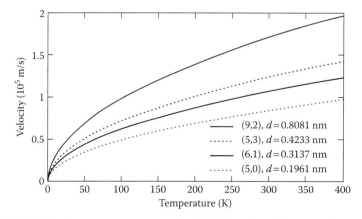

FIGURE 3.12 Velocity versus temperature for CNT, for various diameter values.

3.6.5 Velocity in Nonparabolic Band Structure

In nonparabolic band structure, the band energy is

$$E(k) = \pm \frac{t3a_{C-C}}{2} \sqrt{\left(\frac{2}{3d}\right)^2 + k_t^2} \tag{3.32}$$

For the nonparabolic energy band, $E = \frac{1}{2}mv^2$ is not valid and for the velocity calculation we can employ the wave packet velocity definition as follows (for more details, see Appendix D):

$$v = \frac{1}{\hbar} \frac{dE}{dk} \tag{3.33}$$

Differentiating the energy relationship with respect to the wave vector we can assert

$$|v| = \frac{\left(\frac{E_g}{2}\right)^2 \left(\frac{3d}{2}\right)^2 k}{\hbar E} \tag{3.34}$$

which when solved for k yields

$$k = \frac{2}{E_g} \frac{2}{3d} \sqrt{E^2 - \left(\frac{E_G}{2}\right)^2} \tag{3.35}$$

The average velocity of the carriers

$$v_i = \frac{\int |v| D(E) f(E) dE}{\int D(E) f(E) dE} \tag{3.36}$$

In general, the velocity depends on the temperature and carrier concentration

$$v_i = \frac{\int \left|\frac{\left(\frac{E_g}{2}\right)^2 \left(\frac{3d}{2}\right)^2}{E}\right| \frac{D_0}{\hbar} \frac{\left(\frac{2}{E_g}\right)\left(\frac{2}{3d}\right) \sqrt{E^2 - (E_G/2)^2} E}{\sqrt{E^2 - (E_G/2)^2}} \frac{1}{e^{\frac{E_k - E_{F1}}{k_B T}} + 1} dE}{n_{CN}} \tag{3.37}$$

Using the carrier concentration relationship and simplifying the final equation, we can find the intrinsic velocity

$$v_i = v_{th} \frac{\dot{F}_0(\dot{\eta}_{FCN})}{F_{CN}(\eta_{FCN})} \tag{3.38}$$

where $v_{th\,CNT} = (E_g / 2)(3d / 2\hbar)$ is a CNT thermal velocity that indicates that the ultimate drift velocity depends on temperature.

$$v_i = \left(\frac{E_g}{2}\right)\left(\frac{3d}{2}\right)\frac{D_0}{\hbar} \frac{\displaystyle\int \frac{1}{e^{\frac{E_k - E_{F1}}{k_B T}} + 1} dE}{n_{CN}} \tag{3.39}$$

In Maxwell–Boltzmann (nondegenerate) approximation, the 1 in the denominator is neglected as the Fermi energy E_F is below the conduction band edge (E_{CND}). Since no quantum state exists in the forbidden band gap, the minimum energy of occupation is E_C:

$$v_i = \left(\frac{E_g}{2}\right)\left(\frac{3d}{2}\right)\frac{D_0}{\hbar} \frac{k_B T \displaystyle\int \frac{1}{e^{x-\eta} + 1} dx}{N_{CN} \displaystyle\int_{E_C}^{E_{top}} \frac{x + E_G/2k_B T}{\sqrt{x^2 + xE_G/k_B T}} \frac{1}{e^{x-\eta} + 1} dx} \tag{3.40}$$

$$v_i = \left(\frac{E_g}{2}\right)\left(\frac{3d}{2}\right)\frac{1}{\hbar} \frac{(e^{+\eta})}{(e^{+\eta}) \displaystyle\int_{E_C}^{E_{top}} \frac{x + E_G/2k_B T}{\sqrt{x^2 + xE_G/k_B T}} e^{-x} dx} \tag{3.41}$$

Applying this approximation, we define novel Fermi integrals for CNT as M integrals (see the MATLAB® code for the solution in the appendix c).

$$v_i = v_{th} \frac{e^{+\eta}}{M e^{+\eta}} \tag{3.42}$$

Numerical solution for nondegenerate approximation shows that similar to the 1D device, ultimate drift velocity of carriers in a CNT for a nonparabolic band structure depends on the temperature (see Figure 3.13). Also as expected, similarity in the nondegenerate regime between CNTs for nonparabolic and parabolic band structures is notable.

On the other hand, the carrier behavior in the non-degenerate regime is totally different from that of degenerate regime, similar to other 1D devices. According to our mathematical calculation, the CNT band structure effect on the velocity of carrier in the degenerate regime illustrates different performance compared to the nondegenerate regime.

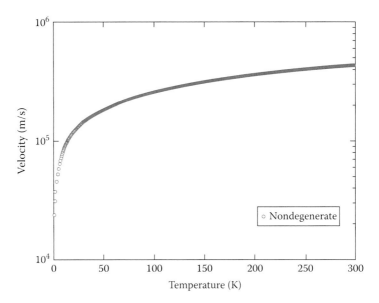

FIGURE 3.13   Intrinsic velocity versus temperature in the nondegenerate limit.

$$v_i = \left(\frac{E_g}{2}\right)\left(\frac{3d}{2}\right)\frac{D_0}{\hbar} \frac{k_B T \int \dfrac{1}{e^{x-\eta}+1}\,dx}{N_{CN}\displaystyle\int_{E_C}^{E_{top}} \dfrac{x+E_G/2k_B T}{\sqrt{x^2+xE_G/k_B T}}\dfrac{1}{e^{x-\eta}+1}\,dx} \tag{3.43}$$

In the degenerate regime, the effect of the exponential part in distribution function is much less than 1; therefore, having compared them, the exponential part can be neglected.

$$v_i = \left(\frac{E_g}{2}\right)\left(\frac{3d}{2}\right)\frac{D_0}{\hbar} \frac{k_B T \int \dfrac{1}{e^{x-\eta}+1}\,dx}{N_{CN}\displaystyle\int_{E_C}^{E_{top}} \dfrac{x+E_G/2k_B T}{\sqrt{x^2+xE_G/k_B T}}\dfrac{1}{e^{x-\eta}+1}\,dx} \tag{3.44}$$

Applying a simplification on the modified form of the ultimate drift velocity of carriers in the degenerate regime indicates that ultimate drift velocity is strongly dependent on carrier concentration and is independent of temperature, as shown in Figure 3.14.

$$v_i = \left(\frac{E_g}{2}\right)\left(\frac{3d}{2}\right)\frac{1}{\hbar}\sqrt{1+\frac{\eta k_B T}{E_G}} \tag{3.45}$$

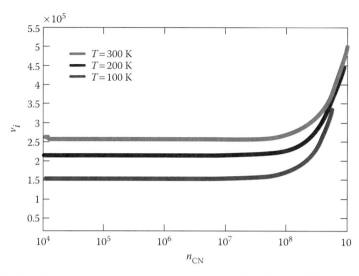

FIGURE 3.14   Velocity versus carrier concentration for $T = 100$, $T = 200$, and $T = 300$ K (room temperature).

3.6.6   High-Field Distribution

Arora (2006) modified the equilibrium distribution function of Fermi–Dirac by replacing the chemical potential with the electrochemical potential. This distribution function can be explained by a tilted band diagram. A channel of CNT-based devices can be assumed as a series of ballistic resistors where the ends of each mean free path can be measured as effective contacts with different Fermi levels with energy difference of $q\varepsilon\ell$. Similar to Büttiker (1986), the inelastic scattering virtual thermalizing is investigated, which can be employed in any regime. Based on this method, carriers are introduced into a "virtual" region where they are thermalized and start their ballistic voyage for the next free path. The carriers that are moving right to left will have their Fermi potential E_{F1} end to the $E_{F1} - q\varepsilon\ell$. Those traveling from left to right will end their voyage with the electrochemical potential $E_{F1} + q\varepsilon\ell$. The gradient of the Fermi energy $E_{F1}(x)$ induces current flow in the presence of an electric field. The electrons drift opposite to the electric field $\bar{\varepsilon}$ applied in the negative x-direction. In a particularly large electric field, all the electrons are moving in the positive x-direction (opposite to the electric field). In other words, random motion will be changed to a streamlined one with ultimate velocity per electron equal to v_i. Therefore, the ultimate drift velocity can be assumed as a ballistic motion, independent of scattering (Arora, 1985a, 2005). The ballistic motion in a mean free path can be broken up by a quantum emission of energy $\hbar\omega_0$, which can be in the form of emitting a phonon, a photon, or spacing with the quantized energy levels. The quantum emission can be initiated by shifting to the higher quantum state; the quantum energy $\hbar\omega_0 = \Delta E_{\perp 1-2}$ as a function of radius on the CNT is seen as

$$\hbar\omega_0 = \frac{3\hbar^2}{2m^*R^2} \tag{3.46}$$

This dependence of the quantum on the radius or diameter of the CNT may give an important clue about the CNT chirality and could be used for characterization of the CNT.

3.6.7 Current–Voltage Characteristics of a CNTFET

Two factors that can be affecting the speed of a device are the transit time and wire delay that is due to finite RC time constants. On the other hand, transit time or gate delay depends on the length of the channel. The two factors are tangled together. In addition, the ultimate saturation velocity has a predominant influence on these factors. The mobility gradient does not have an effect on the saturation velocity except for the case in which higher mobility brings an electron closer to saturation in the presence of a high electric field while having the saturation velocity remain in the same position. The reduction in the conducting channel length of the device results in reduced transit time delay and hence enhanced operational frequency. In any solid state device, it is obvious that the band structure factors, doping profiles, and ambient temperatures play a variety of roles in determining the performance of a device (Arora, 1985, 2005). The conclusion that top mobility leads to higher saturation is not supported by experimental observations (Ahmadi et al., 2009a; Andrieu, 2003). It has been confirmed that low field mobility can be assumed as a function of quantum confinement (Arora, 1985). As devices shrink in all dimensions, the inquisitiveness toward the ballistic nature of the carriers is elevated. Initially, it was in the work of Arora (1983) that the possibility of the ballistic nature of the transport in a very high electric field for a nondegenerate semiconductor was indicated.

3.7 CNTFET

CNTFETs with ballistic motion of carriers in a very high electric field have been predicted, which means that the randomly oriented velocity vectors are streamlined. Therefore, in a CNTFET channel, Ohm's law, which predicts a linear drift velocity response to the applied electric field, $v = \mu_0\varepsilon$, is not suitable. The electric field is now much higher than the channel critical electric field ε_C (Arora, 2006). The velocity saturation effect can be suitably applied in the modeling of a CNT transistor if the empirical relation given in the following is used:

$$v_D = \frac{\mu_0\varepsilon}{1 + \dfrac{\varepsilon}{\varepsilon_C}} \tag{3.47}$$

where
μ_0 is the low field mobility
ε_C is the critical electric field related to the critical applied voltage

Critical velocity V_C by saturation velocity v_{sat} and channel length L is thus given by

$$V_C = \varepsilon_C L = \frac{v_{sat}}{\mu_0} L \tag{3.48}$$

The applied electric field ε is approximated as the gradient of applied voltage V along the channel or as a function of applied drain voltage V_D given by

$$\varepsilon \approx \frac{dV(x)}{dx} \tag{3.49}$$

Based on velocity–field characteristics of Equations 3.47 and 3.48, the current–voltage (I–V) expressions of CNTFET with carrier concentration n_1, length L, and gate capacitance C_G can be obtained as follows for the drain voltage range of $0 \le V \le V_{Dsat}$

$$I_D = \frac{\mu_{lf} C_G}{2L} \frac{\left[2(V_{GS} - V_T)V_D - V_D^2 \right]}{1 + \dfrac{V_D}{V_C}} \tag{3.50}$$

where
C_G is the gate capacitance per unit length
$V_C = v_{sat}L/\mu_{lf}$ is critical voltage, and L is the effective channel length
$V_T = 0.37$ V is the threshold voltage (Fiori and Giuseppe, 2006)

V_{Dsat} is the saturation voltage at the point of current saturation consistent with I_{Dsat} being limited by the drain velocity ($v_D = \alpha v_{Dsat}$). α is always less than 1 as the drain velocity v_D approaches toward saturation v_{Dsat} with the increasing drain field. Saturation current I_{Dsat} for $\alpha < 1$ is evaluated as follows for the drain voltage range of $V_D \ge V_{Dsat}$ (Arora, 2008):

$$I_{Dsat} = \alpha C_G (V_{GT} - V_{Dsat}) v_{sat} \tag{3.51}$$

Reconciling Equations 3.50 and 3.51 at $V_D = V_{Dsat}$ gives (Arora, 2008)

$$V_{Dsat} = \frac{1}{(2\alpha - 1)} \left[(s - \alpha)V_C - (1 - \alpha)V_{GT} \right] \tag{3.52}$$

$$I_{Dsat} = \frac{\alpha}{(2\alpha - 1)} \frac{\mu_{lf} C_G}{2L} V_C \left[\alpha V_{GT} - (s - \alpha)V_C \right] \tag{3.53}$$

with

$$s = \sqrt{ \left(\alpha + (1 - \alpha)\frac{V_{GT}}{V_C} \right)^2 + 2\alpha(2\alpha - 1)\frac{V_{GT}}{V_C} } \tag{3.54}$$

Equations 3.52 and 3.53 simplify considerably when $\alpha = 1$ (the drain velocity is equal to the saturation velocity) (Arora, 2008); these are given by

$$V_{\text{Dsat1}} = V_C \left[\sqrt{1 + \frac{2V_{\text{GT}}}{V_C}} - 1 \right] \tag{3.55}$$

$$I_{\text{Dsat1}} = \frac{1}{2} \frac{C_G \mu_{\text{lf}} W}{L} V_{\text{Dsat1}}^2 \tag{3.56}$$

With the simple geometry of the CNT transistor of Figure 3.15, the gate capacitance is given by C_G (Ilani et al., 2006).

For an SWCNT channel in CNTFETs, the gate-channel capacitance can be expressed as $C_G = (C_e C_Q)/(C_e + C_Q)$, where C_e is the electrostatic gate-coupling capacitance of the gate oxide and C_Q is the quantum capacitance of the gated SWCNT. For a CNT of diameter d placed on a dielectric of thickness h, the electrostatic capacitance comes out to be

$$C \approx \frac{2\pi\varepsilon}{\ln\left[\dfrac{2h}{d}\right]} L \tag{3.57}$$

where ε, h, and d are the dielectric constant, the thickness of silicon dioxide, and the radius of SWCNTs, respectively (Javey et al., 2002; Li et al., 2004; Martel et al., 1998). Furthermore, the quantum capacitance in CNTs is considerably significant. The quantum capacitance can be written as (Burke, 2003) $C_Q = 2e/v_F$, where $v_F \approx 10^6$ m/s is the Fermi velocity of electrons in the CNT (Lin et al., 2005). With $\alpha = 1$, the drain velocity is the ultimate saturation velocity V_{Dsat}. However, when $\alpha < 1$, the drain velocity V_D is smaller than V_{Dsat} due to the presence of the finite electric field at the drain. The origin of V_{Dsat} (as given in Equation 3.55) at the onset of current saturation is normally ascribed to the velocity reaching saturation at the drain end that is only possible if the electric field is infinitely large (Arora and Das, 1990).

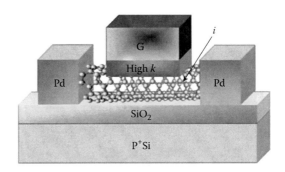

FIGURE 3.15   **(See color insert.)** Illustration of the CNT capacitor device.

To improve this visible disagreement of unachievable infinite drain electric field, the electric field is taken linearly rising from $V_{Dsat}/2L$ at $x = 0$ and reaching the ultimate value $V_{Dsat}/\alpha L$ at $x = L$, which gives an average field of V_{Dsat}/L; however, for $\alpha < 1$, the V_{Dsat1} is replaced with V_{Dsat}, thus leading to the infinite electric field assumption at drain end with the onset of saturation being avoided. The origin of V_{Dsat1} at the onset of current saturation is normally ascribed to the velocity reaching saturation at the drain end that is only possible if the electric field is infinitely large. The drain velocity is a fraction of the saturation velocity. The critical electric field for the onset of velocity saturation is $\varepsilon_C = v_{sat}/\mu_0$. At the onset of current saturation, the drain electric field is $\varepsilon_D = V_{Dsat}/\alpha L$. The electric field is shown to be increasing infinitely with an increase in channel length when $\alpha = 1$. However, in the case of $\alpha < 1$, a finite electric field was observed (Tan et al., 2009), causing the assumption of an infinite electric field not to be attainable at the drain end of a MOSFET due to the presence of device breakdown in a variety of mechanisms. These results validate that α will always be less than 1, thus making the drain velocity v_D smaller than the ultimate saturation velocity v_{sat} due to the presence of the finite electric field at the drain end. The carrier density in the channel is normally degenerate. In this limit, the drain velocity is specified by the Fermi velocity, with n_s given by

$$n_s = \frac{C_G(V_{GT} - V_{Dsat})}{q} \tag{3.58}$$

The drain velocity is always smaller than the one expected from an infinite electric field at the drain. The ultimate carrier concentration appropriate for the infinite electric field at the drain end ($\alpha = 1$) is obtained from Equation 3.58 by replacing V_{Dsat} with V_{Dsat1}. V_{Dsat1} itself with its dependence on v_{sat} requires an iterative solution. In validating the developed Arora model for ballistic transport across CNTFETs, the I–V characteristics of the model have been compared with the published fabrication data (Javey et al., 2005). It is found that in a nanoscale channel, as seen in the comparison of I–V characteristics in Figure 3.16, the channel conductance is nonzero due to an increase in the value of α as the drain voltage is increased. The current at the onset of the saturation region is as given previously in Equation 3.51. In fact, if the binomial expansion of V_{Dsat1} is carried out under the long-channel approximation, the carrier concentration at the drain can be finite. The solid red line is plotted using Equation 3.50 for the range of drain voltage $0 \le V_D \le V_{Dsat}$. The red dotted is from Equation 3.51 and is the extension for the range $V_D \ge V_{Dsat}$ where $\alpha < 1$. The blue-dotted flat curve also represents the same range as red dotted curve but with $\alpha = 1$. In the case of $\alpha = 1$ (blue-dotted line), the velocity-saturation-limited current is being terminated, giving zero transconductance. This is due to the infinite electric field assumed at the drain end making the saturation velocity lower and is almost comparable to the drain velocity v_D that makes the velocity to reach full saturation earlier. In the case of $\alpha < 1$, the drain velocity v_D is revealed to be less than the ultimate saturation velocity v_{sat1} due to a finite electric field at the drain end as prescribed in Figure 3.16. Thus, it can be concluded that the infinite electric field is not achievable in the drain end of a CNTFET device as many researchers have assumed.

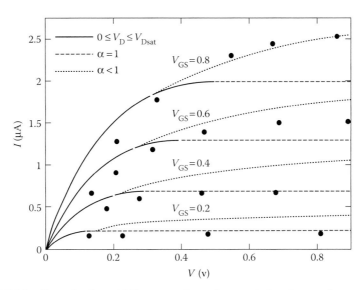

FIGURE 3.16   **(See color insert.)** Current–voltage characteristics of nanowire.

The comparison with the experimental data is not perfect due to experimental geometry. Also, there is a complete ignoring of quantum capacitance that could bring experimental results closer to the modeled one. Considering some of the imperfections, the agreement of the theory and the experimental data are as good as can be expected. For CMOS applications in a planar transistor, normally, *width per length* ratios are matched for equal currents in the n- and the p-channels. In a CNT transistor, similarly, the ratios are matched for the lengths of the n- and p-channels. However, an important factor that appears in the matching of the currents is the critical voltage V_C that itself depends on the saturation velocity. For CMOS applications, it is important to match dimensions proportionally not only to the mobility, but also with the saturation velocity.

3.8   SUMMARY

Working on CNTFETs gives a useful context, in which we learn to treat electronic devices at the molecular scale. The issues related to contact, interfaces, transport, etc. must be addressed for CNTFETs, which are likely to be important for other types of devices as well. The nanotubes are being considered as the best candidates for high-speed applications. The carriers in a CNT, like in any Q1D nanostructure, have analog energy spectrum only in the quasi-free direction, while the other two Cartesian directions are quantum-confined leading to a digital (quantized) energy spectrum. The salient feature of mobility and saturation velocity is the controlling of the charge transport in a semiconducting SWCNT channel. The ultimate drift velocity in SWCNT due to the high-electric-field streaming is based on the asymmetrical distribution function that converts randomness in a zero field to a streamlined one in a very high electric field. Specifically, we show that higher mobility in an SWCNT does not necessarily lead to a higher saturation velocity that is limited by the mean intrinsic velocity depending upon the band parameters. The intrinsic velocity is found to be an appropriate thermal velocity in the nondegenerate regime increasing with the temperature, but independent of carrier concentration. However, in a degenerate regime, this intrinsic velocity is the Fermi velocity that is independent of temperature, but is strongly dependent on the carrier concentration. The velocity that saturates in a high electric field can be lower than the intrinsic velocity, due to the onset of a quantum emission. In an SWCNT, the mobility may also become ballistic if the length of the channel is comparable or less than the mean free path.

APPENDIX A: QUASI 1D MODEL

A.1   DOS FOR 1D

$$g_{1se} = \frac{1}{L}\frac{dN}{dk}\frac{dk}{dE} \tag{3.A.1}$$

Number of electrons stated dN between k and $k + dk$

$$\frac{dN}{dk} = 2\frac{2}{2\pi/L} = \frac{2L}{\pi} \tag{3.A.2}$$

Energy momentum dispersion

$$E = E_{co} + \frac{\hbar^2 k_x^2}{2m^*} + m^2\varepsilon_{oz} + n^2\varepsilon_{oz} \Rightarrow k_x = \sqrt{\frac{2m^*(E - E_{cn})}{\hbar^2}} \tag{3.A.3}$$

$$\frac{dk}{dE} = \frac{1}{2}\left[\frac{2m^*(E - E_{cn})}{\hbar^2}\right]^{-1/2}(2m^*) = m^* \cdot \left[\frac{2m^*(E - E_{cn})}{\hbar^2}\right]^{-1/2} = \frac{m^*}{\hbar^2 k} \tag{3.A.4}$$

DOS is given as

$$g_{1se} = \frac{1}{L}\cdot\frac{dN}{dk}\cdot\frac{dk}{dE} = \frac{1}{L}\cdot\frac{2L}{\pi}\cdot\frac{m^*}{\hbar^2 k} \tag{3.A.5}$$

$$= \frac{2}{\pi}\cdot\frac{m^*}{\hbar^2}\sqrt{\frac{\hbar^2}{2m^*(E - E_{cn})}} \tag{3.A.6}$$

$$= \frac{2}{\pi} \cdot \frac{m^*}{\hbar^2} \frac{\hbar}{\sqrt{2m^*}} (E - E_{cn})^{-1/2} \tag{3.A.7}$$

$$= \frac{2}{\pi} \cdot \frac{m^*}{\hbar} \frac{1}{\sqrt{2m^*}} (E - E_{cn})^{-1/2} \tag{3.A.8}$$

$$= \frac{\sqrt{2m^*}}{\pi \hbar} \cdot (E - E_{cn})^{-1/2} \tag{3.A.9}$$

where m^* is m_1.

A.2 ELECTRON CONCENTRATION FOR 1D

$$n_1 = \int_{E_C}^{top \approx \infty} g_{1se}(E) f(E) dE \tag{3.A.10}$$

$$= \frac{\sqrt{2m^*}}{\pi \hbar} \int_{E_C}^{top \approx \infty} (E - E_{cn})^{-1/2} \cdot \frac{1}{e^{\frac{E_k - E_F}{k_B T}} + 1} dE \tag{3.A.11}$$

$$= \frac{\sqrt{2m^*}}{\pi \hbar} \int_{E_C}^{top \approx \infty} (k_B T x)^{-1/2} \cdot \frac{1}{e^{x - \eta} + 1} k_B T \, dx \tag{3.A.12}$$

$$= \frac{\sqrt{2m^* k_B T}}{\pi \hbar} \int_{E_C}^{top \approx \infty} \frac{\sqrt{\pi}}{\sqrt{\pi}} \frac{x^{-1/2}}{e^{x - \eta} + 1} dx \tag{3.A.13}$$

$$= \frac{\sqrt{2m^* k_B T}}{\hbar} \int_{E_C}^{top \approx \infty} \frac{1}{\sqrt{\pi}} \frac{x^{-1/2}}{e^{x - \eta} + 1} dx \tag{3.A.14}$$

$$= \frac{\sqrt{2m^* k_B T / \pi}}{\hbar} \int_{E_C}^{top \approx \infty} \frac{1}{\sqrt{\pi}} \frac{x^{-1/2}}{e^{x - \eta} + 1} dx \tag{3.A.15}$$

$$= N_{1D} \Im_{-\frac{1}{2}}(\eta_c) \tag{3.A.16}$$

A.3   STRONG DEGENERATE DOPING FOR 1D

$$n_1 = \int\limits_{E_C}^{top \approx \infty} g_{1se}(E)f(E)dE = N_C \mathfrak{I}_{-\frac{1}{2}}(\eta_c) \tag{3.A.17}$$

$$\text{where } \mathfrak{I}_{-\frac{1}{2}}(\eta_c) \approx \frac{2}{\sqrt{\pi}} \eta^{\frac{1}{2}} \Rightarrow n_1 \approx N_c \frac{2}{\sqrt{\pi}} \eta^{\frac{1}{2}}$$

$$E_F - E_c \approx \frac{\pi}{4}\left[\frac{n_1}{N_c}\right]^2 k_B T = \frac{\pi}{4}\left[\sqrt{\left(\frac{\pi\hbar^2}{2m^* k_B T}\right)} \cdot \frac{n_1}{1}\right]^2 k_B T = \frac{\pi}{4}\frac{\pi\hbar^2 n_1^2}{2m^* k_B T}[k_B T] = \frac{\hbar^2 \pi^2 n_1^2}{8m_n^*} \tag{3.A.18}$$

$$k_F \approx \left(\frac{\pi n}{2}\right)_1 \quad v_F \approx \frac{\hbar(\pi n_1)}{2m^*} \tag{3.A.19}$$

A.4   INTRINSIC VELOCITY FOR 1D

$$v_{i1} = \frac{1}{n_1}\int\limits_{E_C}^{top \approx \infty} |v| g_{1se}(E)f(E)dE = \frac{1}{n_1}\int\limits_{E_C}^{\infty} \sqrt{\frac{2(E - E_{cn})}{m}} g_{1se}(E)f(E)dE \tag{3.A.20}$$

$$= \frac{\sqrt{2m^*}}{n_1 \pi \hbar} \cdot (E - E_{cn})^{-1/2} \int\limits_{E_C}^{\infty} \sqrt{\frac{2}{m}} \cdot (E - E_{cn})^{1/2} \cdot \frac{1}{e^{\frac{E_k - E_F}{k_B T}} + 1} dE \tag{3.A.21}$$

$$= \frac{\sqrt{2m^*}}{n_1 \pi \hbar} \cdot \sqrt{\frac{2}{m}} \cdot \int\limits_{E_C}^{\infty} (k_B T x)^0 \cdot \frac{1}{e^{\frac{E_k - E_F}{k_B T}} + 1} k_B T dx \tag{3.A.22}$$

$$= \frac{\sqrt{2m^* k_B T}}{n_1 \pi \hbar} \cdot \sqrt{\frac{2k_B T}{m}} \int\limits_{E_C}^{\infty} \frac{x^0}{e^{\frac{E_k - E_F}{k_B T}} + 1} dx \tag{3.A.23}$$

$$= \frac{N_1}{n_1}\frac{2}{\sqrt{\pi}} \cdot \frac{\sqrt{\pi}}{2}(k_B T)\left(\sqrt{k_B T}\right) \cdot \frac{1}{2\pi^2} \cdot \left(\frac{2m^*}{\hbar^2}\right)^{3/2} v_{th} \cdot \int\limits_{E_C}^{\infty} \frac{x^0}{e^{x - \eta} + 1} k_B T dx \tag{3.A.24}$$

$$= \frac{N_1}{n_1}\frac{2}{\sqrt{\pi}} \cdot \frac{\sqrt{\pi}}{2} \cdot \frac{1}{2\pi^2} \cdot \left(\frac{2m^* k_B T}{\hbar^2}\right)^{3/2} v_{th} \cdot \int\limits_{E_C}^{\infty} \frac{x^0}{e^{x - \eta} + 1} k_B T dx \tag{3.A.25}$$

$$= \frac{1}{\Gamma(3/2)\Im_{\frac{1}{2}}(\eta_c)} \cdot v_{th} \cdot \Gamma(2)\Im_1(\eta_c) \tag{3.A.26}$$

$$= \frac{\Gamma(2)\Im_1(\eta_c)}{\Gamma(3/2)\Im_{\frac{1}{2}}(\eta_c)} \cdot v_{th} \tag{3.A.27}$$

$$= \frac{2}{\sqrt{\pi}} \frac{\Im_1(\eta_c)}{\Im_{\frac{1}{2}}(\eta_c)} \cdot v_{th} \tag{3.A.28}$$

APPENDIX B: DEGENERATE INTRINSIC VELOCITY

B.1   Q3D VELOCITY

$$v_{i3} = \frac{2}{\sqrt{\pi}} \frac{\Im_1(\eta_c)}{\Im_{\frac{1}{2}}(\eta_c)} \cdot v_{th} \tag{3.B.1}$$

$$= \frac{2}{\sqrt{\pi}} \frac{\Im_1(\eta_c)}{\Im_{\frac{1}{2}}(\eta_c)} \cdot v_{th} = \frac{2}{\sqrt{\pi}} \frac{3\sqrt{\pi}}{4\eta_c^{3/2}} \frac{1}{\Gamma(2)} \frac{\eta^2}{2} \cdot v_{th} \tag{3.B.2}$$

$$= \frac{2}{\sqrt{\pi}} \frac{\sqrt{\pi}}{2} \frac{\eta^2 3}{4\eta_c^{3/2}} \cdot v_{th} = \frac{3}{4} \cdot \eta_c^{1/2} \cdot v_{th} = \frac{3}{4}\sqrt{\frac{E-E_c}{k_B T}}\sqrt{\frac{2k_B T}{m^*}} \tag{3.B.3}$$

$$= \frac{3}{4}\sqrt{\frac{2k_B T}{m^*} \cdot \frac{E-E_c}{k_B T}} = \frac{3}{4}\sqrt{\frac{2(E-E_c)}{m^*}} = \frac{3}{4}v_F = \frac{3}{4}\frac{\hbar}{m^*}(3n\pi^2)^{1/3} \tag{3.B.4}$$

B.2   Q2D VELOCITY

$$v_{i2} = \frac{\Gamma(3/2)}{\Gamma(1)} \frac{\Im_{\frac{1}{2}}(\eta_c)}{\Im_0(\eta_c)} \cdot v_{th} \tag{3.B.5}$$

$$= \frac{\sqrt{\pi}}{2} \frac{4\eta_c^{3/2}}{3\sqrt{\pi}} \frac{1}{\eta} v_{th} \tag{3.B.6}$$

$$= \frac{2}{3} \cdot \eta_c^{1/2} \cdot v_{th} = \frac{2}{3}\sqrt{\frac{E-E_c}{k_B T}}\sqrt{\frac{2k_B T}{m^*}} = \frac{2}{3}v_F = \frac{2}{3}\frac{\hbar}{m^*}\sqrt{2\pi n_2} \tag{3.B.7}$$

B.3   Q1D VELOCITY

$$v_{i1} = \frac{\Gamma(1)}{\Gamma(3/2)} \frac{\Im_0(\eta_c)}{\Im_{-\frac{1}{2}}(\eta_c)} \cdot v_{th} \tag{3.B.8}$$

$$= \frac{1}{\sqrt{\pi}} \cdot \frac{\eta}{2} \frac{\sqrt{\pi}}{\eta^{1/2}} v_{th} \tag{3.B.9}$$

$$= \frac{1}{2} \cdot \eta_c^{1/2} \cdot v_{th} = \frac{1}{2} \sqrt{\frac{E - E_c}{k_B T}} \sqrt{\frac{2k_B T}{m^*}} = \frac{1}{2} v_F = \frac{1}{2} \frac{\hbar}{m^*} \frac{n_1 \pi}{2} \tag{3.B.10}$$

$$= \frac{1}{4} \frac{\hbar}{m^*} n_1 \pi \tag{3.B.11}$$

B.4   Q3D NONDEGENERATE VELOCITY LIMIT

$$v_{i3} = \frac{2}{\sqrt{\pi}} \frac{\Im_1(\eta_c)}{\Im_{\frac{1}{2}}(\eta_c)} \cdot v_{th} \tag{3.B.12}$$

$$= \frac{2}{\sqrt{\pi}} \cdot v_{th} \quad \text{where } \frac{\Im_1(\eta_c)}{\Im_{\frac{1}{2}}(\eta_c)} = 1 = \frac{e^\eta}{e^\eta} \tag{3.B.13}$$

$$= \frac{2}{\sqrt{\pi}} \sqrt{\frac{2k_B T}{m^*}} = \sqrt{\frac{8k_B T}{\pi m^*}} \tag{3.B.14}$$

B.5   Q2D NONDEGENERATE VELOCITY LIMIT

$$v_{i2} = \frac{\Gamma(3/2)}{\Gamma(1)} \frac{\Im_{\frac{1}{2}}(\eta_c)}{\Im_0(\eta_c)} \cdot v_{th} \tag{3.B.15}$$

$$= \frac{\sqrt{\pi}}{2} \cdot v_{th} \quad \text{where } \frac{\Im_{\frac{1}{2}}(\eta_c)}{\Im_0(\eta_c)} = 1 = \frac{e^\eta}{e^\eta} \tag{3.B.16}$$

$$= \frac{\sqrt{\pi}}{2} \cdot \sqrt{\frac{2k_B T}{m^*}} \tag{3.B.17}$$

$$= \sqrt{\frac{\pi k_B T}{2m^*}} \tag{3.B.18}$$

B.6   Q1D NONDEGENERATE VELOCITY LIMIT

$$v_{i1} = \frac{\Gamma(1)}{\Gamma(3/2)} \frac{\Im_0(\eta_c)}{\Im_{-\frac{1}{2}}(\eta_c)} \cdot v_{th} \tag{3.B.19}$$

$$= \frac{1}{\sqrt{\pi}} \cdot v_{th} \quad \text{where } \frac{\Im_0(\eta_c)}{\Im_{-\frac{1}{2}}(\eta_c)} = 1 = \frac{e^\eta}{e^\eta} \tag{3.B.20}$$

$$= \frac{1}{\sqrt{\pi}} \cdot \sqrt{\frac{2k_B T}{m^*}} \tag{3.B.21}$$

$$= \sqrt{\frac{2k_B T}{m^* \pi}} \tag{3.B.22}$$

APPENDIX C: GAMMA FUNCTION

	Gamma Function	
I	$\Gamma(i+1) = i!$	$\Gamma(i+1/2) = \sqrt{\pi}\dfrac{(2i)!}{2^{2i} i!}$
-1	Nil	$\Gamma(-1/2) = -2\sqrt{\pi}$
0	$\Gamma(1) = 1$	$\Gamma(1/2) = \sqrt{\pi}$
1	$\Gamma(2) = 1$	$\Gamma(3/2) = \sqrt{\pi}/2$
2	$\Gamma(3) = 2$	$\Gamma(5/2) = 3\sqrt{\pi}/4$
3	$\Gamma(4) = 6$	$\Gamma(7/2) = 15\sqrt{\pi}/8$
4	$\Gamma(5) = 24$	$\Gamma(9/2) = 105\sqrt{\pi}/16$
5	$\Gamma(6) = 120$	$\Gamma(11/2) = 945\sqrt{\pi}/32$
6	$\Gamma(7) = 720$	$\Gamma(-13/2) = 10395\sqrt{\pi}/64$

APPENDIX D: CNT CARRIER VELOCITY IN NONPARABOLIC BAND

$$E(k) = \pm \frac{t3a_{C-C}}{2} \sqrt{\left(\frac{2}{3d}\right)^2 + k_t^2} \tag{3.D.1}$$

$$v = \frac{1}{\hbar} \frac{dE}{dk} \tag{3.D.2}$$

$$|v| = \frac{\left(\dfrac{E_g}{2}\right)^2 \left(\dfrac{3d}{2}\right)^2 k}{\hbar E} \tag{3.D.3}$$

$$k = \frac{2}{E_g} \frac{2}{3d} \sqrt{E^2 - \left(\frac{E_G}{2}\right)^2} \qquad (3.D.4)$$

$$v_i = \frac{\int \left| v \right| D(E) f(E) dE}{\int D(E) f(E) dE} = \frac{\int \left| v \right| D(E) f(E) dE}{n_{CN}} \qquad (3.D.5)$$

$$n_{CN} = \frac{N_{CN}}{2} \int_{E_C}^{E_{top}} \frac{x + E_G / 2k_B T}{\sqrt{x^2 + xE_G / k_B T}} \left(\frac{1}{e^{x-\eta}+1}\right) dx \qquad (3.D.6)$$

$$v_i = \frac{\int \left| \frac{\left(\frac{E_g}{2}\right)^2 \left(\frac{3d}{2}\right)^2}{E} \right| \frac{D_0}{\hbar} \frac{\left(\frac{2}{E_g}\right)\left(\frac{2}{3d}\right)\sqrt{E^2 - (E_G/2)^2} E}{\sqrt{E^2 - (E_G/2)^2}} \frac{1}{e^{\frac{E_k - E_{F1}}{k_B T}}+1} dE}{n_{CN}} \qquad (3.D.7)$$

$$v_i = \frac{\int \left| \frac{\left(\frac{E_g}{2}\right)^2 \left(\frac{3d}{2}\right)^2}{E} \right| \frac{D_0}{\hbar} \frac{\left(\frac{2}{E_g}\right)\left(\frac{2}{3d}\right)\sqrt{E^2 - (E_G/2)^2} E}{\sqrt{E^2 - (E_G/2)^2}} \frac{1}{e^{\frac{E_k - E_{F1}}{k_B T}}+1} dE}{n_{CN}} \qquad (3.D.8)$$

$$v_i = \left(\frac{E_g}{2}\right)\left(\frac{3d}{2}\right)\frac{D_0}{\hbar} \frac{\int \frac{1}{e^{\frac{E_k - E_{F1}}{k_B T}}+1} dE}{n_{CN}} \qquad (3.D.9)$$

$$v_i = v_{th} \frac{\dot{F}_0(\dot{\eta}_{FCN})}{F_{CN}(\eta_{FCN})} \qquad (3.D.10)$$

$$v_i = v_{th} \frac{\dot{F}_0(\dot{\eta}_{FCN})}{F_{CN}(\eta_{FCN})} \qquad (3.D.11)$$

$$F_{CN}(\eta_{FCN}) = M e^{+\eta} \qquad (3.D.12)$$

$$\dot{F}_0(\dot{\eta}_{FCN}) = e^{+\eta} \qquad (3.D.13)$$

$$v_i = v_{th} \frac{e^{+\eta}}{M e^{+\eta}} \qquad (3.D.14)$$

REFERENCES

Ahmadi M. T., M. L. P. Tan, R. Ismail, and V. K. Arora. 2009a. The high-field drift velocity in degenerately-doped silicon nanowires. *Int. J. Nanotechnol.* 6(7/8), 601–617.

Ahmadi M. T., H. H. Lau, R. Ismail, and V. K. Arora. 2009b. Current-voltage characteristics of a SNW transistor. *Microelectronics J.* 40, 547–549.

Andrieu F., et al. 2003. *SiGe Channel p-MOSFET's Scaling Down.* Estoril, Portugal, September 16–18, pp. 267–270.

Appenzeller J., J. Knoch, V. Derycke, R. Martel, S. Wind, and P. Avouris. 2002. Field modulated carrier transport in CNT transistors. *Phys. Rev. Lett.* 89(12), doi 126801.

Appenzeller J. et al. 2005. Comparing CNT transistors the ideal choice: A novel tunneling device design. *IEEE Trans. Electron Devices.* 52(12), 2568–2576.

Arora V. K. 1983. Free-carrier absorption in quasi-two-dimensional semiconducting structures. *Phys. Rev. B.* 28, 971–976.

Arora V. K. 1985a. Quantum well wires: Electrical and optical properties. *J. Phys. C.* 18, 3011.

Arora V. K. 1985b. High field distribution and mobility in semiconductors. *Jpn. J. Appl. Phys.* 24, 537.

Arora V. K. 1985c. Quantum well wires: Electrical and optical properties. *J. Phys. C.* 18, 3011–3016.

Arora V. K. 2000. Quantum engineering of nanoelectronic devices: The role of quantum emission in limiting drift velocity and diffusion coefficient. *Microelectronics J.* 31(11–12), 853–859.

Arora V. K. 2006. Failure of Ohm's law: Its implications on the design of nanoelectronic devices and circuits. *Proceedings of the IEEE International Conference on Microelectronics.* Belgrade, Serbia and Montenegro, May 14–17, pp. 17–24.

Auth C. P. and J. D. Plummer. 1998. A simple model for threshold voltage of surrounding-gate MOSFET's. *IEEE Trans. Electron Devices* 45(11), 2381–2383.

Burke P. J. 2003. An RF circuit model for CNTs. *IEEE Trans. Nanotechnol.* 2(1), 55–58.

Burke P. J. 2004. AC performance of nanoelectronics: Towards a ballistic THz nanotube transistor. *Solid State Electron.* 48, 1981–1986.

Büttiker M. 1986. Role of quantum coherence in series resistors. *Phys. Rev. B. Condens. Matter.* 33, 3020–3026.

Chai G., H. Heinrich, L. Chow, and T. Schenkel. 2007. Electron transport through single CNTs. *Appl. Phys. Lett.* 91, 1031.

Chen J., C. Klinke, A. Afzali, K. Chan, and Ph. Avouris. 2004. Self-aligned CNT transistors with novel chemical doping. *IEDM Technology Digest.* San Francisco, CA, pp. 695–698.

Feynman R. 1959. There's plenty of room at the bottom. Speech, American Physical Society Meeting, Caltech. December 29.

Fiori G. and I. Giuseppe. 2006. Threshold voltage dispersion and impurity scattering limited mobility in CNT field effect transistors with randomly doped reservoirs. *Solid-State Device Research Conference IEEE,* Purdue University, West Lafayette, IN, pp. 202–205.

Guo, J., M. Lundstrom, and S. Datta. 2002. Performance projections for ballistic carbon nanotube field-effect transistors, *Appl. Phys. Lett.* 80, 3192–3194.

Hashempour H. and F. Lombardi. 2008. Device model for ballistic CNTFETs using the first conducting band. *IEEE Design & Test Archive,* 25(2), 178–186.

Heinze S., J. Tersoff, R. Martel, V. Derycke, J. Appenzeller, and Ph. Avouris. 2002. Carbon nanotubes as Schottky barrier transistors. *Phys. Rev. Lett.* 89(10), 106801.

Hernandez E., V. Meunier, B. W. Smith, R. Rurali, H. Terrones, B. N. Nardelli, M. Terrones, D. E. Luzzi, and J. C. Charlier. 2003. Fullerene coalescence in nanopeapods: A path to novel tubular carbon. *Nano Lett.* 3, 1037.

Hirahara K., K. Suenaga, S. Bandow, H. Kato, T. Okazaki, H. Shinohara, and S. Iijima. 2000. One-dimensional metallofullerene crystal generated inside single walled CNTs. *Phys. Rev. Lett.* 85, 5384.

Iijima S. 1991. Helical microtubules of graphitic carbon. *Nature* (London). 354(6348), 56–58.

Ilani S., L. A. K. Donev, M. Kindermann, and P. L. Mceuen. 2006. Measurement of the quantum capacitance of interacting electrons in CNTs. *Nat. Phys.* 2, 687–691.

ITRS. International Technology Roadmap for Semiconductors. 2005 Edition. Semiconductor Industry Association, accessed on 2 August, 2008 at http://public.itrs.net.

Javey A., H. Kim, M. Brink, Q. Wang, A. Ural, J. Guo, P. McIntyre, P. Mceuen, M. Lundstrom, and H. Dai. 2002. Advancements in complementary carbon nanotube. *Nat. Mater.* 1, 241.

Javey A., J. Guo, D. B. Farmer, Q. Wang, D. Wang, R. G. Gordon, M. Lundstorm, and H. Dai. 2004. Carbon nanotube field-effect transistors with integrated ohmic contact and high-k gate dielectric. *Nano Lett.* 4, 447.

Javey A. et al. 2005. High performance n-type CNT field-effect transistors with chemically doped contacts. *Nano Lett.* 5(2), 345–348.

Jin S., T.-W. Tang, and M. V. Fischetti. 2007b. Three-dimensional simulation of one-dimensional transport in SNW transistors. *IEEE Trans. Nanotechnol.* 6(5), 524–529.

Jishi R. A., D. Inomata, K. Nakao, M. S. Dresselhaus, and G. Dresselhaus. 1994. Electronic and lattice properties of CNTs. *J. Phys. Soc.* 63(6), 2252–2260.

Li J., Q. Ye. A. Cassell, H. T. Ng, R. Stevens, J. Han, and M. Meyyappan. 2003. Bottom-up approach for CNT interconnects. *Appl. Phys. Lett.* 82(15), 2491–2493.

Li J. Q., Q. Zhang, D. J. Yang, and J. Z. Tian. 2004. Fabrication of carbon nanotube FET by AC dielectrophoresis method. *Carbon.* 42, 2263–2267.

Lieber C. M. 2003. Nanoscale science and technology: Building a big future from small things. *MRS Bull.* 28, 486–491.

Lin Y. M., J. Appenzeller, and Ph. Avouris. 2004. Novel structures enabling bulk switching in carbon nanotube FETs, *62nd Device Research Technical Conference Digest.* University of Notre Dame, Indiana, pp. 133.

Lin Y. M., J. Appenzeller, Z. Chen, Z.-G. Chen, H.-M. Cheng, and Ph. Avouris. 2005a. Demonstration of a high performance 40-nm-gate CNT field-effect transistor. *63rd Device Research Conference Digest, 2005*, IEEE, Santa Barbara, CA, June 22, vol.1, pp. 113–114.

Lin Y. M. et al. 2005b. High-performance CNT field-effect transistor with tunable polarities. *IEEE Trans. Nanotechnol.* 4(5), 481–489.

Lundstrom M. 1997. Elementary scattering theory of the Si MOSFET. *IEEE Electron. Device Lett.* 18, 361–363.

Lundstrom M. and J. H. Rhew. 2002. A Landauer approach to nanoscale MOSFETs. *J. Comput. Electron.* 1(4), 481–489.

Lundestorm M. S. and J. Guo. 2006. *Nanoscale Transistors: Device Physics, Modeling and* of materials. *Carbon*, 38, 1751.

Martel R., T. Schmidt, H. R. Shea, T. Hertel, and P. Avouris. 1998a. Single- and multi wall carbon nanotube field-effect transistors. *Appl. Phys. Lett.* 73, 2447–2449.

Martel R., T. Schmidt, H. Shea, T. Hertel, and P. Avouris. 1998b. Single- and multi-wall CNT field-effect transistors. *Appl. Phys. Lett.* 73(17), 2447–2449.

McEuen P. L., M. S. Fuhrer, and H. Park. 2002. Single-walled CNT electronics. *IEEE Trans. Nanotechnol.* 1(1), 78–85.

Najari M., S. Frégonèse, C. Maneux, T. Zimmer, H. Mnif, and N. Masmoudi. 2008 Towards compact modelling of Schottky barrier CNTFET. *International Conference on Design and Technology of Integrated Systems in Nanoscale Era.* Tozeur, Tunisia, March 25–27, pp. 1–6.

Ossaimee M. I., S. H. Gamal, K. A. Kirah and O. A. Omar. 2008. Ballistic transport in Schottky barrier CNT FETs. *Electronics Lett.* 44(5), 336–337.

Pierret R. F. 2003. *Advanced Semiconductor Fundamentals.* Prentice Hall, Englewood Cliffs, NJ.

Pourfath M., E. Ungersboeck, A. Gehring, B. H. Cheong, W. J. Park, H. Kosina, and S. Selberherr. 2005. Optimization of Schottky barrier CNT field effect transistors. *Microelectron. Eng.* 81, 428–433.

Pourfath M., et al. 2008. Numerical study of quantum transport in CNT transistors. *Math. Comput. Simul.* 79(4), 1051–1059.

Radosavlijevic M., J. Appenzeller, J. Knoch, and P. Avouris. 2004. High performance of potassium n-doped CNT field effect transistors. *Appl. Phy. Lett.* 84(18), 3693–3695.

Raychowdhury A., S. Mukhopadhyay, and K. Roy. 2004. A circuit-compatible model of ballistic CNT field-effect transistors. *IEEE Trans. Computer-Aided Des. Integr. Circuits Syst.* 23(10), 1411.

Saito R., M. Fujita, G. Dresselhaus, and M. S. Dresselhaus. 1992. Electronic structure of chiral graphene tubules. *Appl. Phys. Lett.* 60(18), 2240–2206.

Svizhenko A. and M. P. Anantram. 2005. Effect of scattering and contacts on current and electrostatics in CNTs. *Phys. Rev. B.* 72, 085430.

Sze S. M. 1981. *Physics of Semiconductor Devices.* Wiley, New York.

Tan M. L. P., V. K. Arora, I. Saad, M. T. Ahmadi, and R. Ismail. 2009. The drain velocity overshoot in an 80 nm metal-oxide-semiconductor field-effect transistor. *J. Appl. Phys.* 105, 074503.

Tans S. J., A. R. M. Verschueren, and C. Dekker. 1998. Room-temperature transistor based on a single CNT. *Nature*, 393(6680), 49–52.

Tasis D., N. Tagmatarchis, A. Bianco, and M. Prato. 2006. Chemistry of CNTs. *Chem. Rev.* 106, 1105.

Wang Z. Y., Z. B. Zhao, and J. S. Qiu. 2006. Development of filling CNTs. *Prog. Chem.* 18, 563.

Wind S. J., J. Appenzeller, and P. Avouris. 2003. Lateral scaling in CNT field effect transistors. *Phys. Rev.* 91(5), 058301.

Winstead B. and U. Ravaioli. 2000. Simulation of Schottky barrier MOSFETs with a coupled quantum injection/Monte Carlo technique. *IEEE Trans. Electron Devices*, 47(6), 1241–1246.

Xia T. S., L. F. Register, and S. K. Banerjee. 2005. Simulation study of the CNT field effect transistors beyond the complex band structure effect. *Solid State Electronics.* 49, 860–864.

Xuemei X. and D. Mohan. 2006. BSIM3 homepage.html accessed on 15 November available at http://www.device.eecs.berkeley.edu/~bsim3/latenews.

4 Carbon Nanotube Circuit Analysis and Simulation

Desmond C. Y. Chek and Razali Ismail

CONTENTS

4.1 INTRODUCTION

Circuit theory and electronic textbooks rely heavily on Ohm's law, which states that the current response to the applied voltage across the length of a resistor is linear [1]. Ohm's law enjoyed its superiority in the performance assessment of all conducting materials until it was discovered that the velocity cannot increase indefinitely with the increase of electric field, and eventually saturates to a value and leads to saturation of current [2]. In this chapter, we discuss the application of nonohmic law [3–6] in a carbon nanotube field effect transistor (CNFET) circuit. One of the impacts of the breakdown of Ohm's law is the resistance blowup, which indeed affects the transient response when a nanoresistor is used in capacitor and inductor circuits [3–6]. In Section 4.2, the RC transient response in CNFET is examined by applying the nonohmic law. Besides, the load capacitor and frequency response of CNFET are discussed in Section 4.4, and the propagation delay in CNFET logic gates is discussed in Section 4.5. The ultimate goal of this chapter is to validate the application of nonohmic law in CNFET circuit by comparing the theoretical results with those of Hspice simulation results. We conclude this chapter with a discussion on the two quality measures, which indeed examine the overall performance of CNFET in logic gates and compare it with that of logic gates made of state-of-the-art metal–oxide–semiconductor field-effect transistor (MOSFET).

4.2   RC TRANSIT DELAY IN CNFET CIRCUIT

It is well known that the rise/fall time is the duration needed to charge/discharge the load capacitor through a transistor. Figure 4.1 shows an example of the corresponding equivalent RC circuit for charging process through a p-type CNFET as an effective resistor.

To model the rise time of the charging process in Figure 4.1, the drain current–drain voltage (I_{DS}–V_{DS}) of a p-type CNFET should be obtained. Then, by selecting the curve with maximum gate voltage V_{GS} (in this case $V_{GS} = -1$ V), an empirical equation that is similar to the Arora's theoretical formalism [6] is used to fit the selected I_{DS}–V_{DS} curve. The empirical equation is given by

$$i = I_{sat} \tanh\left(\frac{V^a}{bV_c}\right) \tag{4.1}$$

where
    a and b are fitting parameters
    i is the charging current
    V is the supply voltage
    V_c is the critical voltage

Figure 4.2 demonstrates the fitting of the empirical equation shown in Equation 4.1 to an established p-type CNFET [7] with $V_{GS} = -1$ V.

As shown in Figure 4.2, the empirical equation with $I_{sat} = 63$ μA, $V_c = 0.2112$ V, $a = 0.65$, and $b = 2.6$ has a very good fitting with the p-type CNFET model [7]. Hence, this empirical equation can be used to represent the established model with $V_{GS} = -1$ V. The total supply voltage for the RC circuit in Figure 4.1 equals the resistance voltage and the capacitor voltage as

$$V = v_R(t) + v_{Cap}(t) = \left[bV_c \tanh^{-1}\left(\frac{i(t)}{I_{sat}}\right) \right]^{1/a} + \frac{q(t)}{C} \tag{4.2}$$

where $v_R(t)$ is the resistance voltage obtained from Equation 4.1. By differentiating Equation 4.2 with respect to time t, it becomes

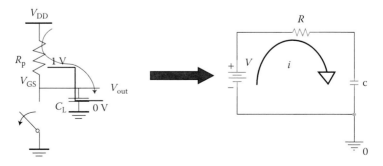

FIGURE 4.1   Equivalent RC circuit for the p-type CNFET charging process.

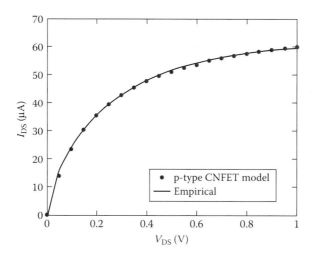

FIGURE 4.2 Fitting curve between the established p-type CNFET and the empirical equation.

$$0 = \frac{1}{a}\left[bV_c \tanh^{-1}\left(\frac{i(t)}{I_{sat}} \right) \right]^{\frac{1}{a}-1} \frac{\partial}{\partial t}\left[bV_c \tanh^{-1}\left(\frac{i(t)}{I_{sat}} \right) \right] + \frac{i(t)}{C} \tag{4.3}$$

By using the approximation

$$\tanh^{-1}\left(\frac{i}{I_{sat}} \right) = \frac{1}{1-\left(\dfrac{i}{I_{sat}} \right)^2} \cdot \frac{i}{I_{sat}} \tag{4.4}$$

Equation 4.3 can be simplified into

$$\frac{1}{a}\left[bV_c \tanh^{-1}\left(\frac{i}{I_{sat}} \right) \right]^{\frac{1}{a}-1} \frac{b}{1-\left(\dfrac{i}{I_{sat}} \right)^2} \frac{di}{i} = -\frac{1}{\tau_{RC_o}} dt \tag{4.5}$$

where
  $\tau_{RC_o} = R_o C$ is the ohmic RC time constant
  $R_o = V_c/I_{sat}$ is the ohmic resistance

By applying the boundary condition and integrating both sides of Equation 4.5 with respect to i and t, respectively, it becomes

$$\int \frac{1}{a}\left[bV_c \tanh^{-1}\left(\frac{i}{I_{sat}}\right)\right]^{\frac{1}{a}-1} \frac{b}{\left(1-\frac{i}{I_{sat}}\right)\left(1+\frac{i}{I_{sat}}\right)i}\,di = -\frac{t}{\tau_{RC_o}} + \ln k \tag{4.6}$$

where k is a constant value. To solve the integration part of Equation 4.6, the numerator of its left-hand side (LHS) is approximated to a cubic polynomial equation:

$$\int \frac{\frac{b}{a}\left[bV_c \tanh^{-1}\left(\frac{i}{I_{sat}}\right)\right]^{\frac{1}{a}-1}}{\left(1-\frac{i}{I_{sat}}\right)\left(1+\frac{i}{I_{sat}}\right)i}\,di = \int \frac{Wi^3 + Xi^2 + Yi + Z}{\left(1-\frac{i}{I_{sat}}\right)\left(1+\frac{i}{I_{sat}}\right)i}\,di \tag{4.7}$$

Then, the right-hand side (RHS) of Equation 4.7 can be rearranged using the partial fraction method to be

$$\frac{Wi^3 + Xi^2 + Yi + Z}{\left(1-\frac{i}{I_{sat}}\right)\left(1+\frac{i}{I_{sat}}\right)i} = \frac{A}{1-\frac{i}{I_{sat}}} + \frac{B}{1+\frac{i}{I_{sat}}} + \frac{C}{i} + D \tag{4.8}$$

Hence, it is necessary to find the coefficients A, B, C, D, W, X, Y, and Z. By plotting Equation 4.9

$$y = \frac{b}{a}\left[bV_c \tanh^{-1}\left(\frac{i}{I_{sat}}\right)\right]^{\frac{1}{a}-1} \tag{4.9}$$

in MATLAB®, a curve fitting cubic expression as depicted in Figure 4.3 can be obtained. In this case, the constants $W = 5.7 \times 10^{13}$, $X = -4.7 \times 10^9$, $Y = 1.52 \times 10^5$, and $Z = 0.0428$ can be obtained. With these constant values, Equation 4.9 can be solved by applying the partial fraction method, where $A = 4.13 \times 10^4$, $B = 3.36 \times 10^5$, $C = 0.0428$, and $D = -2.26 \times 10^5$.

By obtaining all the required constant values, Equation 4.6 can be changed to a simpler equation

$$\int \frac{A}{1-\frac{i}{I_{sat}}} + \frac{B}{1+\frac{i}{I_{sat}}} + \frac{C}{i} + D = -\frac{t}{\tau_{RC_o}} + \ln k \tag{4.10}$$

and ultimately reduced to

$$\frac{(I_{sat}+i)^{I_{sat}B}i^C 2.718^{Di}}{(I_{sat}-i)^{I_{sat}A}} = ke^{-\frac{t}{\tau_{RC_o}}} \tag{4.11}$$

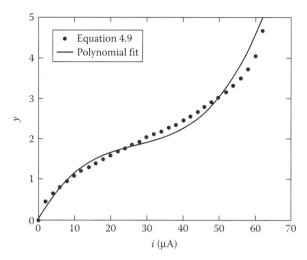

FIGURE 4.3 Approximation in real equation and polynomial equation.

With the given boundary condition

$$i(0) = I_{sat} \tanh\left(\frac{V^a}{bV_c}\right) \tag{4.12}$$

the constant value k is calculated as

$$k = \frac{\left[\left(I_{sat}\left(1+\tanh\left(\frac{V^a}{bV_c}\right)\right)\right)^{I_{sat}B}\right]\left[I_{sat}\tanh\left(\frac{V^a}{bV_c}\right)\right]^C e^{D\left(I_{sat}\tanh\left(\frac{V^a}{bV_c}\right)\right)}}{\left(I_{sat}\left(1-\tanh\left(\frac{V^a}{bV_c}\right)\right)\right)^{I_{sat}A}} \tag{4.13}$$

By substituting $V = V_{DD} = 1$ V and all the constant values in Equation 4.13, we get $k = 3.9753 \times 10^{-13}$. Figure 4.4 illustrates the charging current as a function of time by using Equation 4.11.

The ohmic RC time constant τ_{RC_o} in this case is 4.883 ps with ohmic resistance $R_o = V_c/I_{sat} = 3.352$ kΩ and load capacitance $C = 1.4566$ fF. The details of the load capacitance modeling are discussed in Section 4.3. With the current response as shown in Figure 4.4, the resistor and capacitor voltage response can be obtained easily by using Equation 4.2. Figures 4.5 and 4.6 show, respectively, the resistor and capacitor voltage response as a function of time.

From Figure 4.6, the rise time, which is the time taken to charge the load capacitor from 10% to 90% of its total supply voltage, is computed to be 25.111 ps. It is well known that RC time constant is defined as the time, at which the capacitor potential is $(1 - e^{-1})$ of the step potential V. In other words, when $t = \tau_{RC}$, the capacitor voltage v_{Cap} is $(1-e^{-1})V$. Hence, from Equation 4.2, when $t = \tau_{RC}$,

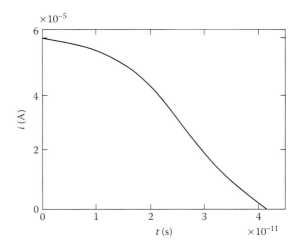

FIGURE 4.4   Current in RC circuit as a response to time.

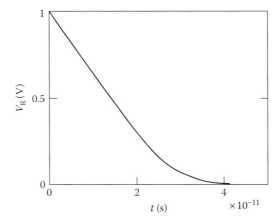

FIGURE 4.5   Resistor voltage in RC circuit as a response to time.

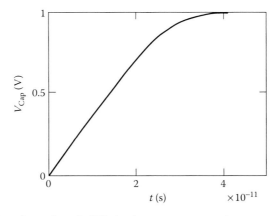

FIGURE 4.6   Capacitor voltage in RC circuit as a response to time.

$$V = v_R(\tau_{RC}) + v_{Cap}(\tau_{RC}) = \left[bV_c \tanh^{-1}\left(\frac{i(\tau_{RC})}{I_{sat}}\right) \right]^{1/a} + (1 - e^{-1})V \qquad (4.14)$$

and it can be solved to obtain the current when $t = \tau_{RC}$.

$$i(\tau_{RC}) = I_{sat} \tanh\left[\frac{(e^{-1}V)^a}{bV_c} \right] \qquad (4.15)$$

By applying Equation 4.15 in Equation 4.11 when $t = \tau_{RC}$, we get

$$\frac{\left(I_{sat} + i(\tau_{RC})\right)^{I_{sat}B} i_{RC}{}^C e^{Di(\tau_{RC})}}{\left(I_{sat} - i(\tau_{RC})\right)^{I_{sat}A}} = ke^{-\frac{\tau_{RC}}{\tau_{RC_0}}} \qquad (4.16)$$

Equation 4.16 is then simulated in MATLAB to obtain the relative RC time constant as a function of normalized voltage as shown in Figure 4.7.

From Figure 4.7, it is obvious that the relative RC time constant increases with the voltage in CNFET RC circuit due to the resistance blowup effect, similar to the results shown in Ref. [8]. By using the ohmic RC constant $0_{RC_0} = 4.883$ ps and critical voltage $V_c = 0.2112$ V, the RC time constant as a function of voltage can be obtained as shown in Figure 4.8.

The equivalent RC circuit shown in Figure 4.9 is an effective resistor for the discharging process through an n-type CNFET.

The established I_{DS}–V_{DS} of an n-type CNFET [7] with $V_{GS} = 1$ V, fitting the empirical Equation 4.1, is depicted in Figure 4.10.

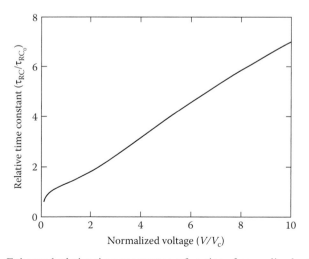

FIGURE 4.7 Enhanced relative time constant as a function of normalized voltage.

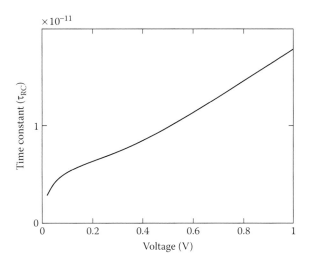

FIGURE 4.8   RC time constant as a function of voltage.

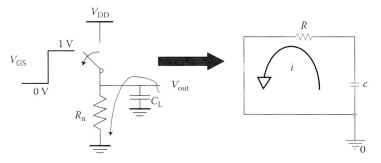

FIGURE 4.9   Equivalent RC circuit for the n-type CNFET discharging process.

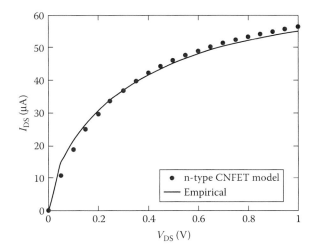

FIGURE 4.10   Fitting curve between the established n-type CNFET and the empirical equation.

In this case, the empirical equation with $I_{sat} = 63$ µA, $V_c = 0.2905$ V, $a = 0.58$, and $b = 2.55$ offers a highly reasonable fitting with the n-type CNFET model [7]. Hence, this empirical equation can be used to represent the established model with $V_{GS} = 1$ V.

The steps shown in Equations 4.2 through 4.12 are carried out having presumed $t_{fall} \approx t_{rise} = 25.111$ ps in order to find the fall time.

4.3 CAPACITANCE AND FREQUENCY RESPONSE OF CNFET

Load capacitance, C_L, plays an important role in determining the frequency response of a CNFET circuit. In general, the load capacitance, similar to the MOSFET circuit, can be divided into three categories—extrinsic or fan-out capacitance (C_{ext}), intrinsic capacitance (C_{int}), and wire capacitance (C_W). Figure 4.11 shows that the load capacitance between two inverters is actually a combination of intrinsic, extrinsic, and wire capacitances [9].

From Figure 4.11, C_W is the wire capacitance connecting the two inverters. Extrinsic capacitance serves as the gate capacitance C_G for the second inverter, and the intrinsic capacitance consists of gate-to-drain capacitance (C_{GD}) and drain-to-bulk capacitance (C_{DB}) for the first inverter. The combination of these capacitances is the so-called load capacitance. It is worth mentioning that this intrinsic capacitance is actually dependent on the input and output of the inverter circuit. The complete set of intrinsic and extrinsic capacitances in CNFET [10] is demonstrated in Figure 4.12.

It is well known that when two conductors are separated by a dielectric or an insulator, a capacitor exists. As shown in Figure 4.12, by applying a voltage on the CNFET, and having the conductors such as source terminal, drain terminal, gate

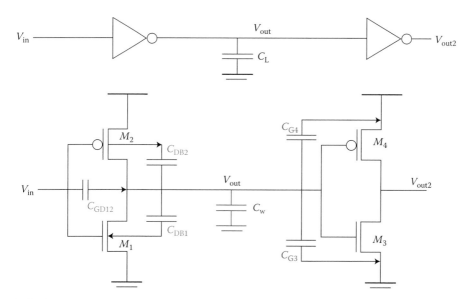

FIGURE 4.11 Equivalent circuit of inverters and load capacitances.

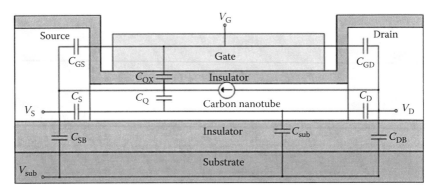

FIGURE 4.12   Intrinsic and extrinsic capacitances in CNFET.

terminal, CNT, and bulks separated by the gate oxide and insulator on the substrate, the capacitances appear. The gate or extrinsic capacitance is a combination of oxide capacitance (C_{ox}) in series with quantum capacitance (C_Q). The intrinsic capacitance in this case consists of gate-to-source capacitance (C_{GS}), gate-to-drain capacitance (C_{GD}), source-to-bulk capacitance (C_{SB}), drain-to-bulk capacitance (C_{DB}), and substrate capacitance (C_{sub}). In general, the wire capacitance is dominating in the load capacitance due to the negligible low values of intrinsic and extrinsic capacitances.

By applying the knowledge of the load capacitance on CNFET, the frequency response of a p-type and an n-type CNFET can be obtained easily. For a p-type model [7], its oxide capacitance per unit length is given by $C'_{ox} = 2\pi\kappa\varepsilon_o / \ln\left[(2t_{ox} + d)/d\right]$, while the quantum capacitance per length is given by

$$C'_Q = \frac{4m^* q^2}{n\pi^2 \hbar^2} \tag{4.17}$$

where n is the degenerate one-dimensional (1D) carrier concentration [7]. By assuming the high gate bias ($V_{GS} = 1$ V), the gate capacitance per length calculated using $C'^{-1}_g = C'^{-1}_Q + C'^{-1}_{ox}$ is 72.7 aF/μm. With the length of 50 nm CNT, the gate capacitance equals 3.635 aF, and the cutoff frequency of CNFET is given by

$$f_T = \frac{1}{2\pi} \frac{g_m}{C_L} \tag{4.18}$$

where

$$g_m = \frac{\partial I_{DS}}{\partial V_{GS}}\bigg|_{V_{DS}} = G\sqrt{\frac{2m^*}{q\left(1 + \dfrac{C'_Q}{C'_{ox}}\right)}} \frac{\mu_{eff}}{L} \frac{V_{DS}}{1 + \dfrac{V_{DS}}{V_c}} \left(\frac{1}{2}(V_{GS} - V_T)^{-\frac{1}{2}}\right) \tag{4.19}$$

is the transconductance of CNFET with I_{DS} shown in Ref. [7]. By assuming the high gate and drain bias ($V_{DS} = V_{GS} = 1$ V) and considering only the gate capacitance ($C_L = C_g$), the cutoff frequency of p-type CNFET is computed to be 1.643 THz with transconductance being 37.5397 µS. However, if the intrinsic capacitances are taken into account, the cutoff frequency is expected to be reduced due to the increasing C_L. As shown in Figure 4.12, the intrinsic capacitance is a combination of C_{GS}, C_{GD}, C_{SB}, and C_{DB}. The gate-to-source and gate-to-drain capacitances are given by [11]

$$C_{GS} = \frac{L}{2} C'_{ox} \left[\frac{C'_Q + 2(1-\beta)C'_C}{C'_{tot} + C'_Q} \right] \tag{4.20}$$

$$C_{GD} = \frac{L}{2} C'_{ox} \left[\frac{C'_Q + 2(\beta)C'_C}{C'_{tot} + C'_Q} \right] \tag{4.21}$$

where
    C'_{tot} is the total capacitance per length
    β and C'_C are fitting parameters, with C'_C being the capacitance per length between the channel and the external drain/source terminal

The total capacitance per length is given by

$$C'_{tot} = C'_{ox} + C'_{sub} + C'_C \tag{4.22}$$

where

$$C'_{sub} = \frac{2\pi\kappa_{sub}\varepsilon_0}{\ln(4H_{sub}/d)} \tag{4.23}$$

is the substrate capacitance per length, κ_{sub} is the relative permittivity of the substrate, and H_{sub} is the thickness of substrate. By considering 10 µm silicon oxide substrate ($\kappa_{sub} = 3.9$ and $C'_C = 0$) [12], the gate-to-source and gate-to-drain capacitances are equal to 1.745 aF. The source-to-bulk and drain-to-bulk capacitances per length can be expressed as [11]

$$C_{SB} = C_{GS} \left(\frac{C'_{sub}}{C'_{ox}} \right) \tag{4.24}$$

$$C_{DB} = C_{GD} \left(\frac{C'_{sub}}{C'_{ox}} \right) \tag{4.25}$$

Hence, the total intrinsic capacitance was computed to be 3.664 aF. By assuming that the load capacitance is a summation of gate and intrinsic capacitances, the cutoff frequency was calculated to be 818.5 GHz. As mentioned earlier, the wire capacitance dominates the load capacitance. Thus, the frequency response is expected to degrade dramatically if the wire capacitance is taken into account. By considering a local wire with width 32 nm [13], the wire capacitance per length for Cu and MWNT are 144.93 and 130.15 aF/µm, respectively. Figure 4.13 illustrates the p-type CNFET cutoff frequency as a function of wire/interconnect length.

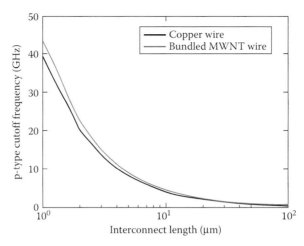

FIGURE 4.13   Cutoff frequency as a function of wire/interconnect length in a p-type CNFET.

As shown in Figure 4.13, when the length of the wire increases, it is obvious that the cutoff frequency will drop linearly. For example, by considering the gate and intrinsic capacitances and using the 1 μm copper wire, the cutoff frequency is 39.24 GHz. However, in case the 1 μm copper wire is replaced by the 100 μm copper wire, the cutoff frequency will be reduced to 411.99 MHz.

For the n-type CNFET, the same equations (Equations 4.17 through 4.25) are utilized to examine its cutoff frequency with different load capacitance. By using $C_g'^{-1} = C_{ox}'^{-1} + C_Q'^{-1}$, the gate capacitance is calculated to be 3.635 aF, which is the same as shown in p-type CNFET. However, due to different materials being used for the electrode, the n-type CNFET has lower mobility that causes the transconductance to be slightly reduced to 35.204 μS. Hence, by considering only the gate capacitance as load capacitance, its cutoff frequency is 1.541 THz. As mentioned earlier, it is expected that the cutoff frequency reduces if the intrinsic capacitance is taken into account. By using exactly the same parameters as in the p-type CNFET and Equations 4.20 through 4.25, the intrinsic capacitance for n-type CNFET is the same as the p-type CNFET, which is 3.664 aF. Thus, the cutoff frequency is 767.6 GHz if the gate and intrinsic capacitances are used as load capacitance. Next, a local wire with width 32 nm [13] is considered to contribute to the load capacitance. Figure 4.14 demonstrates the n-type CNFET cutoff frequency as a function of wire/interconnect length.

By taking the gate and intrinsic capacitances into account and using the 1 μm copper wire, the cutoff frequency is 36.806 GHz. However, if the 1 μm copper wire is replaced by the 100 μm copper wire, the cutoff frequency will be reduced to 386.4 MHz. Therefore, it can be summarized that the wire capacitance dominates the load capacitance, which indeed significantly degrades the performance of the CNFET circuit. This also explains why it is hard to achieve the terahertz frequency response when the CNFET was used in the circuit.

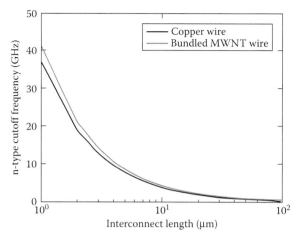

FIGURE 4.14 Cutoff frequency as a function of wire/interconnect length in an n-type CNFET.

4.4 PROPAGATION DELAY IN CNFET LOGIC GATES

4.4.1 CALCULATION USING NONOHMIC LAW

Propagation delay refers to the amount of time starting from the time when input becomes stable and valid to the time when output becomes stable and valid. Figure 4.15 shows the high-to-low propagation delay (t_{pHL}) and low-to-high propagation delay (t_{pLH}) clearly, where t_{pHL} and t_{pLH} are defined as the time from 50% of the input to the time at 50% of the output. The figure also shows the rise time (t_r) and the fall time (t_f), in which t_r is the time rising from 10% to 90% of the output, while t_f is the time falling from 90% to 10% of the output. The total propagation delay (t_p) is the average of t_{pHL} and t_{pLH}.

$$t_p = \frac{t_{pHL} + t_{pLH}}{2} \tag{4.26}$$

In this section, the propagation delays for logic gates are examined by applying the nonohmic law, and then compared with the simulation results obtained from Hspice. It is well known that the propagation delay can be calculated through

$$t_{pLH} = 0.69 R_{eqp} C_L \tag{4.27}$$

$$t_{pHL} = 0.69 R_{eqn} C_L \tag{4.28}$$

where

R_{eqp} is the equivalent resistance for a p-type CNFET to charge the load capacitor from $V_{DD}/2$ to V_{DD}

R_{eqn} is the equivalent resistance for an n-type CNFET to discharge the load capacitor from V_{DD} to $V_{DD}/2$ [9]

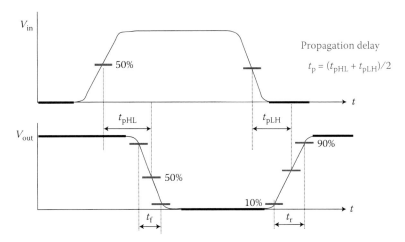

FIGURE 4.15   Propagation delay and rise/fall time.

By incorporating the nonohmic effect into the equivalent resistance, it can be expressed as

$$R_{eqp(n)} = \frac{1}{2}\left[\frac{V_{DD}}{I_{sat}\tanh\left(\dfrac{V_{DD}^{a}}{bV_c}\right)} + \frac{V_{DD}/2}{I_{sat}\tanh\left(\dfrac{(V_{DD}/2)^a}{bV_c}\right)}\right] \tag{4.29}$$

where the constants I_{sat}, V_c, a, and b can be obtained from Section 4.2 with $V_{DD} = 1$ V. Hence, the R_{eqp} in this case is calculated to be 13.196 kΩ, while the R_{eqn} is 14.609 kΩ. As the equivalent resistance is constant for every logic gate, the propagation delay is actually dependent on the load capacitance that is connected to the output of each logic gate. As mentioned in Section 4.3, the wire capacitance dominates the load capacitance. In order to investigate the effect of wire length on the propagation delay, three wire lengths are considered in this case—1, 10, and 100 μm. By using the local 32 nm Cu wire [13], the resistivity and wire capacitances for three wires are determined. Besides, the intrinsic and extrinsic capacitances are also included in the calculation in order to model the logic gate accurately.

As mentioned in Figure 4.11, the load capacitance for an inverter connected to another inverter is given by

$$C_L = C_{int} + C_{ext} + C_W \tag{4.30}$$

where

$$C_{int} = C_{GD1} + C_{GD2} + C_{DB1} + C_{DB2} \tag{4.31}$$

$$C_{ext} = C_{G3} + C_{G4} \tag{4.32}$$

By using the same values for C_{int} and C_{ext} as in Section 4.3, the load capacitance is now dependent on the wire capacitance of different lengths. For a nine-stage oscillator ring, as shown in Figure 4.16, the load capacitances between inverters is exactly the same as given in Equation 4.30. Therefore, the total propagation delay from the first inverter to the final inverter in an oscillator ring can be expressed as

$$t_{pLH} = N \times 0.69 R_{eqp} C_L \tag{4.33}$$

$$t_{pHL} = N \times 0.69 R_{eqn} C_L \tag{4.34}$$

where N is the number of stage.

For a NAND gate, for example, the 4-input NAND gate or NAND4 as shown in Figure 4.17, the load capacitance is a combination of the wire capacitance and the output-surrounding intrinsic and extrinsic capacitances of the next NAND4 gate [9]

$$C_L = C_{DB4} + C_{DB5} + C_{DB6} + C_{DB7} + C_{DB8} + 2C_{GD4}$$

$$+ 2C_{GD5} + 2C_{GD6} + 2C_{GD7} + 2C_{GD8} + C_W + C_{ext} \tag{4.35}$$

where the values of C_{int} can be obtained from Section 4.3, while C_{ext} in this case can be taken from Equation 4.32, as the output of NAND4 is assumed to be connected to one p-type and one n-type CNFET of the next NAND4 gate.

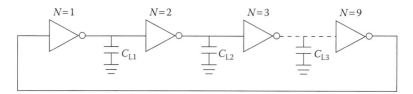

FIGURE 4.16 Nine-stage oscillator ring.

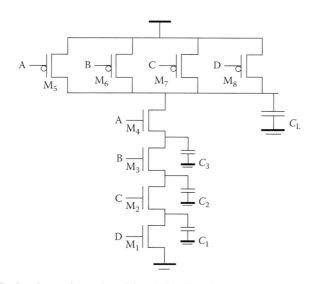

FIGURE 4.17 Load capacitance in a 4-input NAND gate.

Also shown in Figure 4.17 are the intrinsic capacitances at a series of n-type CNFET, which can be expressed as [9]

$$C_1 = C_{DB1} + C_{SB2} + 2C_{GD1} + 2C_{GS2} \tag{4.36}$$

$$C_2 = C_{DB2} + C_{SB3} + 2C_{GD2} + 2C_{GS3} \tag{4.37}$$

$$C_3 = C_{DB3} + C_{SB4} + 2C_{GD3} + 2C_{GS4} \tag{4.38}$$

By applying the Elmore delay model [9] on NAND4 gate, the propagation delay is given by

$$t_{pLH} = 0.69 R_{eqp} C_L \tag{4.39}$$

$$t_{pHL} = 0.69 R_{eqn} (C_1 + 2C_2 + 3C_3 + 4C_L) \tag{4.40}$$

In this case, the worst case is considered where only one p-type CNFET is assumed to be turned ON in one time. The same concepts can be applied on NAND3 and NAND2 gates in order to calculate their load capacitance and propagation delay. For a NOR gate, for example, the 4-input NOR gate or NOR4 as shown in Figure 4.18, the load capacitance is the combination of wire capacitance and output-surrounding intrinsic and extrinsic capacitances of the next NOR4 gate [9], which are exactly the same as the formula given in Equation 4.35. The values of C_{int} and C_{ext} can be again obtained from Section 4.3 and Equation 4.32, respectively, as the output of NOR4 was assumed to be connected to one p-type and one n-type CNFET of the next NOR4 gate.

The intrinsic capacitance at a series of p-type CNFET, as shown in Figure 4.18, can also be obtained by using Equations 4.36 through 4.38. Again, by applying the Elmore delay model [9] on NOR4 gate, the propagation delay is given as

$$t_{pLH} = 0.69 R_{eqp} (C_1 + 2C_2 + 3C_3 + 4C_L) \tag{4.41}$$

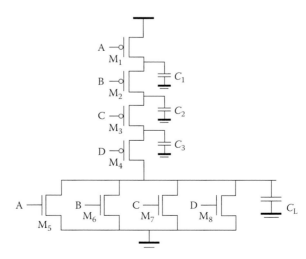

FIGURE 4.18   Load capacitance in a 4-input NOR gate.

$$t_{pHL} = 0.69 R_{eqn} C_L \qquad (4.42)$$

Here the worst case is considered where only one n-type CNFET is assumed to be turned ON in one time. The same concepts can be applied on NOR3 and NOR2 gates in order to calculate their load capacitance and propagation delay. Table 4.1 summarizes the load capacitance for all the aforementioned logic gates, while Table 4.2 shows their propagation delay computation using the theory discussed in this section. The calculation is done for three different lengths of wire.

4.4.2 SIMULATION USING HSPICE

In order to verify the propagation delay theory, a nonohmic law-based theory that has been discussed earlier, the calculation results in Table 4.2 are compared with the

TABLE 4.1

Load Capacitance Calculation of Wires with Three Different Lengths

	C_L		
Logic Gates	1 μm (C_W = 144.93 aF)	10 μm (C_W = 1.4493 fF)	100 μm (C_W = 14.493 fF)
Inverter	155.864 aF	1.460 fF	14.504 fF
Nine-stage oscillator ring	155.864 aF	1.460 fF	14.504 fF
NAND2	162.931 aF	1.467 fF	14.511 fF
NAND3	166.508 aF	1.471 fF	14.515 fF
NAND4	170.085 aF	1.474 fF	14.518 fF
NOR2	162.931 aF	1.467 fF	14.511 fF
NOR3	166.508 aF	1.471 fF	14.515 fF
NOR4	170.085 aF	1.474 fF	14.518 fF

TABLE 4.2

Propagation Delay Calculation of Wires with Three Different Lengths

	Propagation Delay (ps)					
Logic Gates	1 μm		10 μm		100 μm	
	t_{pHL}	t_{pLH}	t_{pHL}	t_{pLH}	t_{pHL}	t_{pLH}
Inverter	1.57	1.42	14.72	13.30	146.20	132.06
Nine-stage oscillator ring	14.14	12.77	132.47	119.66	1315.83	1188.56
NAND2	3.36	1.49	29.47	13.36	292.62	132.13
NAND3	5.25	1.52	44.70	13.39	439.16	132.16
NAND4	7.29	1.55	59.87	13.42	585.81	132.19
NOR2	1.64	3.03	14.79	26.78	146.27	264.32
NOR3	1.68	4.74	14.83	40.38	146.31	396.68
NOR4	1.71	6.585	14.86	54.08	146.34	529.15

Hspice simulation results. In addition, the logic function for every CNFET logic gate is presented in this section to examine the performance of the established CNFET model when used in circuit and logic levels.

First, the CNFET library file (.lib) is set up according to the established CNFET model [7]. Next, an Hspice simulation file (.sp) is used to test and validate the CNFET model that was built in Hspice. The simulation results for p-type and n-type CNFET are shown in Figures 4.19 and 4.20, respectively. Hence, this CNFET library file in Hspice can now be used to represent the established three-terminal CNFET model, and it is ready to be used as a fundamental to develop the CMOS inverter, NOR, NAND, and oscillator ring.

In this section, three set of simulations are carried out for three different lengths of wire and load capacitances. Besides, the logic function for each gate is also determined to further verify the established model in Hspice. Figure 4.21a through g

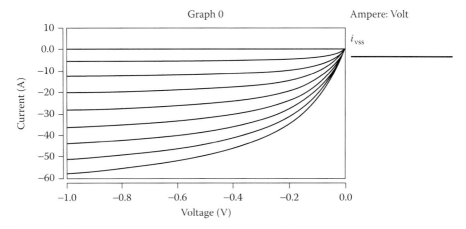

FIGURE 4.19   Simulation result of a p-type CNFET in Hspice.

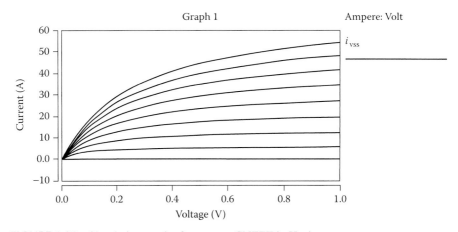

FIGURE 4.20   Simulation result of an n-type CNFET in Hspice.

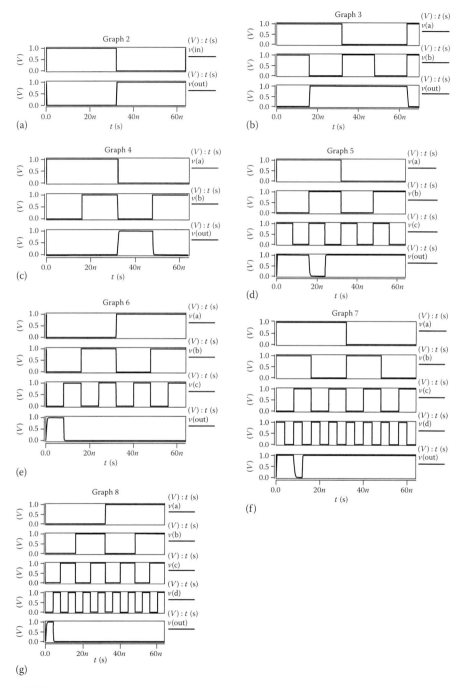

FIGURE 4.21 Simulation results of CNFET logic gates built in Hspice. (a) CMOS inverter, (b) 2-input NAND gate, (c) 2-input NOR gate, (d) 3-input NAND gate, (e) 3-input NOR gate, (f) 4-input NAND gate, and (g) 4-input NOR gate.

present the time simulation results of logic gates in Hspice. It is shown that the logic function for each logic gates below is accurate.

In addition to the logic function verification, the propagation delay, rise/fall time, and average power consumption of each gate can be obtained easily through simulation in Hspice. In order to have a fair comparison with the propagation delay calculated manually in Table 4.2, the load capacitances summarized in Table 4.1 have been used. Tables 4.3 through 4.5 illustrate the simulated propagation delay, rise/fall time, and average power consumption in Hspice for three different lengths of wire.

These Hspice simulated propagation delays (as shown in Tables 4.3 through 4.5) are then compared with the computed propagation delay (as shown in Table 4.2) in order to examine and verify the application of nonohmic law at the CNFET circuit level.

TABLE 4.3

Simulated Propagation Delay, Rise/Fall Time, and Average Power Consumption for CNFET with 1 μm Wire

Logic Gates	Rise Time (ps)	Fall Time (ps)	t_{pLH} (ps)	t_{pHL} (ps)	t_p (ps)	Average Power (nW)
Inverter	2.8288	3.1907	1.4588	1.5886	1.5237	2.3981
Nine-stage oscillator ring	3.3991	3.6024	18.327	18.465	18.396	23.720
NAND2	3.0456	5.3429	1.5895	2.5251	2.0573	4.2103
NAND3	3.1211	7.4240	1.6806	3.5188	2.5997	8.3057
NAND4	7.6842	11.932	3.8179	10.354	7.0858	14.354
NOR2	4.4651	3.3427	2.4255	1.6509	2.0382	2.6306
NOR3	6.0395	3.6482	3.1635	1.8185	2.4910	2.5232
NOR4	84.429	63.897	79.201	23.576	51.389	102.56

TABLE 4.4

Simulated Propagation Delay, Rise/Fall Time, and Average Power Consumption for CNFET with 10 μm Wires

Logic Gates	Rise Time (ps)	Fall Time (ps)	t_{pLH} (ps)	t_{pHL} (ps)	t_p (ps)	Average Power (nW)
Inverter	26.434	29.991	13.389	14.516	13.953	22.440
Nine-stage oscillator ring	31.746	33.901	171.57	172.59	172.08	222.35
NAND2	27.133	48.254	13.716	22.872	18.294	22.940
NAND3	26.482	64.643	13.714	30.347	22.031	45.910
NAND4	70.816	176.83	34.965	95.777	65.371	141.45
NOR2	40.145	29.943	20.219	14.608	17.414	22.890
NOR3	53.071	30.526	25.465	15.036	20.750	23.228
NOR4	138.65	93.807	108.74	41.071	74.905	124.76

TABLE 4.5
Simulated Propagation Delay, Rise/Fall Time, and Average Power Consumption for CNFET with 100 μm Wires

Logic Gates	Rise Time (ps)	Fall Time (ps)	t_{pLH} (ps)	t_{pHL} (ps)	t_p (ps)	Average Power (nW)
Inverter	260.12	295.13	133.66	144.71	139.18	226.52
Nine-stage oscillator ring	309.19	332.05	1.6942	1.7067	1.7005	2225.3
NAND2	260.68	472.02	133.97	225.90	179.94	221.21
NAND3	260.38	636.88	134.03	298.63	216.33	452.95
NAND4	308.75	884.44	165.86	428.12	296.99	547.28
NOR2	395.15	295.27	198.70	144.14	171.42	226.78
NOR3	518.88	295.56	255.17	145.09	200.13	226.72
NOR4	705.48	366.29	386.52	181.08	283.80	328.58

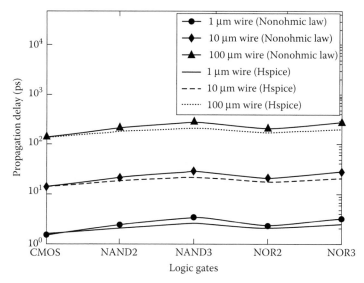

FIGURE 4.22 Comparison results for Hspice simulated propagation delay and the computed propagation delay for CNFET.

Figure 4.22 demonstrates the comparison results for CMOS inverter, NAND2, NAND3, NOR2, and NOR3, where the Hspice simulation results were proved to obey well the nonohmic law calculation. In general, the Hspice simulated propagation delay is slightly lower than the theoretical propagation delay. The comparison results for NAND4 and NOR4 are not shown in Figure 4.22 as they appear to violate the nonohmic law. This is most likely due to the large interval time in Hspice simulation, which results in large propagation delay. The reduction in the time intervals resulted in errors and in the termination of the simulations. In fact, this is a limitation of Hspice software.

4.5 RESULT COMPARISON AND ANALYSIS

In this section, it is worthwhile to compare the future device (established CNFET model) with the current technology (MOSFET) to further examine its performance and capability. The 32 nm MOSFET model considered in this case is one of the latest predictive models [14]. In order to determine the width, W, of the MOSFET model, it is compared with the established CNFET model [7], as shown in Figure 4.23.

By using $W = 120$ nm for PMOS and $W = 45$ nm for NMOS, both the PMOS and NMOS models have comparable on-current with the p-type and n-type CNFET models, respectively. Next, this MOSFET model card is used as a library file in Hspice, while the SPICE simulation file for each logic gates is set up based on the library file to obtain their rise/fall time, propagation delay, and average power consumption. In this case, the load capacitances for each gate are assumed to be equal to their respective wire capacitance due to the negligible intrinsic capacitance in MOSFET. Tables 4.6 through 4.8 illustrate the simulated propagation delay, rise/fall time, and average power consumption in Hspice for three different lengths of wire in MOSFET model.

These propagation delays for MOSFET model (as shown in Tables 4.6 through 4.8) are then compared with the simulated propagation delays for CNFET (as shown in Tables 4.3 through 4.5) in order to examine the performance of CNFET as a

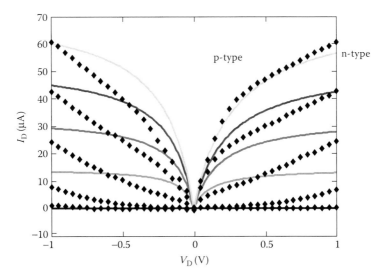

FIGURE 4.23 Comparison results of I_{DS}–V_{DS} characteristics for both n-type and p-type CNFET models (solid line) with the 32 nm MOSFET model (dotted line). p-Type CNFET was asserted with V_{GS} from −0.2 V at the bottom to −1.0 V at the top with −0.2 V step increment, while n-type was asserted with V_{GS} from 0.2 V at the bottom to 1.0 V at the top with 0.2 V step increment.

TABLE 4.6
Simulated Propagation Delay, Rise/Fall Time, and Average Power Consumption for MOSFET with 1 µm Wire

Logic Gates	Rise Time (ps)	Fall Time (ps)	t_{pLH} (ps)	t_{pHL} (ps)	t_p (ps)	Average Power (nW)
Inverter	7.4527	6.8815	4.1192	4.1508	4.1350	11.774
Nine-stage oscillator ring	12.504	10.571	74.193	73.413	73.803	123.50
NAND2	12.662	15.506	6.2568	12.326	9.2912	16.623
NAND3	13.283	28.348	7.7194	22.924	15.322	24.589
NAND4	14.731	45.325	8.9356	36.396	22.666	27.232
NOR2	24.378	7.2378	18.268	4.6819	11.475	12.972
NOR3	48.186	18.037	51.208	7.0935	29.151	26.801
NOR4	79.191	19.815	97.509	6.8562	52.183	36.131

TABLE 4.7
Simulated Propagation Delay, Rise/Fall Time, and Average Power Consumption for MOSFET with 10 µm Wires

Logic Gates	Rise Time (ps)	Fall Time (ps)	t_{pLH} (ps)	t_{pHL} (ps)	t_p (ps)	Average Power (nW)
Inverter	42.730	34.206	17.994	17.952	17.973	33.329
Nine-stage oscillator ring	52.892	45.998	271.13	270.16	270.65	326.02
NAND2	47.305	67.221	20.482	38.128	29.305	38.367
NAND3	49.518	105.08	22.462	61.566	42.014	65.753
NAND4	52.576	147.24	24.085	88.018	56.052	67.833
NOR2	101.45	34.854	54.289	18.213	36.251	32.704
NOR3	169.99	46.182	109.57	23.635	66.603	47.137
NOR4	246.54	49.848	181.63	24.014	102.82	56.585

comparison to latest technology. Figure 4.24 demonstrates the comparison results for CMOS inverter, NAND2, NAND3, NOR2, and NOR3.

In general, the 32 nm MOSFET model has higher propagation delay compared to the CNFET model, proving the extraordinary performance of CNFET in circuit and logic levels. Due to the limitation of Hspice, the NAND4 and NOR4 gates are again neglected in this comparison. However, by sizing the width of NMOS and PMOS in every MOSFET logic gate, the propagation delay for each gate can be improved significantly. The widths of NMOS and PMOS are sized in order to obtain the same t_{pLH} and t_{pHL}. By doing so, the total propagation delay can be reduced. Figure 4.25 shows the comparison of propagation delay for CNFET and MOSFET after sizing.

TABLE 4.8

Simulated Propagation Delay, Rise/Fall Time, and Average Power Consumption for MOSFET with 100 µm Wires

Logic Gates	Rise Time (ps)	Fall Time (ps)	t_{pLH} (ps)	t_{pHL} (ps)	t_p (ps)	Average Power (nW)
Inverter	395.12	316.95	160.05	156.69	158.37	238.01
NAND2	399.63	591.04	162.40	297.03	229.72	242.40
NAND3	401.91	875.40	164.71	446.65	305.68	473.11
NAND4	405.89	1167.3	166.62	602.99	384.80	476.57
NOR2	879.95	317.57	399.92	156.89	278.40	237.43
NOR3	1.3985	327.96	682.22	164.84	423.53	250.08
NOR4	1.9364	333.02	992.81	166.37	579.59	258.03

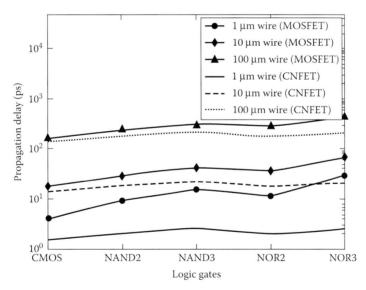

FIGURE 4.24 Comparison results for propagation delay in MOSFET and CNFET. Both models are simulated in Hspice.

As shown in Figure 4.25, the propagation delays for MOSFET logic gates after sizing are reduced compared to the MOSFET logic gates before sizing as shown in Figure 4.24. For a 100 µm wire, the MOSFET logic gates have propagation delay as low as that of the CNFET.

Power delay product (PDP) and energy delay product (EDP) are two quality measures for a logic gate. In general, these two measures determine the average performance of a logic gate from the aspects of propagation delay and power consumption. A good logic gate must have low PDP and EDP. Therefore, PDP and EDP are used

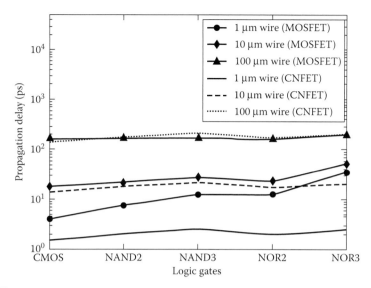

FIGURE 4.25 Comparison results for propagation delay in MOSFET after sizing and CNFET. Both models are simulated in Hspice.

to compare the performance of established CNFET and 32 nm MOSFET. From Tables 4.3 through 4.5, the PDP and EDP can be calculated easily by using

$$\text{PDP} = P_{av}t_p \tag{4.43}$$

$$\text{EDP} = \text{PDP} \times t_p = P_{av}t_p^2 \tag{4.44}$$

Figures 4.26 and 4.27 demonstrate the PDP and EDP of different logic gates made of CNFET as a function of wire length.

The same steps are then carried out on MOSFET by applying the same equations on Tables 4.6 through 4.8. Figures 4.28 and 4.29 show the PDP and EDP for logic gates made of 32 nm MOSFET model as a function of wire length.

By comparing Figures 4.26 and 4.28, the logic gates made of CNFETs have generally lower PDP compared to logic gates made of MOSFETs. On the other hand, the EDP for CNFET logic gates shown in Figure 4.27 is obviously lower than the EDP for MOSFET logic gates shown in Figure 4.29. However, by sizing the width of NMOS and PMOS accordingly, the propagation delay for MOSFET logic gates can be reduced as shown in Figure 4.25. Again, by applying the same steps on these sized MOSFET logic gates, their PDP and EDP can be obtained as shown in Figures 4.30 and 4.31.

It is shown that the sized MOSFETs can produce logic gates with lower PDP and EDP compared to CNFETs and MOSFETs with no sizing, which is due to their lower propagation delay and power consumption. In fact, the performance of CNFET logic gates can also be enhanced by using multichannel CNFET or the so-called CNFET arrays [15–17] instead of single CNFET as a channel, which is not included in the objectives of this chapter.

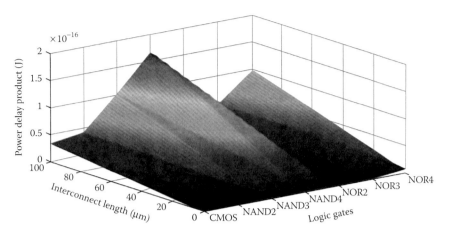

FIGURE 4.26 **(See color insert.)** PDP for CNFET logic gates as a function of wire length.

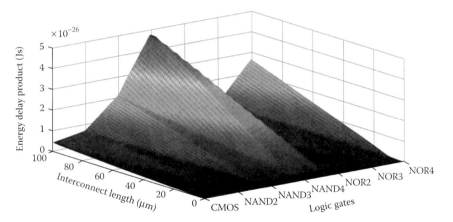

FIGURE 4.27 **(See color insert.)** EDP for CNFET logic gates as a function of wire length.

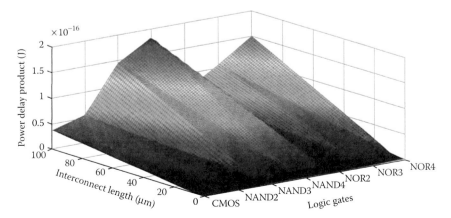

FIGURE 4.28 **(See color insert.)** PDP for MOSFET logic gates as a function of wire length.

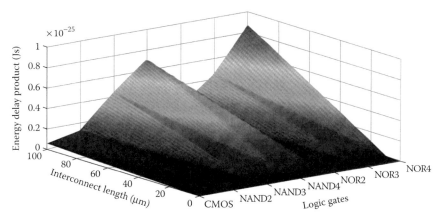

FIGURE 4.29 **(See color insert.)** EDP for MOSFET logic gates as a function of wire length.

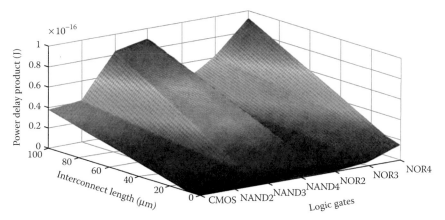

FIGURE 4.30 **(See color insert.)** PDP for sized MOSFET logic gates as a function of wire length.

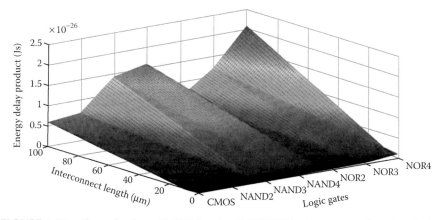

FIGURE 4.31 **(See color insert.)** EDP for sized MOSFET logic gates as a function of wire length.

4.6 CONCLUSION

CNT is a promising material for semiconductor devices due to its excellent electrical properties. This chapter focused on the application of CNFET at circuit- and logic gate levels. The nonohmic law was used to replace the well-known Ohm's law in circuit application due to the velocity saturation effect. The current is no longer increased with the applied voltage, but saturated at the critical voltage. Initially, the transient response of 1D nanoresistor has been investigated by using the nonohmic law. Since one of the main objectives of this chapter was to examine the performance of the established CNFET model at circuit level, the RC transient response in CNFET circuit was examined. By using different wire lengths, the theoretical results were compared with the Hspice simulation results to confirm the application of nonohmic law in CNFET circuit. As the ultimate results, the CNFET was compared with the state-of-the-art MOSFET. The established CNFET model was found to have better performance compared to the MOSFET. However, the finding of this chapter provides evidence to state that having sized the MOSFET leading to the trend of multichannel CNFET facilitates a better performance than the CNFET model. In fact, the logic gates made of CNFET can be enhanced by using multichannel CNFET instead of the single CNT channel. The multichannel CNFET is able to provide higher on-current and better performance compared to the single channel CNFET. Besides, the width of CNFETs can be resized to have lower propagation delay and power consumption in every logic gate. This is a clear illustration of the fact that the CNFET can be used to replace the MOSFET.

REFERENCES

1. A. S. Sedra and K. C. Smith. *Microelectronic Circuits*, Revised Edition. Oxford: Oxford University Press, 2007.
2. D. R. Greenberg and J. A. Del Alamo. Velocity saturation in the extrinsic device: A fundamental limit in HFETs. *IEEE Transactions on Electron Devices*. 41, 1334–1339, 1994.
3. V. K. Arora. High field distribution and mobility in semiconductors. *Japanese Journal of Applied Physics*. 24, 537, 1985.
4. V. K. Arora and M. B. Das. Effect of electric-field-induced mobility degradation on the velocity distribution in a sub-mu m length channel of InGaAs/AlGaAs heterojunction MODFET. *Semiconductor Science and Technology*. 5, 967–973, 1990.
5. V. K. Arora. Quantum engineering of nanoelectronic devices: The role of quantum emission in limiting drift velocity and diffusion coefficient. *Microelectronics Journal*. 31, 853–859, 2000.
6. T. Saxena, D. C. Y. Chek, M. L. P. Tan, and V. K. Arora. Micro-circuit modeling and simulation beyond Ohm's law. *IEEE Transaction on Education*. 54(1), 34–40, 2011.
7. D. C. Y. Chek, M. L. P. Tan, M. T. Ahmadi, R. Ismail, and V. K. Arora. Analytical modeling of high performance single-walled carbon nanotube field-effect-transistor. *Microelectronic Journal*. 41, 579–584, 2010.
8. V. K. Arora, A. M. B. Hashim, D. C. Y. Chek, and T. Saxena. The impact of the breakdown of Ohm's law on switching delay due to reactive elements connected in series with a micro/nano-resistor. In *International Semiconductor Device Research Symposium (ISDRS)*. University of Maryland, College Park, MD, 2009.

9. J. M. Rabaey, A. Chandrakasan, and B. Nikolic. *Digital Integrated Circuit: A Design Perspective.* Upper Saddle River, NJ: Prentice Hall, 2003.
10. T. Kazmierski, D. Zhou, and B. Al-Hashimi. HSPICE implementation of a numerically efficient model of CNT transistor. In *Forum on Specification and Design Languages (FDL 2009).* Austria, 2009.
11. J. Deng and H.-S. P. Wong. A compact SPICE model for carbon-nanotube field-effect transistors including nonidealities and its application—Part I: Model of the intrinsic channel region. *IEEE Transactions on Electron Devices.* 54, 3186–3194, 2007.
12. J. Deng and H.-S. P. Wong. A compact SPICE model for carbon-nanotube field-effect transistors including nonidealities and its application—Part II: Full device model and circuit performance benchmarking. *IEEE Transactions on Electron Devices.* 54, 3195–3205, 2007.
13. H. Li, W.-Y. Yin, K. Banerjee, and J.-F. Mao. Circuit modeling and performance analysis of multi-walled carbon nanotube interconnects. *IEEE Transactions on Electron Devices.* 55, 1328–1337, 2008.
14. Nanoscale Integration and Modeling (NIMO) Group, Ed. Predictive technology model. In *Latest Models*, Vol. 2010. Arizona State University, 2008.
15. P. Avouris, R. Martel, V. Derycke, and J. Appenzeller. Carbon nanotube transistors and logic circuits. *Physica B.* 323, 6–14, 2002.
16. A. Javey, Q. Wang, A. Ural, Y. Li, and H. Dai. Carbon nanotube transistor arrays for multistage complementary logic and ring oscillators. *Nano Letters.* 2, 929–932, 2002.
17. C. Chen, D. Xu, E. S.-W. Kong, and Y. Zhang. Multichannel carbon-nanotube FETs and complementary logic gates with nanowelded contacts. *IEEE Electron Device Letters.* 27, 852–855, 2006.

5 Graphene Nanoribbon Field Effect Transistors

Noraliah Aziziah Amin, Mohammad Taghi Ahmadi, and Razali Ismail

CONTENTS

5.1   INTRODUCTION

Silicon-based electronics is expected to reach its limitation when a three-terminal device known as the field effect transistor (FET) reaches below the 10 nm regime technology as it suffers from numerous drawbacks. Therefore, as the technology progresses, a new channel material plays an important role as a successor of the traditional silicon-based FET for increasing the speed of the device. Among the new materials are nanowires, carbon nanotubes (CNTs), graphene and its derivatives, and graphene nanoribbons (GNRs). The major issue in controlling the CNT's chirality (rolling direction) [1–3] has led to the search for its successor—GNR—which is much easier to deal with. A variety of applications of the graphene nanoribbon field effect transistor (GNRFET) have recently manipulated the advantages of GNR, such as the graphene-based nanoswitch [4], tunneling transistor based on GNR [5], as well as for optical detection [6]. In order to apply GNR as a nanostructure device, the key to this application is to first understand the underlying physics of this material. This chapter reviews and discusses the theoretical physics related to graphene and GNR.

5.2   GRAPHENE AND GRAPHENE NANORIBBON

Graphene, the two-dimensional (2D) allotrope of sp^2-bonded carbon [7,8] arranged in a honeycomb lattice, is a small band gap material. As shown in Figure 5.1, this 2D structure is very useful for describing the properties of its carbon-based siblings including 0D fullerenes (spherical shape of graphene sheets), 1D CNTs (folded graphene sheet in nanometer-sized cylindrical shape), 3D graphite [9] (stacking of several layers of graphene sheets), as well as GNR—a strip of graphene with dimension less than 10 nm [10,11] able to produce band gap due to quantum confinement effect [10,11]. Simply put, GNR originated from graphene and hence has inherited similar properties. There are several ways of obtaining GNRs from CNTs, one of which is by unzipping a single-walled carbon nanotube (SWNT) to form a monolayer GNR [12–15]. However, this method has serious drawbacks related to structural defects and surface contamination that will lead to electrical degradation [16]. Alternatively, GNR can also be obtained through a simpler method by repeated peeling of graphite [17].

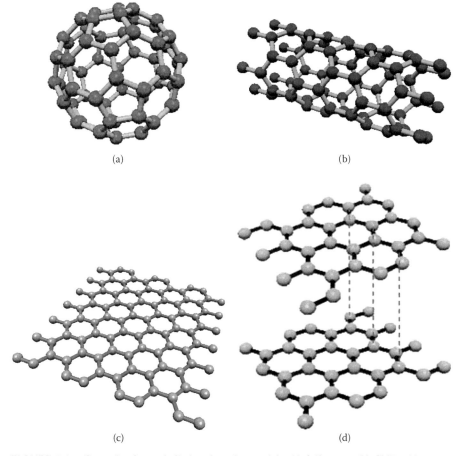

(a)   (b)

(c)   (d)

FIGURE 5.1   **(See color insert.)** Carbon-based materials: (a) fullerenes, (b) CNTs, (c) monolayer graphene, and (d) graphite.

5.3 BAND STRUCTURE OF GRAPHENE AND GRAPHENE NANORIBBON

It could be best to start exploring GNR from the band structure of graphene where the linear band dispersion relation at the Brillouin zone is given most attention. The electronic properties of graphene and thus GNR can be described by its band energy near the K and K' points. The energy band dispersion relation in graphene is given by [18]

$$E(\vec{k}) = \pm t \sqrt{1 + 4\cos\left(\frac{k_x 3 a_{C-C}}{2}\right)\cos\left(\frac{k_y \sqrt{3} a_{C-C}}{2}\right) + 4\cos^2\left(\frac{k_y \sqrt{3} a_{C-C}}{2}\right)} \quad (5.1)$$

where
 $a_{C-C} = 1.42$ Å is the carbon to carbon (C–C) bond length
 $t = 2.7$ eV is the nearest neighbor C–C tight-binding energy
 $k_{x,y,z}$ is the wave vector component [19]

In the graphene band structure, as shown in Figure 5.2, the conduction and valence bands touch at the Dirac point, or K-point, at the six corners of the Brillouin zone. Then, the energy–wave vector, $E(k)$, relationship of the GNR at the low energy limit can be obtained by the approximation of the graphene band structure.

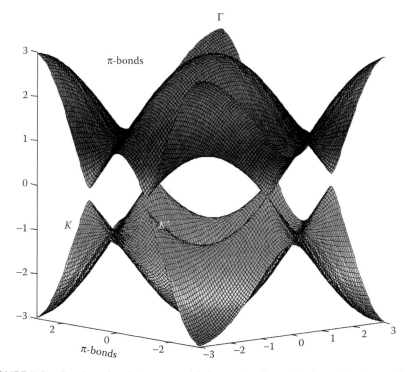

FIGURE 5.2 Graphene band structure with the conduction and valence bands touching at the Dirac point or K-point at the six corners of Brillouin zone.

The $E(k)$ relationship is [19]

$$E(\vec{k}) = \pm \frac{3a_{c-c}t}{2}\sqrt{k_x^2 + \beta^2} \tag{5.2}$$

where

k_x is the wave vector along the length of the nanoribbon
β is the quantized wave vector expressed as [19]

$$\beta = \frac{2\pi}{a_{c-c}\sqrt{3}}\left(\frac{p_i}{N+1} - \frac{2}{3}\right) \tag{5.3}$$

where

N is the number of dimer lines that determines the width of the ribbon [19]
p_i is the subband index

Then, the band energy can be written as

$$E = \pm \frac{E_g}{2}\sqrt{1 + \frac{k_x}{\beta^2}} \tag{5.4}$$

where the equation is not in parabolic form, and the square root approximation needs to be employed for the case of semiconducting GNR.

5.4   CARRIER STATISTICS OF MONOLAYER GRAPHENE NANORIBBON

The density of state (DOS) is a fundamental tool to help describe the electrical properties of the material. The DOS in GNR can be obtained by taking the derivatives of energy over the wave vector, which can be written as [19]

$$\text{DOS}_{\text{GNR}}(E) = \frac{\Delta n}{\Delta EL} = \frac{1}{2\pi}\left(\frac{2m^*}{\hbar^2}\right)^{1/2}\left(E - \frac{E_g}{2}\right)^{-1/2} \tag{5.5}$$

Once the DOS is obtained, the number of occupied conduction band levels (electrons) can be analytically calculated through the carrier concentration equation:

$$n = \int_{E_C}^{E_{\text{top}}} \text{DOS}(E)f(E)\,dE \tag{5.6}$$

while for the occupied valence band (holes) it can be written as

$$p = \int_{E_{\text{bottom}}}^{E_V} \text{DOS}(E)[1 - f(E)]\,dE \tag{5.7}$$

where $f(E)$ is the Fermi–Dirac distribution function defined as $f(E) = (1 + \exp[(E - E_F)/kT])^{-1}$ with k being the Boltzmann constant, T is the absolute temperature, and E_F is the Fermi level. However, when considering the degenerate semiconductor (high doping condition) for the degenerate n-type, the Fermi level lies within the conduction band. The same happens for degenerate p-type semiconductor where the Fermi level lies in the valence band, indicating a high density of holes.

Therefore, the carrier concentration for semiconducting GNR in the nondegenerate limit is derived as [19]

$$n_{GNR(nondegenerate)} = N_C e^{-((E_C - E_F)/k_B T)} \tag{5.8}$$

where N_C is the effective DOS that depends on the material dimension as well as the Fermi–Dirac integral. For the degenerate case [19],

$$n_{GNR(degenerate)} = \left(\frac{8m^*(E_F - E_C)}{\pi^2 \hbar^2} \right)^{1/2} \tag{5.9}$$

5.5 STRUCTURE OF MONOLAYER GRAPHENE NANORIBBON

Interestingly, based on its edge atom alignments, GNR can be classified into two different nomenclatures—armchair (AGNR) and zigzag (ZGNR) [20]. Both types of GNR exhibit semiconducting properties where the band gap increases or decreases proportional to the ribbon width. The GNR is metallic when $m = 3p + 2$ and semiconducting when $m = 3p$ and $m = 3p + 1$ [21].

The ribbon widths for both ZGNR and AGNR are determined by the number of dimer lines denoted by N, as illustrated in Figure 5.3. The equations that represent the width are [20]

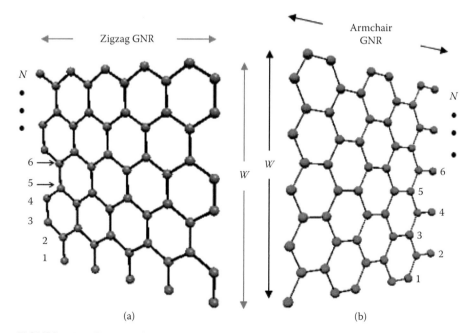

FIGURE 5.3 **(See color insert.)** Structure of GNR: (a) ZGNR and (b) AGNR.

$$W_{zz} = \frac{\sqrt{3}}{2} N \ \text{(ZGNR)} \qquad (5.10)$$

and

$$W_{a} = \frac{\sqrt{3}}{2} a_{c-c}(N-1) \ \text{(AGNR)} \qquad (5.11)$$

Even though the dangling bonds at the edge state of GNR [8,22] do not contribute to the Fermi electronics state [22–24], they do impact the transport behavior of carriers in GNR in terms of scattering, where it is widely termed as edge roughness. The scattering by edge roughness, on the other hand, can be dampened where the dangling bonds can be terminated by hydrogen [8,25,26]. The other scattering mechanisms associated with GNR are not considered in this section.

5.6   CAPACITANCE OF MONOLAYER GRAPHENE NANORIBBON

The capacitance of nanomaterial is an important topic to study, where it provides an understanding of device applications. As for GNR, smaller quantum capacitance could reduce the switching time for graphene-based devices [27]. The capacitance in monolayer GNR is derived based on the definition of capacitance, C, which is given by $C = Q/V$, where Q represents the total free charge confined by the total potential, V. The device capacitance depends on the device geometry as well as on the dielectric constant of the medium. In the case of GNRFET, the capacitance formed between the channel and the gate is taken into account. In addition, as the dimensions shrink to nanometer scale, the quantum effects cannot be simply neglected [28] as they will affect the performance of the device. This quantum capacitance concept was first introduced by Luryi [29]. Shylau et al. [30] then experimentally extracted the quantum capacitance from the total capacitance in which the gate distance is closer to the GNR so that the classical capacitance is comparable to quantum capacitance. Applying a gate voltage V_g to the metallic gate of a GNRFET will induce the charge carriers with density n to the GNR, and hence the chemical potential μ in the nanoribbon is shifted from the neutral point μ_0 analogous to [30]

$$eV_g = \mu - \mu_0 \qquad (5.12)$$

where e is the charge carrier.

The total capacitance, C_{total} derived from the definition of capacitance, $C = e\partial n/\partial V_g$ where the electron density by cooperating the chemical potential is [30]

$$n = \int_{\mu_0}^{\mu_0 + eV_Q} \rho_0(E)\,dE \qquad (5.13)$$

with $\rho_0(E)$ as the DOS for intrinsic ribbon (charge neutral).

Alternatively, the quantum capacitance of GNR also can be reached analytically by combining the electrostatic capacitance formed between gate electrodes together with the electron density in the GNR as a function of gate voltage [31]. John et al. [28] suggested that the quantum capacitance also can be derived by employing the effective mass approximation with parabolic bands for 1D DOS, which yields

$$C_Q = \frac{2vq^2}{hv_F}$$ (5.14)

One of the benefits provided by the "quantum capacitance limit" is that a relatively low quantum capacitance reflects the changes in the local electrostatic potential, which is also related to the changes in gate-to-source voltage.

5.7 CONDUCTANCE OF MONOLAYER GRAPHENE NANORIBBON

Naeemi and Meindl [25] suggested that each conduction channel in GNR contributes one quantum conductance given by

$$G_0 = \frac{2e^2}{h}$$ (5.15)

This conductance will not be affected by the edge scattering existing at the edge of the GNR, provided there are no momentum deviations along the GNR length. Considering Matthiessen's rule for the scattering mean free path (MFP), the total conductance in GNR is given by

$$G = \frac{2e^2}{h} \sum \frac{1}{1 + L(1/l_D + 1/l_n)}$$ (5.16)

where
  L is the GNR length
  l_D is the scattering MFP due to the defects

Interestingly, the work in Ref. [25] shows how the analytical modeling of metallic GNR was carried out. An additional feature is the analytical approximation for the conductance having lengths longer than MFP. Alternatively, the GNR conductance can also be analytically calculated based on the approach used by Ahmadi et al. [32], which was based on the Landauer formula. The Landauer formula was then corporate with the number of subbands and energy dispersion relation given by

$$G = \frac{2q^2}{h} \int_{-\infty}^{+\infty} dE M(E) T(E) \left(-\frac{df}{dE} \right)$$ (5.17)

where
  T represents the transmission probability
  $M(E)$ is the number of subbands
  $\dfrac{df}{dE}$ is the derivative of the Fermi–Dirac distribution function

The simulation result presented in Ref. [32] shows that the conductance in GNR is minimum at the Dirac point, suggesting that the resistance is higher at the point closer to the charge neutrality point.

5.8   MOBILITY OF MONOLAYER GRAPHENE NANORIBBON

The most striking part of graphene is its excellent transport properties, which is high carrier mobility where the electrons can move through it faster—most suitable for modern electronic applications. The high carrier mobility in graphene was reported to be in the order of 10^4 cm^2/Vs at room temperature [33] and even higher at 4 K with 6×10^4 cm^2/Vs [7]. Besides, graphene was also found to have long MFP at low bias due to the weak acoustic phonon (AP) scattering [7]. On the other hand, the room temperature mobility of GNR can reach up to 200,000 cm^2/Vs for the suspended GNR with typical carrier concentration of 2×10^{11} cm^{-2} [34], which provides near-ballistic transport [7] as well as high switching speed [35,36]. Obradovic et al. [37] suggested that nearly ballistic mobilities in GNR are achievable provided the GNR band gap is reduced to below 0.25 eV, where the mobilities are predicted to be over 10,000 cm^2/Vs.

However, there are several kinds of scattering mechanisms that will possibly impede the carriers in graphene and GNR. Among the possible factors that may degrade the transport in graphene are phonons, defects, and line edge roughness (LER) [38,39], where LER is the dominant factor that limits the carrier mobility in GNR, especially the narrow one [38,40]. LER is caused by the dangling bonds at the edges of GNR, which enhance the scattering process in the GNR and hence degrade the carrier mobility. The carrier mobility will be reduced further with the downsizing of the GNR, because the narrower the GNR, the greater the scatterings. In addition, the narrow GNR provides a smaller transport channel for the carriers compared to the larger size GNR. However, this kind of scattering can be reduced with a few conditions as suggested by Fang et al. [38].

The high carrier mobility in graphene is dampened by the strong impurity scattering, which caused the MFP of the carriers to decrease and hence reduced the carrier mobility [34]. Besides, the scattering mechanisms also change with the GNR width, temperature, and the Fermi level location (carrier concentration) [38]. At low-field mobility, scatterings by optical and acoustic phonons and the existence of charged impurity scatterings are among the dominant scattering mechanisms in these materials. Optical phonon (OP) scattering is only possible if the kinetic energy of the carriers exceeds the boundary phonon energy of ~160 meV [38]. This OP scattering limits the carrier mobility even at room temperature. On the other hand, AP scattering tends to dominate above 200 K when the GNRs are subjected to temperature effect.

Carrier mobility in GNRs can be divided into several parts, which are low-field mobility, effective mobility, and field-effect mobility. Bresciani et al. [41], for example, have carried out an analytical model to calculate the low-field mobility of GNRs based on their band structures in the presence of edge roughness. Besides, Fang et al. also worked on the analytical modeling of low-field mobility in GNRs. Their research was carried out by comparing the scattering effects of phonon, impurity, and LER [38]. Obradovic et al., however, evaluated the low-field mobility in GNRs by using the Kubo–Greenwood formula, and used Monte Carlo simulation in the computation of high-field mobility [37]. Interestingly, Obradovic et al. also found that mobility was higher in GNR than in CNT due to the higher group velocity in GNR.

Calculation of high carrier mobility is a complicated task due to the short-range scatterers as outlined in Ref. [42]. This high carrier mobility reflects the lower resistivity in the sample [42]. The most common ways to determine the mobility in graphene are the field-effect mobility and the effective mobility [43]. The effective mobility extracted from the channel resistance is given by [43]

$$\mu_{\text{eff}} = \frac{L_{\text{g}}}{R_{\text{eff}} C_{\text{g}} (V_{\text{gs}} - V_{\text{th}}) W} \qquad (5.18)$$

where
L_{g} is the gate length
C_{g} is the gate capacitance
V_{th} is the threshold voltage
V_{gs} is the gate-to-source voltage
W is the gate width

The field-effect mobility is, however, determined based on the transconductance, g_{m}, and is given by [43]

$$\mu_{\text{FE}} = \frac{L_{\text{g}}}{W} \frac{g_{\text{m}}}{C_{\text{g}} (V_{\text{ds}} - IR_{\text{C}})} \qquad (5.19)$$

where
V_{ds} is the drain-to-source voltage
R_{C} is the sum of source and drain resistances

The low-field mobility in GNRFET has also been analytically investigated in Ref. [44], where it is defined as field independent but temperature dependent [45]. Besides, Pennington and Goldsman [45] also suggested that low-field mobility corresponds to the mobility in low electric field and therefore low-field mobility is assumed to occur in very low electric field and at almost equilibrium condition. This low-field mobility is related to the thermal velocity and intrinsic velocity given by [44]

$$\mu_{\text{0GNR}} = \frac{q\ell_0}{m^* v_{\text{th}}^2} v_i \qquad (5.20)$$

The low-field mobility can be further categorized into two types, which are the degenerate and nondegenerate regimes. The low-field mobility in the degenerate regime is expressed by [44]

$$\mu_{\text{0D\_GNR}} = \frac{q\ell_0 \hbar \left(\sqrt{\pi n_1} \right)}{4m^* k_{\text{B}} T \sqrt{2}} \qquad (5.21)$$

where
$\hbar$ is Planck's constant
n_1 is the carrier concentration in the 1D device

The low-field mobility equation in the degenerate regime shows that the low-field mobility in GNRFET directly depends on the carrier concentration. However, the low-field mobility of GNRFET differs in the nondegenerate regime where the mobility is dependent on the temperature through the thermal velocity, as can be seen through the analytical model given by [44]

$$\mu_{0\text{ND\_GNR}} = v_{\text{th}} \frac{q\ell_0}{m^* v_{\text{th}} \sqrt{\pi}} \tag{5.22}$$

5.9 DRIFT VELOCITY OF GRAPHENE NANORIBBON IN HIGH ELECTRIC FIELD

Drift velocity is an important parameter that helps predict the carrier mobility and thus device characteristics. Besides, the carrier velocity control in high electric field is very essential, especially when dealing with radio frequency operation [46]. Some existing works on high field transport [46–48] discussed the OP influence on the velocity when subjected to high electric field. However, our works [49] expand the drift velocity of GNR in both the degenerate and nondegenerate regimes. In this case, the analytical model of the low-field mobility is utilized in order to demonstrate the drift velocity, v_d, in a high electric field, E, which is [50,51]

$$v_d(E) = \frac{\mu_0 E}{1 + \mu_0 E / v_{\text{sat}}} \tag{5.23}$$

where
μ_0 is the low-field mobility
v_{sat} is the saturation velocity

The saturation velocity in the nondegenerate regime refers to the thermal velocity, whereas the Fermi velocity is considered for the degenerate regime [52]. The low-field mobility in the degenerate and nondegenerate regimes is then substituted in Equation 5.23 when considering the respective cases. The horizontal electric field due to the source–drain voltage is considered, as the drift velocity depends on this electric field where [53]

$$E = \frac{V_{\text{DS}}}{L_{\text{CH}}} \tag{5.24}$$

where
V_{DS} is the source–drain voltage
L_{CH} is the channel length

5.10 GNRFET

A GNRFET is an FET device with GNR as the channel material. The GNR sandwiched between the source and drain electrodes controlled by the gate through the gate voltage applied is illustrated in Figure 5.4.

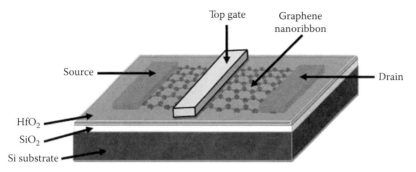

FIGURE 5.4 **(See color insert.)** Structure of a GNRFET with GNR as the channel material.

One of the advantages of utilizing GNR as the channel material in GNRFET is its powerful ability to control the electrostatics and hence is expected to reduce the short channel effect [36], which depends on the device electrostatics. GNRFET performance is usually evaluated through current–voltage (I–V) characteristics. Liang et al. [36], for example, calculated the I–V characteristics for ballistic GNRFET by employing the "top-of-barrier" approach. The I–V characteristics of GNRFET show better performance than those of the silicon MOSFET (Si MOSFET). This is due to the higher average velocity in the GNR. Pei et al. [21] evaluated GNRFET through the analytical modeling of the current at ballistic limit. The computation of the transmission coefficient is also included in the current model, taking into account the edge and OP scatterings. The fabrication of GNRFET conducted by Wang et al. [10] found that gate material of palladium (Pd) yields a higher ON-current (I_{ON}) than titanium (Ti) and gold (Au). The GNR width effects on the I–V characteristics were studied by Choudhury et al. [54], assuming a Schottky-barrier FET. The results suggested that the current is larger for GNR with higher width and vice versa. Yan et al. [55] worked on the I–V characteristics of GNRFET by implementing first-principles transport calculations. The work carried out showed that GNRFET can achieve high performance levels similar to other FETs made of SWNTs.

5.11 CONCLUSION

GNR derived from 2D graphene has great impact on the nanoscale transistor even though it is a new material. Among the issues highlighted for GNR are the dangling bonds at the edges of the GNR that contribute to the carrier scattering and thus are expected to impede the carrier mobility. Even though this problem can be overcome by terminating the dangling bonds with hydrogen atoms, the possibility is still wide open to explore this issue using other chemical species apart from hydrogen. In addition, the nanoscale dimension of GNR itself opens up room for research as it reaches the quantum realm. The quantum effects of GNR on device performance can then be evaluated through conductance, capacitance, carrier mobility, I–V characteristics, and so on. However, these are just a few suggestions out of several other potential doors to be opened concerning GNR.

5.12 SUMMARY

The excellent transport properties of GNR provide a paradigm shift in the material evolution for the advancement of electronics, especially in nanoscale devices. It also motivates and thus enhances the research in this field in terms of modeling and fabrication, as nowadays the market demands electronic devices with high speed, low dimension, and, of course, high energy efficiency. Overall, there is still much research to be conducted with GNR that will drive the material evolution to a higher echelon.

REFERENCES

1. Harris, G. L., P. Zhou, et al. Semiconductor and photoconductive GaN nanowires and nanotubes. *Summaries of Papers Presented at the Conference on Lasers and Electro-Optics, 2001 (CLEO '01) Technical Digest.* San Francisco, May 6–11, p. 239, 2001.
2. Liang, G. Widths effects in ballistics graphene nanoribbon FETs. *2nd IEEE International Nanoelectronics Conference (INEC).* Shanghai, March 24–27, pp. 1187–1188, 2008.
3. Dumlich, H. and S. Reich. Chirality-dependent growth rate of carbon nanotubes: A theoretical study. *Physical Review B.* 82(8): 085421, 2010.
4. Cresti, A. Proposal for a graphene-based current nanoswitch. *Nanotechnology.* 19(26): 265401, 2008.
5. Jyotsna, C. and G. Jing. Atomistic simulation of graphene nanoribbon tunneling transistors. *2010 3rd International Nanoelectronics Conference (INEC).* Hong Kong, January 3–8, 200–201, 2010.
6. Sheikhi, M. H., M. Berahman, et al. The numerical modeling for electrical behavior of graphene nanoribbon in the present of optical detection. *Proceedings of the 2009 Second International Conference on Computer and Electrical Engineering, IEEE Computer Society.* 2: 20–22, 2009.
7. Novoselov, K. S., A. K. Geim, et al. Electric field effect in atomically thin carbon films. *Science.* 306(5696): 666–669, 2004.
8. Dubois, S. M. M., Z. Zanolli, et al. Electronic properties and quantum transport in graphene-based nanostructures. *European Physical Journal B.* 72(1): 1–24, 2009.
9. Castro Neto, A. H., F. Guinea, et al. The electronic properties of graphene. *Reviews of Modern Physics.* 81(1): 109–162, 2009.
10. Wang, X. R., Y. J. Ouyang, et al. Room-temperature all-semiconducting sub-10-nm graphene nanoribbon field-effect transistors. *Physical Review Letters.* 100(20): 206803, 2008.
11. Li, X. L., X. R. Wang, et al. Chemically derived, ultrasmooth graphene nanoribbon semiconductors. *Science.* 319(5867): 1229–1232, 2008.
12. Jiao, L., L. Zhang, et al. Narrow graphene nanoribbons from carbon nanotubes. *Nature.* 458(7240): 877–880, 2009.
13. Kosynkin, D. V., A. L. Higginbotham, et al. Longitudinal unzipping of carbon nanotubes to form graphene nanoribbons. *Nature.* 458(7240): 872–875, 2009.
14. Hirsch, A. Unzipping carbon nanotubes: A peeling method for the formation of graphene nanoribbons. *Angewandte Chemie, International Edition.* 48(36): 6594–6596, 2009.
15. Rangel, N. L., J. C. Sotelo, et al. Mechanism of carbon nanotubes unzipping into graphene ribbons. *Journal of Chemical Physics.* 131(3): 031105, 2009.
16. Kim, K., A. Sussman, and A. Zettl. Graphene nanoribbons obtained by electrically unwrapping carbon nanotubes. *ACS Nano.* 4(3): 1362–1366, 2010.
17. Moriki, T., A. Kanda, et al. Electron transport in thin graphite films: Influence of microfabrication processes. *Physica E: Low-Dimensional Systems & Nanostructures.* 40(2): 241–244, 2007.

18. Lundstrom, M. and J. Guo. *Nanoscale Transistors: Device Physics, Modeling and Simulation.* Springer, New York, 2006.
19. Johari, Z., M. T. Ahmadi, et al. Modelling of graphene nanoribbon Fermi energy. *Journal of Nanomaterials.* 2010: 909347, 2010. doi:10.1155/2010/909347.
20. Wakabayashi, K., M. Fujita, et al. Electronic and magnetic properties of nanographite ribbons. *Physical Review B.* 59(12): 8271, 1999.
21. Pei, Z., M. Choudhury, et al. Analytical theory of graphene nanoribbon transistors. *Proceedings of the 2008 IEEE International Workshop on Design and Test of Nano Devices, Circuits and Systems.* Cambridge, MA, September 29–30, 3–6, 2008.
22. Nakada, K., M. Fujita, et al. Edge state in graphene ribbons: Nanometer size effect and edge shape dependence. *Physical Review B.* 54(24): 17954, 1996.
23. Areshkin, D. A., D. Gunlycke, et al. Ballistic transport in graphene nanostrips in the presence of disorder: Importance of edge effects. *Nano Letters.* 7(1): 204–210, 2006.
24. Cresti, A. and S. Roche. Range and correlation effects in edge disordered graphene nanoribbons. *New Journal of Physics.* 11(9): 095004, 2009.
25. Naeemi, A. and J. D. Meindl. Conductance modeling for graphene nanoribbon (GNR) interconnects. *IEEE Electron Device Letters.* 28(5): 428–431, 2007.
26. Hod, O., J. E. Peralta, et al. Edge effects in finite elongated graphene nanoribbons. *Physical Review B.* 76(23): 4, 2007.
27. Mišković, Z. and N. Upadhyaya. Modeling electrolytically top-gated graphene. *Nanoscale Research Letters.* 5(3): 505–511, 2010.
28. John, D. L., L. C. Castro, et al. Quantum capacitance in nanoscale device modeling. *Journal of Applied Physics.* 96(9): 5180–5184, 2004.
29. Luryi S. Quantum capacitance devices. *Applied Physics Letters.* 52: 501, 1988. doi:10.1063/1.99649.
30. Shylau, A. A., J. W. Klos, et al. Capacitance of graphene nanoribbons. *Physical Review B.* 80(20): 9, 2009.
31. Fang, T., A. Konar, et al. Carrier statistics and quantum capacitance of graphene sheets and ribbons. *Applied Physics Letters.* 91: 092109, 2007.
32. Ahmadi, M. T., Z. Johari, et al. Graphene nanoribbon conductance model in parabolic band structure. *Journal of Nanomaterials.* 2010: 753738, 2010. doi:10.1155/2010/753738.
33. Guo, J., Y. Yoon, et al. Gate electrostatics and quantum capacitance of graphene nanoribbons. *Nano Letters.* 7(7): 1935–1940, 2007.
34. Bolotin, K. I., K. J. Sikes, et al. Ultrahigh electron mobility in suspended graphene. *Solid State Communications.* 146(9–10): 351–355, 2008.
35. Grassi, R., A. Gnudi, et al. Graphene nanoribbons FETs for high-performance logic applications: Perspectives and challenges. *2008 9th International Conference on Solid-State and Integrated-Circuit Technology.* 1–4: 365–368, 2008.
36. Liang, G. C., N. Neophytou, et al. Performance projections for ballistic graphene nanoribbon field-effect transistors. *IEEE Transactions on Electron Devices.* 54(4): 677–682, 2007.
37. Obradovic, B., R. Kotlyar, et al. Analysis of graphene nanoribbons as a channel material for field-effect transistors. *Applied Physics Letters.* 88(14): 142102, 2006.
38. Fang, T., A. Konar, et al. Mobility in semiconducting graphene nanoribbons: Phonon, impurity, and edge roughness scattering. *Physical Review B.* 78(20): 205403, 2008.
39. Yang, Y. X. and R. Murali. Impact of size effect on graphene nanoribbon transport. *IEEE Electron Device Letters.* 31(3): 237–239, 2010.
40. Betti, A., G. Fiori, et al. Physical insights on graphene nanoribbon mobility through atomistic simulations. *2009 IEEE International Electron Devices Meeting (IEDM).* Baltimore, MD, December 7–9, 1–4, 2009.
41. Bresciani, M., P. Palestri, et al. A better understanding of the low-field mobility in graphene nanoribbons. *Proceedings of the European Solid State Device Research Conference (ESSDERC).* Athens, Greece, September 14–18, 480–483, 2009.

42. Adam, S. and S. Das Sarma. Transport in suspended graphene. *Solid State Communications* 146(9–10): 356–360, 2008.

43. Rumyantsev, S., G. Lui, et al. Electrical and noise characteristics of graphene field-effect transistors: Ambient effects, noise sources and physical mechanisms. *Journal of Physics: Condensed Matter.* 22(39): 395302, 2010.

44. Amin, N. A., Z. Johari, et al. Low-field mobility model on parabolic band energy of graphene nanoribbon. *Modern Physics Letters B.* 25(4): 281–290, 2011.

45. Pennington, G. and N. Goldsman. Low-field semiclassical carrier transport in semiconducting carbon nanotubes. *Physical Review B.* 71(20): 204318, 2005.

46. Da Silva, A. M., K. Zou, et al. Mechanism for current saturation and energy dissipation in graphene transistors. *Physical Review Letters.* 104(23): 236601, 2010.

47. Jena, D. A theory for the high-field current-carrying capacity of one-dimensional semiconductors. *Journal of Applied Physics.* 105(12): 123701–123705, 2009.

48. Tse, W. K., E. H. Hwang, et al. Ballistic hot electron transport in graphene. *Applied Physics Letters.* 93(2): 023128, 2008.

49. Amin, N. A., M. T. Ahmadi, et al. Drift velocity and mobility of a graphene nanoribbon in a high magnitude electric field. *Proceedings of the Fourth Global Conference on Power Control and Optimization.* 1337: 177–179, 2011.

50. Dorgan, V. E., M.-H. Bae, et al. Mobility and saturation velocity in graphene on SiO_2. *Applied Physics Letters.* 97(8): 082112–082113, 2010.

51. Scott, B. W. and J. Leburton. High-field carrier velocity and current saturation in graphene field-effect transistors. *2010 10th IEEE Conference on Nanotechnology (IEEENANO).* Seoul, South Korea, August 17–20, 655–658, 2010.

52. Saad, I., M. L. P. Tan, et al. Ballistic mobility and saturation velocity in low-dimensional nanostructures. *Microelectronics Journal.* 40(3): 540–542, 2009.

53. Gray, P. R., P. J. Hurst, et al. *Analysis and Design of Analog Integrated Circuits* (5th Edn.). John Wiley & Sons, Hoboken, NJ, 2010.

54. Choudhury, M., Y. Yoon, et al. Technology exploration for graphene nanoribbon FETs. *2008 45th ACM/IEEE Design Automation Conference.* 1–2: 272–277, 2008.

55. Yan, Q., B. Huang, et al. Intrinsic current–voltage characteristics of graphene nanoribbon transistors and effect of edge doping. *Nano Letters.* 7(6): 1469–1473, 2007.

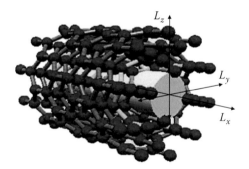

FIGURE 2.2 In a bulk semiconductor all three directions are more than the de Broglie λ_D wavelength.

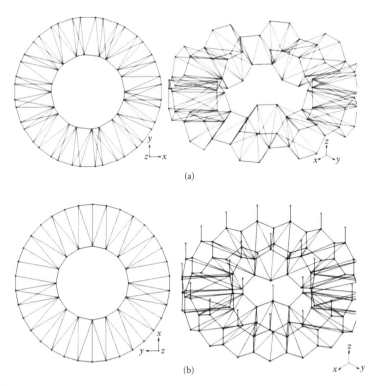

FIGURE 2.12 Top- and side-view of a DWCNT's first ring for a (a) (10, 10), (5, 5) DWCT and (b) (17, 0), (8, 0) DWCNT.

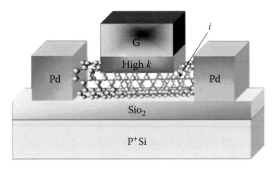

FIGURE 3.15 Illustration of the CNT capacitor device.

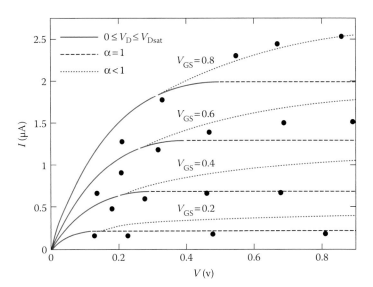

FIGURE 3.16 Current–voltage characteristics of nanowire.

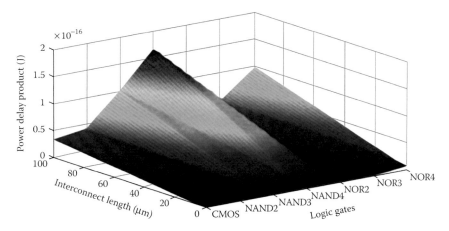

FIGURE 4.26 PDP for CNFET logic gates as a function of wire length.

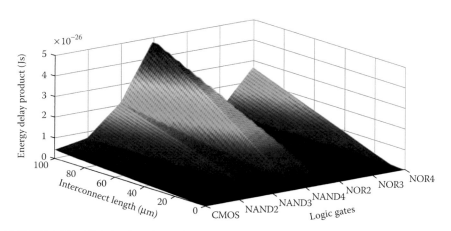

FIGURE 4.27 EDP for CNFET logic gates as a function of wire length.

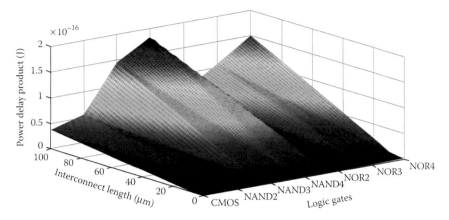

FIGURE 4.28 PDP for MOSFET logic gates as a function of wire length.

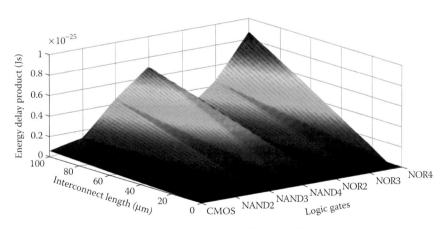

FIGURE 4.29 EDP for MOSFET logic gates as a function of wire length.

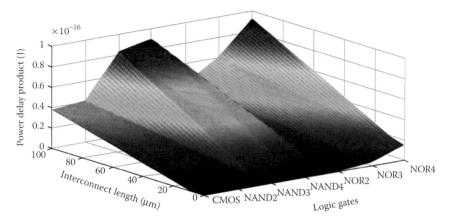

FIGURE 4.30 PDP for sized MOSFET logic gates as a function of wire length.

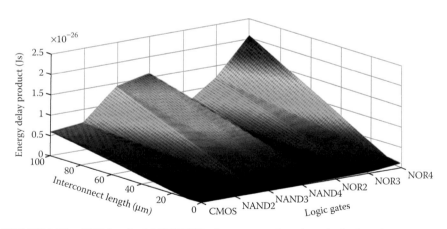

FIGURE 4.31 EDP for sized MOSFET logic gates as a function of wire length.

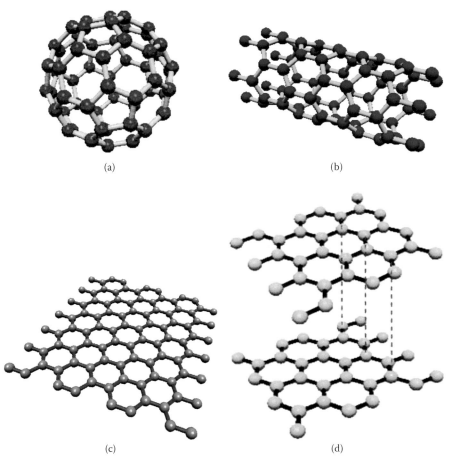

(a)

(b)

(c)

(d)

FIGURE 5.1 Carbon-based materials: (a) fullerenes, (b) CNTs, (c) monolayer graphene, and (d) graphite.

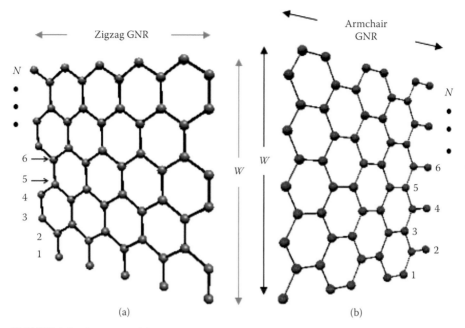

FIGURE 5.3 Structure of GNR: (a) ZGNR and (b) AGNR.

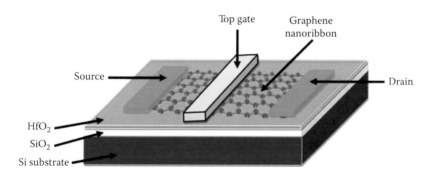

FIGURE 5.4 Structure of a GNRFET with GNR as the channel material.

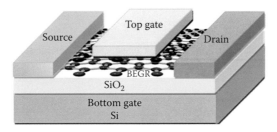

FIGURE 6.6 The schematic of a BGN FET is shown as a function of the top gate voltage (V_{top}) for different bottom gate voltages (V_{bottom}).

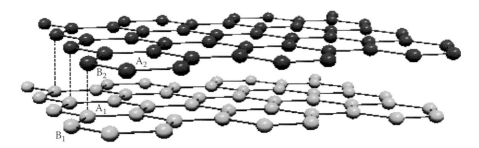

FIGURE 6.7 Schematic of the AB stacking of BGN.

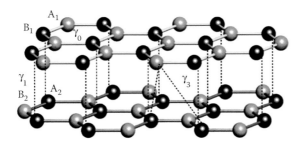

FIGURE 6.10 Structure of BGN with AB stacking.

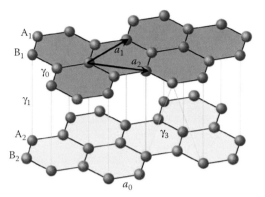

FIGURE 7.1 Bilayer graphene configuration. Atomic structure of AB-stacked bilayer graphene. In AB-stacking BGN, A_1 atoms of layer 1 are below B_2 atoms of layer 2 and vice versa.

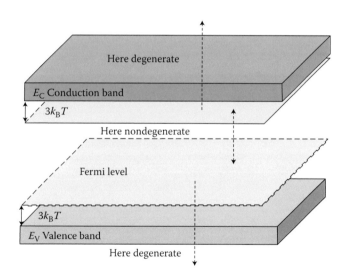

FIGURE 7.4 Definition of the degenerate and the nondegenerate regions. Fermi–Dirac distribution function can be approximated by either $f(E) = \exp((E_F - E)/k_B T)$ in the nondegenerate limit or $f(E) = 1$ in the degenerate limit.

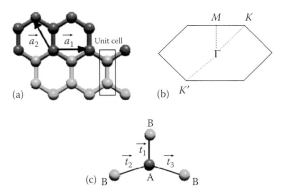

FIGURE 8.2 (a) Lattice of graphene. Unit cell vectors $\vec{a}_1$ and $\vec{a}_2$ allocate the unit cell. (b) The reciprocal lattice and the high symmetry points Γ, K, and M. (c) Each A atom is surrounded by three nearest neighbor B atoms. Vectors $\vec{t}_1$, $\vec{t}_2$, and $\vec{t}_3$ connect the A atoms to the three B atoms.

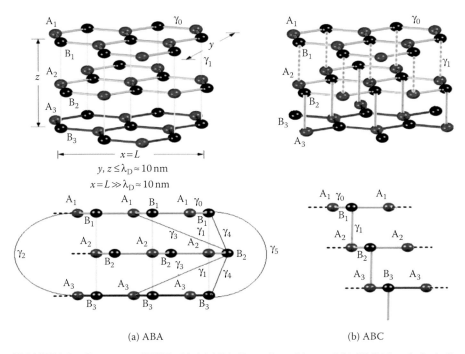

FIGURE 8.5 Structures of TGN with (a) ABA (Bernal) stacking and (b) ABC (rhombohedral) stacking.

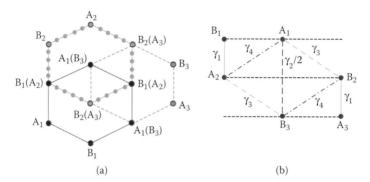

FIGURE 8.6 (a) Lattice structure of ABC-stacked TGN and (b) schematic of the unit cell of ABC-stacked TGN.

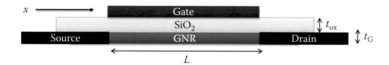

FIGURE 9.1 Schematic cross section of a top-gated GNRFET [9]. Typical device parameters are doping concentration $N = 5 \times 10^{16}$ m^{-2}, $t_{OX} = 1$ nm, $L = 20$ nm, and $w_G = 5$ nm.

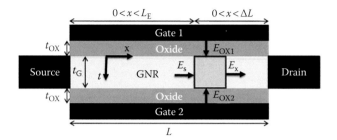

FIGURE 9.6 Schematic cross section of a DG-GNRFET [9]. Typical device parameters are doping concentration $N_d = 5 \times 10^{16}$ m^{-2}, $t_{OX} = 1$ nm, $L = 20$ nm, and $w_G = 5$ nm.

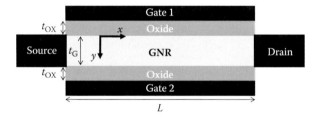

FIGURE 9.11 Schematic cross section of a DG-GNRFET.

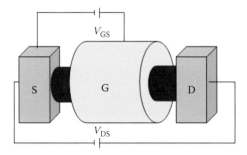

FIGURE 10.6 Schematic of a nanowire transistor with gate dielectric.

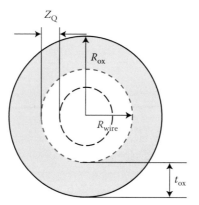

FIGURE 10.7 Schematic of a nanowire cross section with gate dielectric.

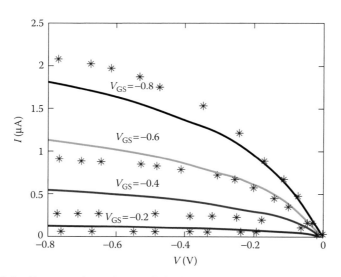

FIGURE 10.8 Current–voltage characteristics in a nanowire compared with the published data taken from Ref. [49].

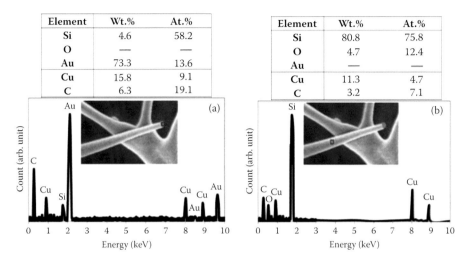

Element	Wt.%	At.%
Si	4.6	58.2
O	—	—
Au	73.3	13.6
Cu	15.8	9.1
C	6.3	19.1

Element	Wt.%	At.%
Si	80.8	75.8
O	4.7	12.4
Au	—	—
Cu	11.3	4.7
C	3.2	7.1

FIGURE 11.15 The corresponding EDX spectra of Au-catalyzed SiNWs grown by VSS mechanism at 320°C (a) tip and (b) stem of the nanowire.

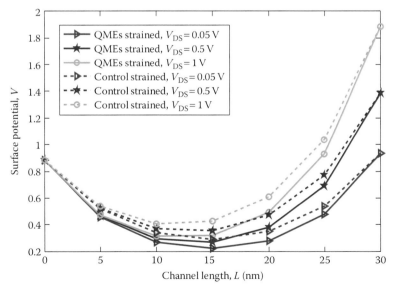

FIGURE 13.12 Surface potential variation along the channel for 30 nm channel length with 20% germanium content, respectively, for conventional and quantum tensile strained Si. The solid curves obtained by the quantum approach and the dashed represent the conventional strained Si.

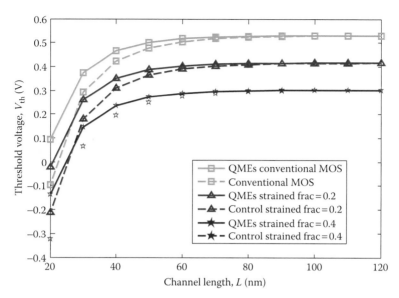

FIGURE 13.13 Threshold voltage variation along the channel, comparing conventional MOSFETs, and conventional and quantum strained Si model, for germanium content 0.2 and 0.4.

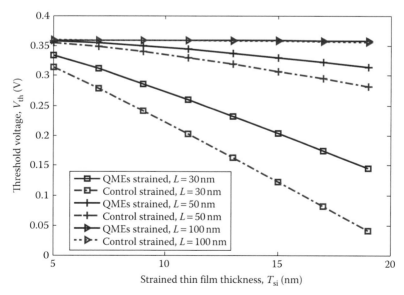

FIGURE 13.15 Plot of threshold voltage performance as the function of strained Si thickness for $L = 30$, 50, and 100 nm for 30% germanium content in $Si_{1-x}Ge_x$ substrate.

6 Carrier Transport, Current–Voltage Characteristics of BGN

Seyed Mahdi Mousavi, Meisam Rahmani,
Hatef Sadeghi, Mohammad Taghi Ahmadi,
and Razali Ismail

CONTENTS

6.1 INTRODUCTION

One of the most attractive properties of graphene is the density of charge particles (electrons and holes) in a graphene material, which is considered in matter physics. The movement of electrons and holes in graphene generates current. The band structure of bilayer graphene nanoribbon (BGN), which is very different from that of single-layer graphene (SLG), is due to applying external electric fields, which is discussed in the next chapter. In the next chapter we also consider the carrier concentration and ballistic conductance behavior in valence and conduction bands, respectively. In this discussion, we describe the movement of electron's transport in BGN, which is due to the electric field. It is found that temperature is an effective factor in carrier mobilities of BGN, leading to movement of carriers in a BGN FET channel. Accordingly, the carrier transport in BGN is an essential phenomenon that helps determine current–voltage characteristics (Cheng 2010).

BGN as a new material with interesting physical properties has attracted the attention of many research groups (Novoselov et al. 2006, 2007). A band gap in BGN can be created by applying a perpendicular electric field and incorporating the inversion symmetry breaking between double layers in the atomic structure. Two different stacking shapes of BGN (AA, AB) in layers result from interlayer coupling effects in low energy, which show different band structure (Xu 2010). AB-stacked is stable and used in graphite, whereas AA-stacked's instability kept it from being investigated experimentally (Figure 6.1).

Among several available material configurations of Bernal AB-stacked bilayer, graphene presents unique electrical flexibility with energy band gap, which is tunable up to a few hundreds of millielectron volts by applying a perpendicular electric field resulting in the electrically induced breakdown behavior of AB-stacked BGN (Yu et al. 2010).

The different stacking structures of BGN in intercoupling between the top and bottom layers are weaker than those of intracoupling within layers leading to electrical conduction. The electrical conduction through the layers is shown by the carrier hopping between the π orbits, with the van der Waals forces contributing to the coupling interaction between the layers' structure (Xu and Ding 2008; Sun and Chang 2008). Then, the transport conduction channels are essentially created by passing through the layers of BGN rather than by passing between those layers (Sun and Chang 2008; Xu 2010). The different configuration of AB stacking compared to AA stacking is due to the shift in the distance of lattice in the armchair edges (Xu and Ding 2008). Figure 6.2 shows the position of different atoms laid on top of each other in a honeycomb lattice. In BGN, with the two Cartesian directions being less than the de Broglie wavelength (λ_D) with a typical value of 10 nm, in which the carriers are confined to the y- and z-directions, the carriers can move only in the x-direction (Ahmadi 2010).

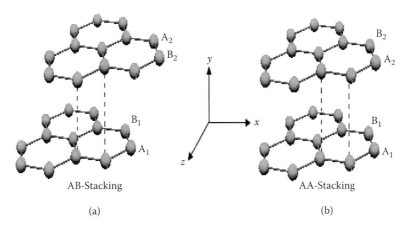

AB-Stacking (a) AA-Stacking (b)

FIGURE 6.1 Configuration of (a) the AB-stacking and (b) the AA-stacking BGNs.

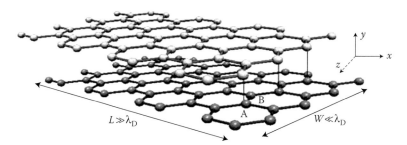

FIGURE 6.2   Structure of BGN in honeycomb lattice with width ($W \ll \lambda_D$) and length ($L \gg \lambda_D$).

6.2   MODELING OF MOBILITY IN BGN

Carrier transport physics is considered essential in bilayer graphene (Das Sarma et al. 2010). The carriers move ballistically where the distances are very low (submicron), and can be produced by being applied under an external electric field (Morozov et al. 2008; Castro 2010; Das Sarma et al. 2010). The carriers' transport on quantum mechanical tunneling between two layers plays an important role in improving the parabolic band dispersion with effective mass in the conduction and valence bands (Das Sarma et al. 2010). They can be measured in field effect transistors (FETs) with Bernal stacking indicating a gate-induced insulating state to find a large band gap of BGN (Xia et al. 2010). In theory, carriers' transport in BGN has a lower amount of minimum conductivity around the Fermi level in comparison with SLG nanoribbon (Das Sarma et al. 2010), showing a lower amount of mobility in BGN compared with SLG (Nagashio et al. 2009; Das Sarma et al. 2010). The minimum conductance of BGN is a function of temperature and carrier density n. In spite of the zero-carrier density's closeness to the Dirac point, graphene reveals a minimum conductivity and integer quantum effect on the order of e^2/h (Dragoman 2010). This conductance value can change roughly with increasing gate voltage (V_g) linearly in momentum space. In this study, we express a computational model of mobility in BGN based on the systematic exposition, which is used for the calculation of mobility by the conductance limit effect. The Landauer formula on the definition of normalized conductance of BGN through the central sample at finite temperature is written as (Xu and Ding 2008)

$$G(E) = \frac{2q^2}{h} \left[\int_{-\infty}^{+\infty} dE(E)T(E)\left(-\frac{df}{dE}\right) \right] \tag{6.1}$$

where
   $T(E)$, the transmission probability of carriers or injected electrons at one channel side to the other end, is equal to one in ballistic limit
   q, the electron charge in ballistic channel
   h, Planck's constant

The energy band structure of BGN (E–k) being close to the Fermi level versus the wave vector can be explained with the tight binding (TB) method as (Castro 2010)

$$E(k) \approx \Delta - \alpha k^2 + \beta k^4 \tag{6.2}$$

where we assume

$\alpha = (V/t_\perp^2) v_F^2$

$\beta = (1/V t_\perp^2) v_F^4$

$\Delta = V/2$

$v_F = 10^6$ m/s is the Fermi velocity

$t_\perp = 0.35$ eV is the effective interlayer hopping energy, in which the nearest neighbor carbon–carbon atoms distance is $a_{c-c} = 1.42$ Å (Castro 2010)

By taking the derivatives of the electron wave vector (k) over the energy $E(dk/dE)$ (Castro 2010; Mousavi 2011; Russo 2011), the number of the modes (subbands) $M(E)$ that are above the cutoff point at energy E in the transmission channel can be obtained as

$$M(E) = \frac{1}{L} \frac{dE}{dk} = \frac{k(-2\alpha + 4\beta k^2)}{L} \tag{6.3}$$

Considering the amount of parameters $M(E)$ and dE used in Equation 6.1, the conductance function of BGN is obtained as

$$G_{BGN} = \frac{2q^2}{hL} \int_{-\infty}^{+\infty} \left[\frac{\left(\alpha \pm \sqrt{\alpha^2 + 4\beta x k_B T}\right)^{3/2}}{\alpha\sqrt{2\beta}} - \frac{\left(\alpha \pm \sqrt{\alpha^2 + 4\beta x k_B T}\right)^{1/2}}{\sqrt{2\beta}} \right].$$

$$\left[-d\left(\frac{1}{1 + \exp\left((E - E_F)/k_B T\right)} \right) \right] \tag{6.4}$$

where k_B is the Boltzmann constant. This equation explains conductance with a normalized Fermi energy function $f(E) = 1/1 + \exp(E - E_F)/k_B T$ in mathematical symbols, where E_F is the 1D Fermi energy, the probability of occupation is half, and T is the ambient temperature. The off-state conductivity can be decreased by increasing the channel length. Moreover, the minimum conductivity value increases when the temperature rises, which indicates the characteristic of an insulating system. The conductance model is considered to point out the mobility of BGN (Gnani et al. 2010; Sadeghi 2011), which relates the mobility to the measurable conductance and the carrier density.

The channel lengths of graphene nanoribbons in quasi-1D structures with narrow widths (<~10 nm) are predicted to display a high carrier mobility in ballistic transport (Li et al. 2008b; Betti et al. 2009). In the Boltzmann point, the intrinsic mobility of a graphene layer is calculated where the low-field Coulomb scattering time (τ) between collisions is considered for a charge, and also the effective mass (m^*) approximation of the carriers in room temperature is used (Tang 2011):

$$\mu = \frac{q\tau}{m^*} \tag{6.5}$$

The quasi-1D intrinsic mobility at low field and short channel is a function of electron conductivity and carrier density (Morozov et al. 2008):

$$\mu = \frac{\sigma}{nq} \tag{6.6}$$

where
σ is the electrical conductivity as a function of conductance and channel length $\sigma = GL$, which is given in units of $1\ \Omega^{-1}\ cm^{-1}$ in 1D systems

the unit of μ is cm^2/Vs (Du 2008; Morozov et al. 2008; Liao 2010; Amin et al. 2011)

The high carrier density $n = 7.1 \times 10^{12}\ cm^{-2}$ per unit length is one of the most important advantages of a bilayer graphene device that can be controlled effectively through a gate voltage, in which an external electric field between the layers is applied (Dong et al. 2009; Mao et al. 2010).

$$\mu_{BGN} = \frac{2q}{hn} \int_{-\infty}^{+\infty} \left[\frac{\left(\alpha \pm \sqrt{\alpha^2 + 4\beta x k_B T}\right)^{3/2}}{\alpha\sqrt{2\beta}} - \frac{\left(\alpha \pm \sqrt{\alpha^2 + 4\beta x k_B T}\right)^{1/2}}{\sqrt{2\beta}} \right] \cdot \left[-d\left(\frac{1}{1 + e^{x-\eta}}\right) \right] \tag{6.7}$$

To simplify the mobility equation, the derivation of the Fermi–Dirac distribution function is employed, and the limit of integral is changed accordingly. Thus the mobility is

$$\mu_{BGN} = -\frac{2q}{nh} \int_{-v_g}^{+v_g} \left(\left[\frac{1}{1 + \exp\left[x - \frac{\left(v_g - V\right)q}{k_B T} \right]} \right] \left[\frac{3}{4}\sqrt{\frac{\alpha + \sqrt{\alpha^2 + 4x\beta k_B T}}{2\beta\left(\alpha^2 + 4x\beta k_B T\right)}} - \frac{\alpha}{8\beta\sqrt{\alpha^2 + 4x\beta k_B T}\sqrt{\frac{\alpha + \sqrt{\alpha^2 + 4x\beta k_B T}}{2\beta}}} \right] \right) dx \tag{6.8}$$

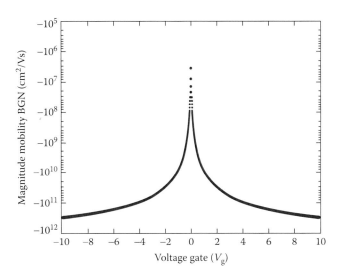

FIGURE 6.3   Mobility model for BGN with $T = 300$ K in Dirac point.

where $x = (E - \Delta)/k_{\mathrm{B}}T$. In spite of the integrals being assessed numerically, the normalized Fermi energy is defined as $\eta = (E_{\mathrm{F}} - \Delta)/k_{\mathrm{B}}T$. The general mathematical model of a BGN carrier mobility depends on the carrier density, which can be solved numerically as shown in Figure 6.3, where interlayer biased voltage is V and v_{g} is the gate voltage.

Figure 6.3 explains that the values of mobility magnitude will be increased by increasing the gate voltage. At low field effect, since filling of the empty bands reduces the phase space for scattering, the mobility will be increased. Mobility on the first band at low field, which is independent of the occupation of the second band, vanishes by increasing the gate voltage. Figure 6.4 also shows the field effect mobility with a linear peak for low voltage, which is similar to the conventional metal oxide semiconductor field effect transistor. These values decrease rapidly as the band filling emerges at the second band with higher charge density. Considering the fact that the gate voltage is changeable in device operation, the effect of voltage variation on mobility in nanoribbon transistors, especially at high fields, needs to be explored.

The temperature effect on carrier mobility of bilayer graphene is presented by the Coulomb impurity scattering within the random phase approximation method (Lv 2009). As shown in Figure 6.5, this model explains the relationship between temperature and the magnitude of mobility in combination with the voltage, which indicates an incremental effect on mobility at different temperatures (88, 260, and 340 K).

A significant shift to the mobility model at the Dirac point illustrates the effect of carrier mobility reduction in BGN (Morozov et al. 2008; Zou et al. 2011).

In summary, BGN with a stable AB structure can be used in a FET channel. Therefore, the carrier mobility model based on the quantum confinement effect indicates that BGN can be assumed as a 1D device in a FET channel. In this study, the carrier density and the temperature effect on the mobility of carriers in BGN are explored. Also, variation of gate voltage during the device operation and its

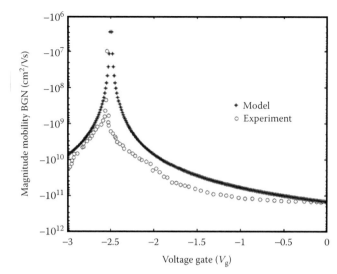

FIGURE 6.4 Comparison of carrier ballistic mobilities of a BGN model (stars) with experimental data (spots) at $T = 300$ K for different voltage gates (Oostinga et al. 2008). The experimental data extracted from conductance were presented by Oostinga et al. (2008) and Zou et al. (2011).

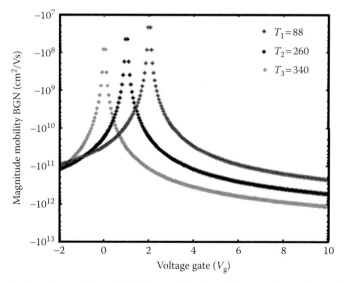

FIGURE 6.5 Carrier mobility of BGN as a function of gate voltage plotted at various temperatures at the charge neutrality point.

effect on mobility are discussed, and a numerical mobility model is presented based on the channel conductance. A comparison between the presented model and the published experimental data is reported, which points out an acceptable agreement between them.

6.3 CURRENT AND VOLTAGE MODEL OF BGN

There are certain characteristics of the electron and hole energy spectra in graphene nanoribbon (GNR) FETs that predicate the exceptional properties of graphene-based devices. The operation of GNR FETs is accompanied by the configuration of the pnp or npn junction under the controlling top gate and the relevant energy barrier. The current across this barrier can be associated with the processes of thin striped GNR (Ryzhii et al. 2008). The device characteristics of GNR FETs, in use near ballistic and drift-diffusion regimes, being considered analogously with those of nanowire FETs and carbon nanotube FETs which shows their superiority. Yet, the unique properties of BGN compared to GNR make it promising for different nanoelectronic device applications. The gapless energy spectrum of graphene layers allows us to use them in terahertz and mid-infrared detectors and lasers. However, the gapless energy spectrum of BGN is a barrier for making digital electronics based on GNR FETs because of the existence of strong interband tunneling in the FET off-state (Ryzhii et al. 2011). This means that the band gap in the channel in an off state is not adequately large, and also the source barrier for the electron is small. Therefore, the barrier for the holes at the drain part is smaller because of the existence of the drain bias V (Majumdar et al. 2010). The width of graphene as the channel material for FETs has very poor I_{on}/I_{off} ratios, even smaller than 10, and seems to have limited capability for digital electronics applications (Fiori and Iannaccone 2009a). However, the restoration of the energy gap in graphene-based structures like GNR and BGN appears to be unavoidable when fabricating an FET (Ryzhii et al. 2011). Nevertheless, in the limited set of device structures considered, the band gap does not permit a proper on and off switching of the transistor. Recently, theoretical and experimental works have demonstrated that bilayer graphene has the attractive property of an energy gap tunable with an applied vertical electric field with few hundreds of milli-electron volts. Therefore, it can be used as a channel material for BGN FETs as long as the FET is in the off state. This means that it can develop the possibility of patterning a bilayer graphene layer with lithographic techniques and nonprohibitive feature sizes (Fiori and Iannaccone 2009b).

Principally, BGN FET characteristics can be investigated using the same approaches as those understood previously for more customary FETs with a 2D electron system in the channel. However, certain important characteristics of BGN FETs, such as the dependence of both the electron density and the energy gap in different parts of the channel on the gate and drain voltages, should be considered along with the "short-gate" effect and drain-induced barrier lowering. We use a considerable sweeping version of the BGN FET analytical device model to evaluate the characteristics of BGN FETs, such as the threshold voltages, current–voltage characteristics, and transconductance in different regimes. Besides, we consider the alternative of creating a significant development in the eventual performance of these FETs by shortening the gate and reducing the gate layer thickness. The device model under consideration that presents the BGN FET characteristics in closed analytical form allows a simple and obvious assessment of the ultimate performance of BGN FETs and their comparison (Ryzhii et al. 2011).

6.3.1 Modeling of BGN FET

First, we investigate a BGN FET with a sandwich structure in the channel as shown in Figure 6.6. It is assumed that the bottom gate, which is positively biased by the relevant voltage $V > 0$, supplies the configuration of the electron channel in the BGN between the drain and source parts. The short top gate moderately controls the source–drain current by structuring the potential barrier for the electrons spreading between the parts. Accordingly, it can be shown that the BGN FETs are degenerate on the condition that the electron structures in the source and drain parts are spread. This means that the bottom gate voltage is adequately high to induce the essential electron density in the source and drain parts. The bottom gate has an important role in electrically inducing the energy gap and the electron channel in the BGN FET. As the applying external electric field is directed toward the BGN sheet in the channel part underneath, the top gate is determined by both V_b and V_t, the energy gap can be diverse in different parts of the BGN channel (Ryzhii et al. 2011). In order to increase the vertical electric field between the two layers of graphene, we have driven the device by imposing a high voltage between the two gates, $V_{driff} = V_{top} - V_{bottom}$ (Fiori and Iannaccone 2009a).

Figure 6.6 shows a schematic view of a double-gate BGN FET, which is made of metal gates, and also a Si substrate that is a gate dielectric (bottom gate) with 285 nm thickness under a SiO$_2$ sheet (Zhang 2009). The distance between ribbons is 0.35 nm, which is located between the source and drain electrodes with a 5 nm length (Xia et al. 2010). The source and drain extensions are 10 nm long and are doped with an equivalent molar fraction of fully ionized donors.

We present a detailed description of the developed model that is calculated theoretically on the current model of BGN. In particular, we assume the ballistic transport and the effective mass approximation, whose main electrical quantities, including the effective mass and the energy gap, have been extracted from the energy bands achieved in the TB method. Regarding the more atomistic models, the mentioned approach may underestimate the actual carrier concentration inside

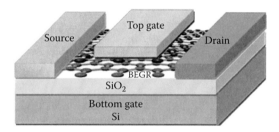

FIGURE 6.6   (See color insert.) The schematic of a BGN FET is shown as a function of the top gate voltage (V_{top}) for different bottom gate voltages (V_{bottom}).

the channel, especially for large drain-to-source (V_{DS}) and gate-to-source (V_{GS}) voltages, when the parabolic band misses to match the exact dispersion relation. In 1D bilayer graphene, with the conventional device equations, the drain current under drain bias can be expressed as

$$I_D = G_{BGN}V_{DS} \tag{6.9}$$

where
  G is defined as conductance
  L is the gate length

The current along the channel can be calculated from Landauer's formula assuming a 1D ballistic channel between contacts that are connected to external reservoirs:

$$G_{BGN} = \frac{nq}{L}\mu_{BGN} \tag{6.10}$$

where
  q is the sheet electron density at the beginning of the channel and other symbols
       have their common meanings
  μ is the mobility of electrons in the layers

The energy band structure of BGN close to the Fermi level versus wave vector with the TB method can be explained in Equation 6.2 (Castro 2010).

In these models, we consider the sizable band gap only in AB-stacking BGN with broken inversion symmetry. The carbon atoms A_1–B_1 lie on the top layer of the honeycomb lattices, while the A_2–B_2 atoms lie on the bottom layer (see Figure 6.7) (Castro 2010; Cheli et al. 2010).

By taking the derivatives of the wave vector over the energy $E(dk/dE)$ (Castro 2010; Mousavi 2011; Russo 2011) from Equation 6.2,

$$dE = \left(-2\alpha k + 4\beta k^3\right)dk \tag{6.11}$$

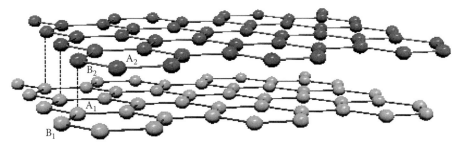

FIGURE 6.7 **(See color insert.)** Schematic of the AB stacking of BGN.

The current along the channel length can be investigated theoretically from the Landauer formula where ballistic conductance in a limit of channel is defined as in Equation 6.1 (Jimenez 2008). By using the Landauer formula that is necessary for investigating the smaller size devices, we can develop the conductance of BGN in the number of modes $M(E)$, Equation 6.3 (Katsnelson et al. 2006; Zhang, 2010). By replacing the derivative of k from Equations 6.2 and 6.3, the conductance explains (Koshino, 2010)

$$G_{BLG} = \frac{2q^2}{hL} \int_{-\infty}^{+\infty} \left(\frac{2\beta}{\alpha} \left(\frac{\alpha \pm \sqrt{\alpha^2 - 4\beta x k_B T}}{2\beta} \right)^{3/2} - \left(\frac{\alpha \pm \sqrt{\alpha^2 - 4\beta x k_B T}}{2\beta} \right)^{1/2} \right) \left(-d\left(\frac{1}{1 + e^{x-\eta}} \right) \right) \quad (6.12)$$

where
  k_B is the Boltzmann constant
  T is the room temperature

The charge mobility of BGN can be defined from channel conductance as a function of gate voltage where this means $\mu = GL/nq$. Quasi 1D intrinsic mobility at low field and short channel is a function of electron conductivity and carrier density.

$$\mu_{BLG} = \frac{2q}{hn} \int_{-\infty}^{+\infty} \left[\frac{\left(\alpha \pm \sqrt{\alpha^2 + 4\beta x k_B T} \right)^{3/2}}{\alpha \sqrt{2\beta}} - \frac{\left(\alpha \pm \sqrt{\alpha^2 + 4\beta x k_B T} \right)^{1/2}}{\sqrt{2\beta}} \right] \cdot \left[-d\left(\frac{1}{1 + e^{x-\eta}} \right) \right] \quad (6.13)$$

Unfortunately, this equation cannot be solved analytically; therefore, a numerical solution is presented. The normalized Fermi energy is $\eta = E_F - \Delta/k_B T$, where E_F is the source Fermi level and $x = \Delta - E/k_B T$,

$$\mu_{BLG} = -\frac{2q}{nh} \int_{-v_g}^{+v_g} \left(\left[\frac{1}{1 + \exp\left[x - \frac{(v_g - V)q}{k_B T} \right]} \right] \left[\frac{3}{4} \sqrt{\frac{\alpha + \sqrt{\alpha^2 + 4x\beta k_B T}}{2\beta(\alpha^2 + 4x\beta k_B T)}} - \frac{\alpha}{8\beta \sqrt{\alpha^2 + 4x\beta k_B T} \sqrt{\frac{\alpha + \sqrt{\alpha^2 + 4x\beta k_B T}}{2\beta}}} \right] \right) dx \quad (6.14)$$

where

    q is the sheet electron density at the beginning of the channel, and other symbols have their common meanings

    μ is the mobility of electrons in the layers

Therefore, current can be achieved also from the formula $I = \mu w/L$. A graphene channel width ($w = 53$ nm) and a channel length ($L = 6.1$ μm) are used in this model (Sidorov 2010).

Following Equation 6.10, the drain current I_d as a function of drain–source voltage is obtained by conducting Equation 6.12. The current along the channel length can be investigated from Landauer's formula assuming a 1D ballistic channel between the drain and source parts (Jimenez 2008).

$$I_{\text{BLG}} = \left| -\frac{4q^2}{hl} \left| \int_0^{+v_g} \left(\left(\frac{1}{1+e^{x - \frac{(v_g - V)q}{k_B T}}} \right) \left(\frac{3}{4}\sqrt{\frac{\alpha + \sqrt{\alpha^2 + 4\beta x k_B T}}{2\beta(\alpha^2 + 4\beta x k_B T)}} - \frac{\alpha}{8\beta\sqrt{\alpha^2 + 4\beta x k_B T}\sqrt{\frac{\alpha + \sqrt{\alpha^2 + 4\beta x k_B T}}{2\beta}}} \right) \right) dx \right| V_{\text{DS}} \right| \quad (6.15)$$

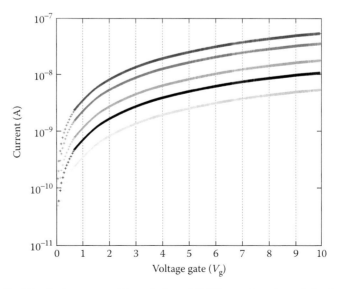

FIGURE 6.8 The drain current (I_D) variation of BGN as a function of the drain–source voltage (V_{DS}).

The drain current in BGN FETs depends on the gate voltages. The source–drain current is due to the applied voltage dependences of the potential distributions along the channel (Ryzhii et al. 2011). In order to validate our model, we have compared the analytical results with the experimental data; when voltage gate changes, the drain current versus V_{ds} curve will change. The obtained analytical formulas for the source–drain current can be used for BGN FET optimization by the proper choice of the thicknesses of gate layers, the top-gate length, and the bias voltages (Ryzhii et al. 2011).

We presented theoretically a model for the current–voltage characteristics of BGN FETs, in which the bottom and the top gates have an important role for achieving the band gap and the current–voltage as shown in Figure 6.8.

6.4 CURRENT–VOLTAGE MODELING OF BGN SCHOTTKY DIODE

6.4.1 INTRODUCTION TO BGN SCHOTTKY BARRIER DIODE

A BGN-based Schottky diode is presented for the first time. Analytical model and numerical solution of the relationship between the current–voltage are proposed. The current–voltage ($I–V$) characteristics of a BGN Schottky diode is dependent on physical parameters, such as effective mass (m^*), GNR width (w), gate insulator thickness (t_{ins}), and electrical parameters such as the Schottky barrier (SB) height (φ_{SB}).

The Schottky diode has turned out to be of huge interest for the p–n junction diode due to the fact that it presents better performance over the conventional semiconductor p–n junction (Jena et al. 2008). The graphene p–n junction diode possesses characteristics that can be applied in some novel tools such as graphene terahertz oscillators, graphene lenses, and filters (Katsnelson et al. 2006; Cheianov et al. 2007; Low et al. 2009; Ryzhii et al. 2009). There has been a universal recognition of the use of the SB diode because of its high operating frequencies and low forward voltage drop (Sankaran and Kenneth 2005). RF signal rectification/detection, mixing, and imaging have been found to be some of its wide applications. By raising a slim molecular ray epitaxial layer over an n layer, there exists a possibility of high-sensitivity silicon substrate of Schottky diode with cutoff frequencies of 1 THz (Sankaran and Kenneth 2005). This study analyzed a model of a BGN Schottky diode. To obtain an analytical relation for the channel current, an analytical equation for potential distribution along the BGN is presented. The derived analytical current well describes the rectifying behavior of the presented BGN Schottky diode. The channel current is analytically derived as a function of various physical and electrical parameters, including gate insulator thickness, SB height, drain bias voltage, gate bias voltage, GNR width, and subband number.

6.4.2 GNR SB DIODE

Currently, there is significant research interest in GNR that is used in SB diodes. There exists a finite width that opens up a band gap in GNR, which allows the structure to become a useful and competent channel material for these diodes (Alam 2009).

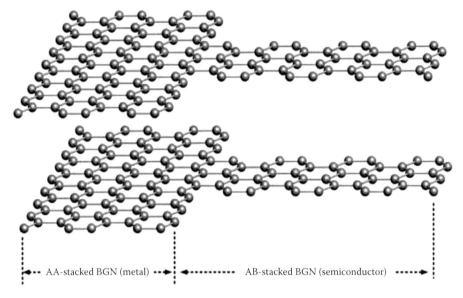

FIGURE 6.9   Graphene nanoribbon Schottky diode.

The contact between GNR and a metal is too inappropriate to make an SB GNR. But it can be used in the manufacturing of a Schottky diode, a transistor, and others. A semiconductor diode is produced as a result of contact between a semiconductor layer and a metal covering. It has nonlinear rectifying characteristics such as hot carriers (electrons for n-type material or holes for p-type material) being broadcast from the SB of the semiconductor (Neamen 2003; Kargar and Lee 2009). As shown in Figure 6.9, a p-type GNR Schottky diode can be used as a channel between the drain and source contacts.

There are two different metals at the two ends of the channel: Ag and Al are used for the drain and source contacts, respectively. The work function of Ag is higher than that of GNR, and it produces a small SB at the drain contact (near ohmic contact), while Al metal has lower work function compared to GNR and it makes a large SB at the source contacts (Kargar and Lee 2009).

There exists a universal applicability of the SB diode, which is associated with other components being dependent on it. Due to the low forward voltage drop, the Schottky diode was found to be very relevant in power applications as a rectifier, resulting in lower levels of power reduction compared to the traditional p–n junction diode. In most cases, it is referred to as the Schottky diode, but has lately been referred to as the hot carrier diode, hot electron diode, and surface barrier diode; it takes its name from Schottky. The Schottky diode can be applied in several applications in which other diodes will not function effectively. This is because of its advantages such as low injunction capacitance, low turn on voltage, and fast recovery time (Neamen 2003; Jimenez 2008; Kargar and Lee 2009).

The ability of the Schottky diode to rectify an alternating current is considered as one of its main relevant characteristics. For the rectification attitude to be derived for the presented device, there is a consideration of metals of varying qualities at the two ends of a p-type semiconducting GNR leading to asymmetric linkage. The impact of different parameters like the gate bias voltage, GNR width, and contact metals on the rectification behavior has been studied (Kargar and Wang 2010), which indicates that there is a possibility of adjusting the rectification properties, threshold voltage, reverse saturation current, and reverse turn-on voltage by adopting these parameters (Kargar and Wang 2010). The method is believed not to be affected by the thermal equilibrium, while the Maxwell–Boltzmann approximation can be taken in case the barrier height is larger than KT. Thus, the thermionic emission characteristics can be described through this method (Neamen 2003).

Switching speeds cannot be reduced because of the preponderance of majority carriers and the minority carriers not being essentially injected or stored. A hot carrier diode, which is also referred to as the Schottky diode, serves as a semiconductor. It is usually a single-crystal silicon, on which independent slim metal layers are embedded to produce electrical contact. The SB to the silicon is formed using either of the layers of the metal. Ohmic contact is formed by a very low resistance. The SB is defined as an electron or hole barrier that is caused by an electric dipole charge distribution related to the contact and future difference created between a metal and semiconductor under an equilibrium condition. The barrier is found to be very abrupt at the top of the metal due to the charge being mostly on the surface (Neamen 2003; Jimenez 2008).

The potential slowly varies across a small distance because the charge is spread over this distance in the semiconductor. The current has a significant transmission through the diode in one direction only. In an n-type semiconductor, electrons can move from the semiconductor to the metal for a polarity of the applied voltage. If the applied voltage is reversed, electrons are blocked from moving into the semiconductor from the metal by a potential barrier. In case of a p-type semiconductor, since holes are positively charged, the polarities are reversed. No reverse bias polarity is experienced in both cases where the applied voltage of one polarity (forward bias) can limit the potential barrier for charge carriers leaving the semiconductor. The SB height can be as a function of the electric field in the semiconductor because of the effect of barrier lowering. The barrier height is also dependent on the surface states in the semiconductor (Neamen 2003).

6.4.3 Proposed Model for Current–Voltage Characteristics of BGN Schottky Diode

There are two single layers stacked in two atomic structures including the hexagonal (AA) and Bernal (AB) stacking, among which the AB-stacked semiconducting with a band gap of 0.02 eV is investigated here, while the hexagonal (AA) being energetically unstable and metallic makes it irrelevant to be studied (Latil and Henrard 2006; Koshino 2008, 2009). As shown in Figure 6.10, in terms of AB configuration, atoms on the top and bottom layers of BGN are called (A_1, B_1) and (A_2, B_2) with hexagonal carbon lattice, respectively (Aoki and Amawashi 2007; Li et al. 2008a). By applying a normal electric field on the BGN's surface, there is the possibility of opening a gap on the BGN's band energy that can be controlled by value of applied voltage (Castro et al. 2008).

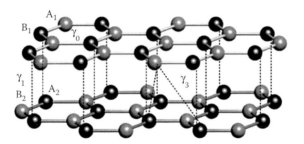

FIGURE 6.10   **(See color insert.)** Structure of BGN with AB stacking.

By using the TB technique, the energy band structure of BGN with AB stacking has been calculated in (Castro et al. 2010)

$$E_k(V) = \pm \sqrt{\varepsilon_k^2 + \frac{t_\perp^2}{2} + \frac{V^2}{4} \pm \sqrt{\left(t_\perp^2 + V^2\right)\varepsilon_k^2 + \frac{t_\perp^4}{4}}} \tag{6.16}$$

where

$t_\perp$ is the interlayer hopping
V is the biased voltage
$\varepsilon_k^2 = \left(V^4/4 + V^2 t_\perp^2/2\right)/\left(V^2 + t_\perp^2\right)$ (Castro et al. 2010)

So, Equation 6.16 can be simplified into energy Equation 6.2, which was earlier mentioned (Stauber et al. 2007; Castro et al. 2010).

The density of state (DOS) shows the number of available states at each energy level that can be occupied. Therefore, the DOS for BGNR can be modeled as

$$D(k) = \frac{1}{2\pi(4\beta k^3 - 2\alpha k)} \tag{6.17}$$

By integrating the distribution function over energy band, carrier concentration as an important parameter can be obtained. Derivatives n over the energy (dn/dE) with respect to

$$dn = \frac{1}{2\pi(4\beta k^3 - 2\alpha k)} \frac{1}{e^{\frac{E-E_F}{k_B T}} + 1} dE \tag{6.18}$$

From Equation 6.2

$$(E - \Delta) + \alpha k^2 - \beta k^4 = 0 \tag{6.19}$$

where $k = \left[\alpha \pm \sqrt{\alpha^2 - 4(E-\Delta)\beta}/2\beta\right]^{1/2}$, which is calculated from band energy equation.

The thermionic emission theory is reported to be the most important step in rectifying into order the contact with an n-type semiconductor through transmission of electrons above the anticipated barrier; there can be electrons in the metal directly adjacent to the empty space in the semiconductor (Eriksson et al. 2003). Assuming that an electron from the valence band of the semiconductor was to flow into the metal, the effect would be equivalent to holes being injected into the semiconductor. There is an establishment of an excess minority carrier hole in the vicinity. However, the dimensions and calcula-tions have clearly illustrated that the proportion of the minority carrier hole current to the overall current is far less in a majority of the situations (Neamen 2003).

The drain current is made up of both a dual component-thermal and a tunnel. The tunneling current plays a very relevant manner in a Schottky contact device (Alam 2009). An efficient functionality of the transistor with a doped nanoribbon has been noticed in terms of on/off current ration, intrinsic switching delay, and intrinsic cutoff frequency. A source-channel flat band condition is achieved at a gate bias exactly the same as half of the band gap in a Schottky source–drain contact device (Alam 2009). Further than this bias gate, the drain current gets almost saturated when it is affected by the gate potential modulates, resulting in limits in the tunnel barrier and on-state current in the Schottky contact device. Contrary to this, the gate modulates the dual component consisting of the thermal and tunnel barrier in a doped contact device beyond the source–gate band voltage (Alam 2009).

When electrons move from the metal into the semiconductor, it can be defined by electron current density $J_{m \to s}$, whereas $J_{s \to m}$ is the electron current density as a result of pouring of electrons from the semiconductor into the metal. The trend of flow of elec-trons depends on the subscripts of the current. In order words, the conventional current direction inverses to the electron pour. $J_{s \to m}$ is relative to the concentration of electrons with x-directed velocities enough to subdue the barrier (Neamen 2003). We can write this as

$$J_{s \to m} = e \int_{-\infty}^{+\infty} v_x \, dn \tag{6.20}$$

where
  e is the magnitude of the electronic charge
  v_x is the carrier velocity in the direction of transport

If all of the electron energy over E_{co} is specified to be kinetic energy, we have

$$E - E_{co} = \frac{1}{2} m^* v_x^2 \tag{6.21}$$

where m^* is the effective mass. The current density of the BGN SB diode given by Equations 6.20 and 6.21 is

$$J_{s \to m} = e \int_{-\infty}^{+\infty} \frac{\sqrt{2}(E - E_{co})^{\frac{1}{2}}}{\sqrt{m^*}} \frac{1}{2\pi(4\beta k^3 - 2\alpha k)} \frac{dE}{\left(e^{\frac{E - E_F}{k_B T}} + 1 \right)} \tag{6.22}$$

Therefore, by replacing K in Equation 6.22 we get

$$
J_{s \to m} = \frac{\sqrt{2}e}{2\pi\sqrt{m^*}}
$$

$$
\int_{-\infty}^{+\infty} \frac{(E - E_{co})^{\frac{1}{2}}}{\left(4\beta \left[\frac{\alpha \pm \sqrt{\alpha^2 - 4(E - \Delta)\beta}}{2\beta} \right]^{\frac{3}{2}} - 2\alpha \left[\frac{\alpha \pm \sqrt{\alpha^2 - 4(E - \Delta)\beta}}{2\beta} \right]^{\frac{1}{2}} \right)} \frac{dE}{\left(e^{\frac{E - E_F}{k_B T}} + 1 \right)} \quad (6.23)
$$

where
$E_{co} = \Delta$, which is the applied voltage.

The current density of the BGN Schottky diode can be gained from Equation 6.23

$$
J_{s \to m} = \frac{\sqrt{2}e}{2\pi\sqrt{m^*}}
$$

$$
\int_{-\infty}^{+\infty} \frac{(k_B T)^{\frac{1}{2}} x^{\frac{1}{2}}}{\left(4\beta \left[\frac{\alpha \pm \sqrt{\alpha^2 - 4(k_B T)\beta x}}{2\beta} \right]^{\frac{3}{2}} - 2\alpha \left[\frac{\alpha \pm \sqrt{\alpha^2 - 4(k_B T)\beta x}}{2\beta} \right]^{\frac{1}{2}} \right)} \frac{(k_B T)dx}{(1 + e^{x - \eta})} \quad (6.24)
$$

where
Normalized Fermi energy $\eta = \frac{E_F - E_{co}}{k_B T}$

$$
x = \frac{E - E_{co}}{k_B T}
$$

The study of the current–voltage response of an electronic device is one of the main topics in electronics. By replacing $\eta \approx \frac{V_{GS} - V_T}{k_B T/q}$ in Equation 6.24, the current–voltage relationship of the BGN Schottky diode can be gained from Equation 6.25,

$$
J_{s \to m} = \frac{\sqrt{2}e}{2\pi\sqrt{m^*}}
$$

$$
\int_{-\infty}^{+\infty} \frac{(k_B T)^{\frac{1}{2}} x^{\frac{1}{2}}}{\left(4\beta \left[\frac{\alpha \pm \sqrt{\alpha^2 - 4(k_B T)\beta x}}{2\beta} \right]^{\frac{3}{2}} - 2\alpha \left[\frac{\alpha \pm \sqrt{\alpha^2 - 4(k_B T)\beta x}}{2\beta} \right]^{\frac{1}{2}} \right)} \frac{(k_B T)dx}{1 + e^{x - \left(\frac{V_{GS} - V_T}{k_B T/q} \right)}} \quad (6.25)
$$

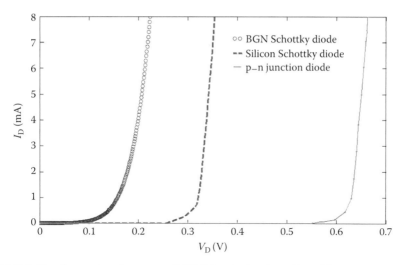

FIGURE 6.11 Comparison between a BGN Schottky barrier with a silicon Schottky diode and a p–n junction diode.

Based on this model, the current–voltage curve of the BGN Schottky diode is plotted in Figure 6.11.

6.4.4 COMPARISON BETWEEN **BGN** SCHOTTKY DIODE AND **BGN** P–N JUNCTION DIODE

The current mechanism differs between a BGN Schottky diode and a BGN p–n junction diode. In the case of the SB diode, it is determined by the thermionic emission of majority carriers over a potential barrier, while for the p–n junction, it is specified by the diffusion of minority carriers (Neamen 2003). The real difference between the turn-on voltages is a function of the barrier height of the metal–semiconductor contact and doping concentrations in the p–n junction, but the comparatively large difference is always achieved. The switching properties or frequency response is regarded as an alternatively relevant variation between an SB diode and a p–n junction diode (Eriksson et al. 2003; Neamen 2003).

There is no relation between the diffusion capacitance and a forward-biased Schottky diode, because the SB diode is known as a majority carrier device. The removal of the diffusion capacitance makes the SB diodes have a higher-frequency than the p–n junction diodes. In a Schottky diode switching from forward to reverse bias, it has no minority carrier stored charge to remove, as in the case with a p–n junction diode. The Schottky diode can be applied in fast-switching applications since there is no minority carrier storage time. A typical switching-time for a Schottky diode is in picoseconds range, while for a p–n junction it is in the nanoseconds range (Neamen 2003; Sankaran and Kenneth 2005; Yoon et al. 2008).

There are differences between a Schottky diode and a p–n junction diode in two main areas. The former deals with the magnitudes of the reverse saturation current densities, whereas the latter is associated with the switching characteristics.

Although the real difference between the turn-on voltages will be affected by the barrier height of the metal–semiconductor contact and doping concentrations in the p–n junction, a comparatively large difference is always achieved. In fact, the switching properties or frequency response is taken into consideration as a range of difference between an SB diode and a p–n junction diode (Eriksson et al. 2003; Neamen 2003). Figure 6.11 shows a comparative study of the presented model, the typical I–V characteristics of a silicon SB diode and a p–n junction diode. It gives a clear illustration of the fact that the effective turn-on voltage of the BGN Schottky diode is smaller than that of the silicon Schottky diode and the p–n junction diode. The silicon SB diode possesses a turn-on voltage that is about 0.37 V smaller than the turn-on voltage of the p–n junction diode.

6.5 CONCLUSION

BGN with different stacking arrangements is used as metal and semiconductor contacts in a Schottky diode junction. Based on this assumption, an analytical model of junction current–voltage characteristic is presented. Simulated results indicate that rectification I–V characteristic of a BGN Schottky diode is a function of effective mass (m^*), temperature, and applied bias voltage. We conclude that by decreasing the temperature, the turn-on voltage shifts rightward to larger values, which shows better performance in comparison with typical rectification behavior of a Schottky diode. A comparative study of the presented model with the conventional diodes is presented. It shows that the effective turn-on voltage of the BGN Schottky diode is smaller compared to that of the silicon Schottky diode and the p–n junction diode. The silicon SB diode possesses a turn-on voltage that is smaller than the turn-on voltage of the p–n junction diode. The presented model can be employed on the BGN Schottky transistor optimization.

REFERENCES

Ahmadi, M. T., Z. Johari, et al. 2010. Graphene nanoribbon conductance model in parabolic band structure. *Journal of Nanomaterials.* 2010: 4 p. doi:10.1155/2010/753738.

Alam, K. 2009. Transport and performance of a zero-Schottky barrier and doped contacts graphene nanoribbon transistors. *Semiconductor Science and Technology.* 24(1): 015007.

Amin, N. A., M. T. Ahmadi, et al. 2011. Effective mobility model of graphene nanoribbon in parabolic band energy. *Modern Physics Letters B.* 25: 739–745.

Aoki, M. and H. Amawashi. 2007. Dependence of band structures on stacking and field in layered graphene. *Solid State Communications.* 142(3): 123–127.

Betti, A., G. Fiori, et al. 2009. Physical insights on graphene nanoribbon mobility through atomistic simulations. *IEEE International Electron Devices Meeting.* Baltimore. MD, 7–9 December, pp. 5424276–5424280.

Castro, E. V. 2010. Electronic properties of a biased graphene bilayer. *Journal of Physics: Condensed Matter.* 22(17): 175503.

Castro, E. V., K. S. Novoselov, et al. 2010. Electronic properties of a biased graphene bilayer. *Journal of Physics: Condensed Matter.* 22(17): 175503–175514.

Castro, E. V., N. M. R. Peres, et al. 2008. Bilayer graphene: gap tunability and edge properties. *International Conference on Theoretical Physics, Dubna-Nano.* 129(1).

Cheianov, V. V., V. Fal'ko, et al. 2007. The focusing of electron flow and a veselago lens in graphene p–n junctions. *Science.* 315: 1252–1255.

Cheng, S.-G. 2010. Transport properties of monolayer and bilayer graphene p–n junctions with charge puddles in the quantum Hall regime. *Journal of Physics: Condensed Matter.* 22: 465301–465308. doi:465310.461088.

Cheli, M., P. Michetti, et al. 2010. Model and performance evaluation of field-effect transistors based on epitaxial graphene on SiC. *IEEE Transactions on Electron Devices.* 57(8): 1936–1941.

Das Sarma, S., S. Adam, et al. 2010. Theory of carrier transport in bilayer graphene. *Physical Review B.* 81(16): 161407.

Dong, H. M., J. Zhang, et al. 2009. Optical conductance and transmission in bilayer graphene. *Journal of Applied Physics.* 106(4): 043103–043106.

Dragoman, D. 2010. Low-energy conductivity of single- and double-layer graphene from the uncertainty principle. *Physica Scripta.* 81(3).

Du, X. 2008. Suspended graphene: A bridge to the Dirac point. *Nature Nanotechnology.* 3: 491–495.

Eriksson, J., N. Rorsman, et al. 2003. 4H-Silicon carbide Schottky barrier diodes for microwave applications. *IEEE Transactions on Microwave Theory and Techniques.* 51(3): 796–804.

Fiori, G. and G. Iannaccone. 2009a. On the possibility of tunable-gap bilayer graphene FET. *IEEE Electron Device Letters.* 30(3): 261–264.

Fiori, G. and G. Iannaccone. 2009b. Performance analysis of graphene bilayer transistors through tight-binding simulations. *IWCE-13: 2009 13th International Workshop on Computational Electronics*, Beijing, 27–29 May, pp. 301–304.

Gnani, E., A. Gnudi, et al. 2010. Effective mobility in nanowire FETs under quasi-ballistic conditions. *IEEE Transactions on Electron Devices.* 57(1): 336–344.

Jena, D., T. Fang, et al. 2008. Zener tunneling in semiconducting nanotube and graphene nanoribbon p–n junctions. *Applied Physics Letters.* 93(11): 112106–112103.

Jimenez, D. 2008. A current-voltage model for Schottky-barrier graphene based transistors. *Nanotechnology.* 19(34): 345204.

Kargar, A. and C. Lee. 2009. Graphene nanoribbon Schottky diodes using asymmetric contacts. *Nanotechnology, 9th IEEE-NANO 2009*, Genoa, 26–30 July, pp. 978–981.

Kargar, A. and D. L. Wang. 2010. Analytical modeling of graphene nanoribbon Schottky diodes. *Proceedings of the SPIE7761 Carbon Nanotubes, Graphene, and Associated Devices III.* 7761: 77610U-1–77610U-8. ISBN: 9780819482570.

Katsnelson, M. I., K. S. Novoselov, et al. 2006. Chiral tunnelling and the Klein paradox in graphene. *Nature Physics.* 2: 620–625.

Koshino, M. 2008. Electron delocalization in bilayer graphene induced by an electric field. *Physical Review B.* 78(155411–155415).

Koshino, M. 2009. Electronic transport in bilayer graphene. *New Journal of Physics.* 11(9): 195406–195415.

Koshino, M. 2010. *Physical Review B.* 81: 125304.

Latil, S. and L. Henrard. 2006. Charge carriers in few-layer graphene films. *Physical Review Letters.* 97(3): 036803–036807.

Li, T. S., Y. C. Huang, et al. (2008a). Transport properties of AB-stacked bilayer graphene nanoribbons in an electric field. *European Physical Journal B.* 64: 73–80.

Li, X. L., X. R. Wang, et al. (2008b). Chemically derived, ultrasmooth graphene nanoribbon semiconductors. *Science.* 319(5867): 1229–1232.

Liao, Z.-M. 2010. Current regulation of universal conductance fluctuations in bilayer graphene. *New Journal of Physics.* 12: 083016–083019.

Low, T., S. Hong, et al. 2009. Conductance asymmetry of graphene p–n junction. *IEEE Transaction on Electron Devices.* 56(6): 1292–1299.

Lv, M. 2009. Screening-induced transport at finite temperature in bilayer graphene. *Physical Review B*. 81: 195409.

Majumdar, K., K. Murali, et al. 2010. High on-off ratio bilayer graphene complementary field effect transistors. *2010 International Electron Devices Meeting, Technical Digest*, San Francisco, CA, December 6–8, pp. 736–739.

Mao, Y. L., G. M. Stocks, et al. 2010. First-principles study of the doping effects in bilayer graphene. *New Journal of Physics*. 12: 033046.

Morozov, S. V., K. S. Novoselov, et al. 2008. Giant intrinsic carrier mobilities in graphene and its bilayer. *Physical Review Letters*. 100(1): 016602.

Mousavi, S. M. 2011. Bilayer graphene nanoribbon carrier statistic in degenerate and non degenerate limit. *Journal of Computational and Theoretical Nanosceince*. 8(10): 2029–2032.

Nagashio, K., T. Nishimura, et al. 2009. Mobility variations in mono- and multi-layer graphene films. *Applied Physics Express*. 2(2): 025003.

Neamen, D. A. 2003. *Semiconductor Physics and Devices*. Third Edition. McGraw-Hill, University of New Mexico, Albuquerque.

Novoselov, K. S., E. McCann, et al. 2006. Unconventional quantum Hall effect and Berry's phase of 2. *Nature Physics*. 2: 177–180. doi:10.1038/nphys245.

Novoselov, K. S., S. V. Morozov, et al. 2007. Electronic properties of graphene. *Physica Status Solidi B*. 244(11): 4106–4111.

Oostinga, J. B., H. B. Heersche, et al. 2008. Gate-induced insulating state in bilayer graphene devices. *Nature Materials*. 7(2): 151–157.

Russo, S. 2011. Electronic transport properties of few-layer graphene materials. *Graphene Times*. arXiv:1105.1479v1 [cond-mat.mes-hall].

Ryzhii, V., M. Ryzhii, et al. 2008. Current-voltage characteristics of a graphene-nanoribbon field-effect transistor. *Journal of Applied Physics*, 103(9): 094510–094517.

Ryzhii, V., M. Ryzhii, et al. 2009. Graphene tunneling transit-time terahertz oscillator based on electrically induced p-i-n junction. *Infrared, Millimeter, and Terahertz Wave, IRMMW-THz 2009. 34th International Conference*, University of Aizu, Aizu-Wakamatsu, Japan. 21–25 September, pp. 1–2.

Ryzhii, V., M. Ryzhii, et al. 2011. Analytical device model for graphene bilayer field-effect transistors using weak nonlocality approximation. *Journal of Applied Physics*. 109(6): 064508–064517.

Sadeghi, H., M. T. Ahmadi, et al. 2011. Ballistic conductance model of bilayer graphene nanoribbon (BGN). *Journal of Computational and Theoretical Nanoscience*. 8(10): 1993–1998.

Sankaran, S. and K. O. Kenneth. 2005. Schottky barrier diodes for millimeter wave detection in a foundry CMOS process. *IEEE Electron Device Letters*. 26(7): 492–494.

Sidorov, A. N., T. Bansal, et al. 2010. Graphene nanoribbons exfoliated from graphite surface dislocation bands by electrostatic force. *Nanotechnology*. 21(19): 195704–195723.

Stauber, T., N. M. R. Peres, et al. 2007. The optical conductivity of graphene in the visible region of the spectrum. *Physical Review B*. 75(11).

Sun, S. J. and C. P. Chang. 2008. Ballistic transport in bilayer nano-graphite ribbons under gate and magnetic fields. *European Physical Journal B*. 64(2): 249–255.

Tang, K. 2011. Electric field induced energy gap in few layer graphene. *The Journal of Physical Chemistry C*. 115: 9458–9464. dx.doi.org/10.1021/jp201761p.

Yu, T., E.-K. Lee, et al. 2010. Bilayer graphene system: Current-induced reliability limit. *IEEE Electron Device Letters*. 31(10): 1155–1157.

Xia, F. N., D. B. Farmer, et al. 2010. Graphene field-effect transistors with high on/off current ratio and large transport band gap at room temperature. *Nano Letters*. 10(2): 715–718.

Xu, N. and J. W. Ding. 2008. Conductance growth in metallic bilayer graphene nanoribbons with disorder and contact scattering. *Journal of Physics: Condensed Matter.* 20: 485213. doi: 485210.481088/480953-488984/485220/485248/485213.

Xu, Y. 2010. Infrared and Raman spectra of AA-stacking bilayer graphene. *Nanotechnology* 21(6): 065711–065716.

Yoon, Y., G. Fiori, et al. 2008. Performance comparison of graphene nanoribbon FETs with Schottky contacts and doped reservoirs. *IEEE Transaction Electron Devices.* 55(9): 2314–2323.

Zhang, Y., T. T. Tang, et al. 2009. Direct observation of a widely tunable bandgap in bilayer graphene. *Nature.* 459(7248): 820–823.

Zhang, Y. 2010. Dimensions and elements of people's mental model of an information-rich web space. *Journal of the American Society for Information Science and Technology.* 61(11): 2206–2218.

Zou, K., X. Hong, et al. 2011. Electron-electron interaction and electron-hole asymmetry in bilayer graphene. *Physical Review B.* 84: 085408. *arXiv:1103.1663v1 [cond-mat.str-el].*

7 Bilayer Graphene Nanoribbon Transport Model

*Hatef Sadeghi, Seyed Mahdi Mousavi,
Meisam Rahmani, Mohammad Taghi
Ahmadi, and Razali Ismail*

CONTENTS

7.1   INTRODUCTION

The silicon-based semiconductor industry is facing a grave problem because of the performance limits imposed on semiconductor devices when it is miniaturized to nano scale [1–3]. To counter this problem, several semiconductor designers have relied on some innovations by utilizing the new computational state variables (spintronic). Many semiconductor designers use new materials such as graphene and carbon-based materials [4]. Research on the application of graphene has increased tremendously. This two-dimensional (2D) material represents a new class of materials [5]. Graphene physics is currently one of the most active research areas. The possibility of changing the band gap in the bilayer graphene using external electric field suggests that this system can be an alternative to silicon-based integrated system technology [6]. Graphene results from stacked layers of graphite carbon atoms. It was discovered by Andre Geim and Kostya Novoselov (Russian) from the University of Manchester in 2004 [1]. They were awarded the Nobel Prize in Physics in 2010 [7].

Graphene can be produced by the process of exfoliation, drawing, hydrazine reduction, sodium reduction of ethanol, and epitaxial growth on silicon carbide or metal substrates [5].

A 2D honeycomb lattice of carbon is called graphene, which has a single atomic layer sheet [8–10]. Until the experimental discovery of single-layer graphene, it was believed that a 2D crystal may be thermodynamically unstable. Further investigations proved that the stability of the graphene is due to the gentle crumples in the third dimension [11].

Exciting electronic, thermal, and photoelectronic properties of graphene as 2D electron gas (2DEG) has attracted huge scientific interest [8,12]. Fabrication of new high-speed graphene transistors [13] realized this hypothesis that graphene can be the best alternative to silicon-based nanoelectronic devices. After the discovery of graphene in 2004 [11], different aspects of electronic properties of mono-, bi-, and tri-layer graphene led the investigations to find out the differences between them. Carrier statistics and conductance of monolayer graphene have been studied [14,15]. This study focuses on the electronic properties of bilayer graphene nanoribbon (BGN) and attempts at uncovering some electronic differences between mono- and bilayer configurations. Electronic properties of BGN and the differences between the properties of monolayer and bilayer graphenes were investigated. Energy band structure, carrier concentration, and conductance of biased BGN were studied, and the mathematical models of carrier concentration and ballistic conductance of BGN were derived. The properties of the proposed conductance model show a good agreement with the published experimental data.

7.2   ELECTRONIC PROPERTIES OF BGN

Graphene nanoribbon (GNR) exhibits excellent electronic and thermal characteristics [11,16]. It is stable at temperatures up to 3000°C [17]. It has high carrier mobility [4] similar to that of carbon nanotubes [1]. Although GNR has an extremely high mobility, it is not appropriate for use in digital electronics because of its zero band energy gap. It is most suitable for use in analog electronics, ultrahigh-speed radio frequency (RF) applications, and interconnects [8,13,18]. BGN can be employed in digital electronics because its band gap can be varied by means of an external perpendicular electric field and can induce a significant band gap between the valence and conduction bands, turning it from a zero-gap semiconductor to an insulator [4,18–20]. Carrier mobility values as high as 200,000 cm^2 (Vs)$^{-1}$ [8,13,21,22] (200 times higher than in silicon [23]) can be achieved by graphene, which is increased by increasing carrier density [4]. The high mobility and the possibility of low-cost mass production of graphene prove its potential for use as in the future high-speed integrated electronic circuits [23]. Recent studies have established graphene transistors operating in the gigahertz regime with a record of f_T = 100–300 GHz [13]. It is notable that almost-ballistic transport in BGN field effect transistors (BGNFETs) is induced by its very high mobility [24]. This means that electrons can travel long distances (~1 μm) in graphene without being scattered [1,11].

BGN is expected to show a different type of behavior compared to GNR, which is not only due to the presence of a gap in the energy spectrum of the latter [25], but also because of each layer's electrons being affected by the electrons at the same layer and also the electrons in the second layer [18]. Recent studies have shown several different properties of BGN compared to GNR in many respects such as minimal conductivity, the quantum Hall effect [26,27], the edge states, and the weak localization [18].

7.3 CONFIGURATION OF BILAYER GRAPHENE

An atomic layer of carbon atoms arranged in a benzene-ring structure is termed GNR [8,17]. Each two nearest carbon atoms (referred to as A and B) with a distance of ~1.41 Å is called a unit cell [23]. BGN consists of two monolayers stacked in two possible configurations. One of them is the simple hexagonal (AA) stacking, which is energetically unstable [28] and the other is Bernal (AB) stacking. A_2 atoms on layer 2 are located directly below B_1 atoms on layer 1 in AB Bernal stacking (Figure 7.1), and the layer spacing is almost 0.34 nm [16–18,29,30]. A unit cell of four atoms in layers 1 and 2 of Bernal stacking BGN induces four energy bands, where a pair of them is very far from each other and the other pair is closed, playing significant roles in BGN's conductivity [23].

Referring to Figure 7.1, γ_0 = 3.16 eV is the value of in-plane hopping energy, γ_1 = 0.39 eV is the interlayer hopping energy of two carbon atoms of double layer that are located directly, and γ_3 = 0.315 eV is the interlayer hopping energy of two atoms

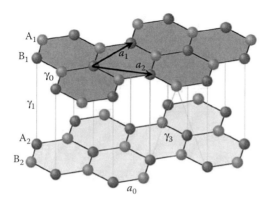

FIGURE 7.1 (See color insert.) Bilayer graphene configuration. Atomic structure of AB-stacked bilayer graphene. In AB-stacking BGN, A_1 atoms of layer 1 are below B_2 atoms of layer 2 and vice versa.

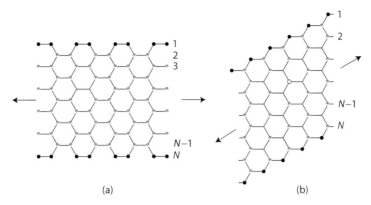

FIGURE 7.2 Edge shape of graphene sheet. The network structure of (a) an armchair GNR and (b) a zigzag GNR.

that are not facing each other directly [32,33]. In BGN, a relatively weak interlayer interaction induces a complicated electronic conduction [26].

A graphene sheet can be cut along a straight line to make two kinds of edge shapes, namely, the armchair edge (Figure 7.2a) and the zigzag edge (Figure 7.2b) [34]. With regard to graphene sheet fabrication process, micromechanical cleavage of bulk graphite was used to produce the first single-layer graphene, and thermal decomposition of graphene on silicon carbide (SiC) wafers was employed for industrial quality production of single- and multilayer graphene sheets [11,19,26]. As shown in Figure 7.1, size of the graphene ribbon is defined as $\vec{c} = m\vec{a}_1 + n\vec{a}_2$, where m and n are 1, 2, 3, …. Vectors $\vec{a}_1$ and $\vec{a}_2$ can be defined as $\vec{a}_1 \equiv \hat{x}a + \hat{y}b$ and $\vec{a}_2 \equiv \hat{x}a - \hat{y}b$, respectively, where $a \equiv 3a_0/2$, $b \equiv \sqrt{3}a_0/2$, and $a_0 = 1.41$ Å.

7.4 BAND STRUCTURE OF BGN

Understanding of the electronic structure of bilayer graphene starts with examining its band structure. It has been shown that the band gap of BGN can be varied by means of an external perpendicular electric field and induced significant band gap between the valence and conduction bands from a zero-gap semiconductor to an insulator [4,18–20]. Spectrum of full tight-binding Hamiltonian of AB-stacked BGN gives electronic structure of the biased BGN as shown in Equation 7.1. It appears that the gap is controlled by V [35] as:

$$E_k^{\pm\pm}(V) = \pm\sqrt{\varepsilon_k^2 + \frac{t_\perp^2}{2} + \frac{V^2}{4} \pm \sqrt{\frac{t_\perp^4}{4} + \left(t_\perp^2 + V^2\right)\varepsilon_k^2}} \tag{7.1}$$

where $\varepsilon_k^2 = (V^4/4 + V^2 t_\perp^2/2)/(V^2 + t_\perp^2)$ [35]. The experimental value of the interlayer hopping ($t_\perp$) energy for graphene is approximately equal to 0.39 eV [36]. Equation 7.1 can readily be rearranged as

$$E^{\pm}(k) \approx \pm\Delta \mp \alpha k^2 \pm \beta k^4 \tag{7.2}$$

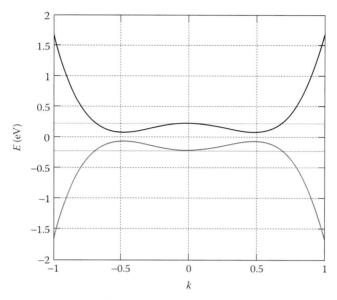

FIGURE 7.3 Band structure of BGN for different values of k in perpendicular external electric fields V = 0.45 V. Upper line represents conduction band while the lower line yields valance band of BGN.

where

$$\Delta = V/2$$

$$\alpha = V v_F^2 / t_\perp^2$$

$$\beta = v_F^4 / V t_\perp^2$$

$v_F = 3t \times a_{c-c} / 2\hbar$, where $a_{c-c} = 1.41$ Å is the lattice-spacing [37] and $t \sim 3.1$ eV is the in-plan hopping [38]

$k = 2\pi n/\lambda$ is the momentum, where λ is the wavelength

It has been demonstrated by some experiments that a perpendicular electric field applied on a BGN is a valuable tool to change its energy band structure [20]. A band energy gap of 250 meV has been reported for a BGN at room temperature [31].

Figure 7.3 shows the closed pair of four energy bands induced in biased BGN based on Equation 7.2, which plays a significant role in BGN conductivity [18]. In general, GNR and BGN are both gapless semimetal materials [27]. Biased BGN is the only known semiconductor with a tunable energy band gap. This type of material may open a way for developing photodetectors [32].

7.5 DEGENERATE AND NONDEGENERATE APPROXIMATION

It has been shown that the Fermi–Dirac distribution function can be approximated by the Maxwell–Boltzmann distribution function if the Fermi level lies in the band gap more than $3k_B T$ from either band edge where $(E - E_F) \geq 3k_B T$. For this positioning of

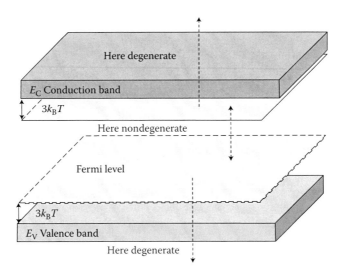

FIGURE 7.4 **(See color insert.)** Definition of the degenerate and the nondegenerate regions. Fermi–Dirac distribution function can be approximated by either $f(E) = \exp((E_F - E)/k_B T)$ in the nondegenerate limit or $f(E) = 1$ in the degenerate limit.

the Fermi level, which is called nondegenerate region (Figure 7.4), the energy states are mostly empty of electrons and $f(E) = \exp((E_F - E)/k_B T)$.

Conversely, in the degenerate region, the Fermi level either exists within $3k_B T$ of band edge or lies inside a band (Figure 7.4). In this region, the Fermi–Dirac distribution function can be approximated by 1 ($f(E) = 1$) and the energy states are mostly filled with electrons (holes) [39–41].

7.6 DENSITY OF STATES IN BGN

One of the fundamental parameters of BGN is the density of state (DOS), which indicates the available energy states [42]. It can be defined as

$$\text{DOS} = \frac{\Delta n}{l \Delta E} \tag{7.3}$$

where
    Δn is the quantum number
    l is the length
    ΔE is the energy

The gradient of DOS near the Dirac point changes the electrical property of the material, which demonstrates either metallic or semiconducting behavior. Derivatives of k over the energy (dk/dE) with respect to Equation 7.2 shows

$$dE = (-2\alpha k + 4\beta k^3)dk \tag{7.4}$$

where

$k = 2\pi n/\lambda$ is the quantized wave vector

l is the length of the nanoribbon, which can be assumed equal to λ, in ballistic limit

$$dE = (-2\alpha k + 4\beta k^3)\frac{2\pi}{l}\,dn \tag{7.5}$$

Since the DOS is the number of states at each energy level that are available to be occupied, DOS can be modeled as a function of wave vector as

$$DOS = \frac{1}{2\pi\left(4\beta k^3 - 2\alpha k\right)} \tag{7.6}$$

The DOS is useful in the understanding of optical excitations [43].

7.7 GENERAL MODEL OF BGN CARRIER CONCENTRATION

Carrier concentration in a band is achieved by integrating the Fermi–Dirac distribution function over energy band [42] as

$$n = \int DOS\, f(E)\, dE \tag{7.7}$$

where

$f(E) = 1/[1 + \exp(E - E_F/k_B T)]$ is the Fermi–Dirac distribution function

k_B is the Boltzmann constant

E_F is the one-dimensional (1D) Fermi energy

T is the temperature

Equation 7.6 with respect to Equation 7.7 indicates

$$n = \int_0^\infty \frac{dE}{2\pi\left(4\beta\left[\frac{\alpha\pm\sqrt{\alpha^2 - 4(E-\Delta)\beta}}{2\beta}\right]^{\frac{3}{2}} - 2\alpha\left[\frac{\alpha\pm\sqrt{\alpha^2 - 4(E-\Delta)\beta}}{2\beta}\right]^{\frac{1}{2}}\right)\left[1 + e^{E - E_F/k_B T}\right]} \tag{7.8}$$

where $k = \left[\alpha\pm\sqrt{\alpha^2 - 4(E-\Delta)\beta}/2\beta\right]^{1/2}$, which is easily obtained from Equation 7.2. Unfortunately, Equation 7.8 cannot be solved analytically. We can replace effective DOS $N_c = k_B T/4\pi$, and normalize the Fermi energy $\eta = E_F - \Delta/k_B T$ and $x = \Delta - E/k_B T$, which can be simplified as

$$n = \pm N_c \int_0^\eta \frac{\left[\alpha\pm\sqrt{\alpha^2 + 4\beta x k_B T}\right]^{\frac{-1}{2}}}{\left[\sqrt{\frac{\alpha^2}{2\beta} + 2x k_B T}\right]\left[1 + e^{x - \eta}\right]}\,dx \tag{7.9}$$

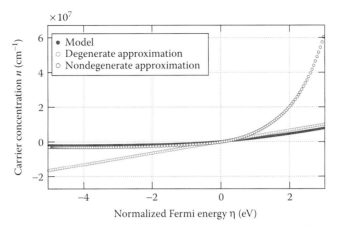

FIGURE 7.5   Carrier concentration of BGN. Normalized Fermi energy as a function of temperature and carrier concentration. Solid line represents model (Equation 7.8); light dark circles represents nondegenerate approximation (Equation 7.10); and dark circles represents degenerate approximation (Equation 7.12).

Equation 7.9 is a BGN carrier concentration, which is shown in Figure 7.5 as a function of normalized Fermi energy (η). Notice that normalized Fermi energy is commonly considered as a function of temperature.

7.8   CARRIER CONCENTRATION OF BGN IN THE NONDEGENERATE REGIME

In the nondegenerate regime, carrier concentration of BGN is defined where $(E - E_F) \gg 3k_B T$ satisfies the nondegenerate condition, meaning that the Fermi level lies in the band gap at more than $3k_B T$ away from the band edge. In other words, nondegenerate approximation can be used to simplify this equation as

$$n = N_c M e^{\eta} \tag{7.10}$$

where M is called the BGN Fermi integral that can be solved numerically by

$$M = \int_0^{\eta} \frac{\left[\alpha \pm \sqrt{\alpha^2 + 4\beta x k_B T}\right]^{-\frac{1}{2}} e^{-x}}{\left[\sqrt{\dfrac{\alpha^2}{2\beta} + 2 x k_B T}\right]} dx \tag{7.11}$$

As shown in Figure 7.6, the nondegenerate approximation can be used instead of the common equation 7.9 in a lower number of carrier concentration. The filled circles represents the model and the light dark circles represents the nondegenerate

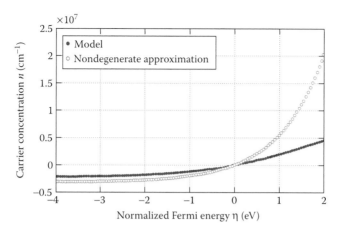

FIGURE 7.6 Carrier concentration of BGN in the nondegenerate regime. Comparison between model and presented nondegenerate approximation. The filled circles represents the model and the light dark circles represents its approximation in nondegenerate regime.

approximation. In the nondegenerate regime, the carrier concentration of BGN is a function of temperature, particularly in a lower number of carrier concentration.

Figure 7.6 shows that BGN in the nondegenerate regime is closely approximated by the model where carrier concentration is small; thus, BGN is a function of normalized Fermi energy (η) for the nondegenerate limit.

7.9 CARRIER CONCENTRATION OF BGN IN THE DEGENERATE REGIME

Nowadays, most of the devices that are scaled down to the nanoscale operate on a degenerate regime; therefore, the degenerate approximation is going to be more dominant. However, when Fermi level subsist inside a valance or conduction band or within $3k_BT$ away from band edge, the probability is one, indicating that we can neglect exponential function in the degenerate limit. Therefore, BGN carrier concentration model could be obtained as:

$$n = N_c \int\limits_0^\eta \frac{\left[\alpha \pm \sqrt{\alpha^2 + 4\beta x k_B T}\right]^{\frac{-1}{2}}}{\left[\sqrt{\dfrac{\alpha^2}{2\beta} + 2x k_B T}\right]} \, dx \qquad (7.12)$$

As shown in Figure 7.7, carrier concentration in the degenerate limit is unlike the Maxwell–Boltzmann approximation, as it is dependent on the normalized Fermi energy and independent of the temperature T.

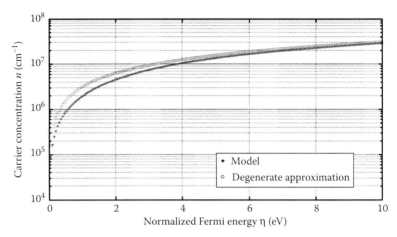

FIGURE 7.7 Carrier concentration of BGN in the degenerate regime. BGN carrier concentration as a function of normalized Fermi energy in the degenerate limit. The filled circles represents model and the light dark circles represents its approximation in the degenerate regime.

7.10 CONDUCTANCE THEORY IN GRAPHENE

7.10.1 MONOLAYER GRAPHENE

It was shown that despite the zigzag-shaped edge of GNR is metallic, in the case of armchair-shaped edge GNRs, depending on their widths, they can be either metallic or semiconducting [26]. The electronic properties of armchair GNR depends sensitively on the ribbon width, which is metallic when $N = 3M - 1$, where M is an integer; otherwise, there is an energy gap between the conduction and valence bands [18,34]. Strongly decreasing the temperature affects the conductance of GNR at the charge neutrality point [44], where minimum conductance of GNR appears. GNR minimum conductance decreases by decreasing the temperature [15]. Based on the quantum confinement effect in GNRs, conductance is a function of the Fermi–Dirac integral, which is based on Maxwell approximation in the nondegenerate limit. Outside of the Dirac point, the conductance balances with the width of the GNR [15]. Elastic scattering strongly affects the variation of total conductance fluctuations in graphene [25,45].

7.10.2 BILAYER GRAPHENE

The number of propagating modes at the Fermi energy is defined as the conductance of the clean (disorder-free) ideal nanoribbons [46]. Each spin-degenerate propagating mode contributes to the total conductance by the conductance unit $G_0 = 2e^2/h$ [18]. The conductance of clean zigzag and armchair BGNs is respectively quantized as $2(n + 1)G_0$ and nG_0, where n is an integer [18]. The minimum conductance of BGN increases dramatically as the temperature increases [4]. The insulating state occurs only in BGN, and it is not the case in GNR, and insulating temperature dependence of the resistance becomes stronger when BGN is highly biased [23].

7.10.3 QUANTUM HALL EFFECT

A remarkable anomalous quantum Hall effect (QHE) appears in GNR, which is $\sigma_{xy} = \pm 4(N + 1/2)e^2/h$, where N is the integer "Landau level" index [5,32,47]. BGN also shows the anomalous QHE, but with the standard sequence of $\sigma_{xy} = \pm 4Ne^2/h$. Since the first plateau at $N = 0$ is absent in the unbiased case, BGN stays metallic at the neutrality point. Biased BGN exhibits a plateau at zero Hall conductivity $\sigma_{xy} = 0$, which may only be due to the opening of a sizable gap between the valence and conduction bands [5,32,48]. Additionally, it is notable that both GNR and BGN have a zeroth Landau level located at the charge-neutrality point, which are eightfold degenerate in BGN and fourfold degenerate in GNR [49], and the Berry phase for carriers in BGN is 2π while it is π in GNR [27].

7.10.4 DISORDER EFFECT IN BILAYER GRAPHENE

Various types of disorder may exist for a BGN, including edge disorder and single-layer disorder. It has been shown that the type of disorder and contact scattering play an important role in conductance [26]. As in the case of GNR, a relatively small-edge disorder strongly suppresses the conductance of the BGN; no difference in conductance is expected for realistic zigzag and armchair BGNs, though. Some reports have established that there are interesting differences in the disorder effect between GNR and BGN [18].

7.11 GENERAL CONDUCTANCE MODEL OF BGNs

Considering the fact that charge carriers in graphene can travel ballistically over submicron distances [22], the ballistic conductance of BGN is the focus of this section. The quantized ballistic conductance (G) is an unimpeded flow of charge or energy carrying particles over relatively long distances in a material [46]. Referring to the point that flat materials with width less than the de Broglie wavelength (almost 10 nm) can be understood as a 1D device, a BGN corresponding to this condition (BGNFET) can be assumed as a 1D device. For a 1D BGNFET where the channel is assumed to be ballistic, the current from source (S) to drain (D) (Figure 7.8) given by the Boltzmann transport equation is [42,46]

$$I = \frac{2q}{\hbar} \int_{E_{F2}}^{E_{F1}} D(E)T(E)\left(-\frac{df}{dE}\right)dE \tag{7.13}$$

where
 q is the electron charge
 $\hbar$ is Planck's constant

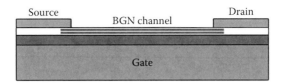

FIGURE 7.8 Simple 1D BGNFET biased by voltage V. Length of BGN is denoted by L and W is width, which is less than 10 nm. The current from S to D is given by the Boltzmann transport Equation 7.13.

E_{F1} and E_{F2} are the Fermi levels of S and D
$D(E)$ is DOS
$T(E)$ is the transmission probability

Based on the definition of conductance $G = I/V$, the Boltzmann transport equation can be written as the Landauer formula with

$$G = \frac{2q^2}{h} \int_{-\infty}^{+\infty} M(E)T(E)\left(-\frac{df}{dE}\right)dE \tag{7.14}$$

where

$$M(E) = \frac{dE}{ldk} = \frac{(-2\alpha k + 4\beta k^3)dk}{ldk} = \frac{-2\alpha k + 4\beta k^3}{l} \tag{7.15}$$

is the number of modes which are above the cutoff at energy E in the transmission channel

f is the Fermi–Dirac distribution function
$T(E)$ is transmission probability

Transport is here assumed to be completely ballistic, that is, no self-energies accounting for inelastic scattering are considered. Note that in a ballistic channel, $T(E)$ is equal to 1 [46]. With regard to band energy structure of BGN in Equation 7.2, momentum (k) is

$$k = \sqrt{\frac{\alpha \pm \sqrt{\alpha^2 - 4\beta(\Delta - E)}}{2\beta}} \tag{7.16}$$

Substituting Equations 7.15 and 7.16 into Equation 7.14, we can achieve the so-called general mathematical model of BGN conductance as

$$G = \frac{2q^2}{hl} \int_{-v_g}^{+v_g} \left[\frac{2\beta}{\alpha}\left(\frac{\alpha \pm \sqrt{\alpha^2 + 4\beta x k_B T}}{2\beta}\right)^{\frac{3}{2}} - \left(\frac{\alpha \pm \sqrt{\alpha^2 + 4\beta x k_B T}}{2\beta}\right)^{\frac{1}{2}} \right] \left(-d\left(\frac{1}{1+e^{x-\eta}}\right)\right) \tag{7.17}$$

where
$$x = (E - \Delta) / k_B T$$
$$\eta = (E_F - \Delta)/k_B T$$

Equation 7.17 can be written as $G = G_0 + G_{BGN}$ where $G_0 = G_0'(v_g) - G_0'(-v_g)$ and

$$G_0'(x) = -\frac{4q^2}{hl} \frac{1}{1 + e^{x - \eta}}$$

$$\left(2\beta \left(\frac{\alpha \pm \sqrt{\alpha^2 + 4\beta x k_B T}}{2\beta} \right)^{\frac{3}{2}} - \alpha \left(\frac{\alpha \pm \sqrt{\alpha^2 + 4\beta x k_B T}}{2\beta} \right)^{\frac{1}{2}} \right) \qquad (7.18)$$

and

$$G_{BGN} = -\frac{4q^2}{hl} \int\limits_{-v_g}^{+v_g} \left(\frac{1}{1 + e^{x - \eta}} \right)$$

$$\left(d \left(2\beta \left(\frac{\alpha \pm \sqrt{\alpha^2 + 4\beta x k_B T}}{2\beta} \right)^{\frac{3}{2}} - \alpha \left(\frac{\alpha \pm \sqrt{\alpha^2 + 4\beta x k_B T}}{2\beta} \right)^{\frac{1}{2}} \right) \right) \qquad (7.19)$$

Since $G_0 \simeq 0$, $G = G_{BGN}$. The general mathematical model of BGN conductance, which can be solved numerically, is presented by Equation 7.20 where V is the biased voltage and vg is the gate voltage.

$$G_{BGN} = -\frac{4q^2}{hl} \int\limits_{-v_g}^{+v_g} \left(\frac{1}{1 + e^{x - \frac{(v_g - V)q}{k_B T}}} \right)$$

$$\left(\frac{3}{4} \sqrt{\frac{\alpha + \sqrt{\alpha^2 + 4\beta x k_B T}}{2\beta (\alpha^2 + 4\beta x k_B T)}} - \frac{\alpha}{8\beta\sqrt{\alpha^2 + 4\beta x k_B T} \sqrt{\frac{\alpha + \sqrt{\alpha^2 + 4\beta x k_B T}}{2\beta}}} \right) dx \qquad (7.20)$$

Figure 7.9 is the numerical solution of Equation 7.20. This conductance model is then expected to provide a clear understanding of BGN field effect transistor applications.

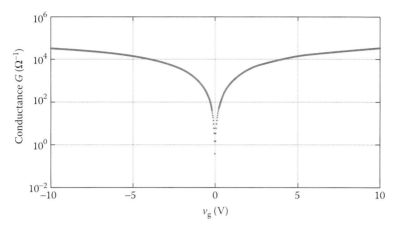

FIGURE 7.9 Numerical solution of BGN conductance model.

7.12 CONDUCTANCE MODEL OF BGN IN NONDEGENERATE REGIME

In the nondegenerate regime, Fermi function could be modified. This modification could be applied by subtracting one from denominator of equation 7.20. The calculation based on this modification shows that the nondegenerate approximation can be used instead of the general model in neutrality point.

Figure 7.10 shows a BGN conductance model in the nondegenerate regime based on Equation 7.20, where the Fermi–Dirac distribution function has been approximated by the Maxwell–Boltzmann distribution function. Green dots represent the nondegenerate approximation, and red dots are the general model.

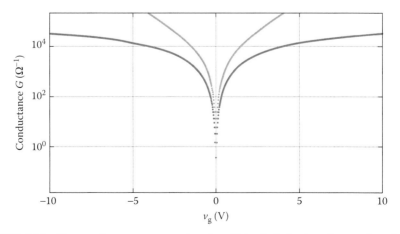

FIGURE 7.10 The nondegenerate approximation (light dark dots) and general model of BGN conductance (dark dots) in $T = 55$ K.

7.13 CONDUCTANCE MODEL OF BGN IN DEGENERATE REGIME

In the degenerate limit where Fermi–Dirac distribution function has been approximated by $f(E) = 1$, Equation 7.20 can be rearranged as

$$G_{BGN}^{D} = -\frac{4q^2}{hl} \int_{-v_g}^{+v_g} \left(\frac{3}{4}\sqrt{\frac{\alpha+\sqrt{\alpha^2+4\beta xk_BT}}{2\beta(\alpha^2+4\beta xk_BT)}} - \frac{\alpha}{8\beta\sqrt{\alpha^2+4\beta xk_BT}\sqrt{\frac{\alpha+\sqrt{\alpha^2+4\beta xk_BT}}{2\beta}}} \right) dx \tag{7.21}$$

Equation 7.21 can be analytically solved with the analytical solution shown in Equation 7.22:

$$G_{BGN}^{D} \simeq -\frac{q^2}{\sqrt{2}hlk_BT} \left(\sqrt{\left(\frac{\alpha+\sqrt{\alpha^2+4\beta xk_BT}}{\beta}\right)^3} - \frac{\alpha}{\beta}\sqrt{\frac{\alpha+\sqrt{\alpha^2+4\beta xk_BT}}{\beta}} \right)\Bigg|_{-v_g}^{v_g} \tag{7.22}$$

Figure 7.11 shows the general model of BGN conductance (dark dots) and analytical solution in the degenerate approximation (light dark dots) from Equation 7.22. It is clear that the analytical degenerate approximation can be used instead of the general model out of the neutrality point.

7.14 TEMPERATURE DEPENDENCE OF BGN

The conductance in BGN is expected to be affected by temperature [22]. Regarding the published BGN conductance experiments, it has been shown that the minimum conductance value depends on the value of gate voltage [23].

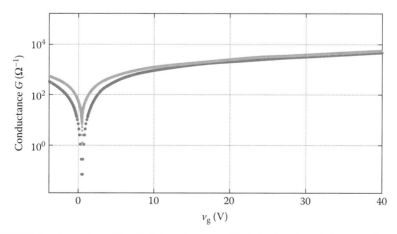

FIGURE 7.11 General model of BGN conductance (dark dots) and analytical solution of the degenerate approximation (light dark dots) in $T = 55$ K.

Based on the zero-gap semiconductor characteristic for small gate voltages, the conductance of GNR located near the charge-neutrality peak is basically temperature independent. The conductance temperature dependence observed in BGN is noticeably different from the one measured in GNR [23]. In BGN, temperature dependence of conductance is strongly affected by the lower applied perpendicular electric field (E_{ex}), and weak temperature dependency has been reported in higher E_{ex} [20]. On the other hand, away from the neutrality point, there has not been any observation of increase in BGN conductivity corresponding to the increase in temperature (T) [22].

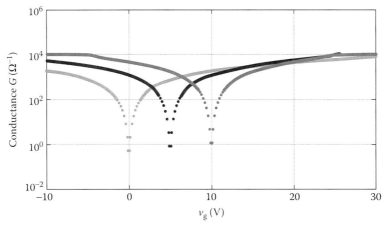

FIGURE 7.12   Temperature dependence of BGN. BGN conductance model in $T = 220$ (moderate dark dots), $T = 160$ (dark dots), and $T = 100$ K (light dark dots). Minimum conductance increases with increasing temperature, but conductance is not affected by temperature out of neutrality point.

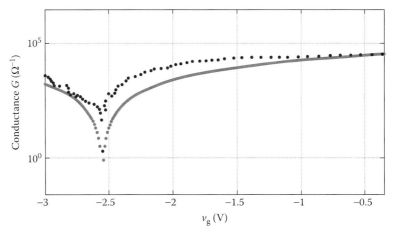

FIGURE 7.13   Comparison of presented model with experimental data. Dark dots are experimental data for BGN conductance from [23] in $T = 55$ mk and solid line are solution of BGN conductance model presented by Equation 7.20.

Figure 7.12 shows the temperature dependence of BGN conductance at 220, 160, and 100 K concerned with our model. The minimum conductance of BGN is increased by increasing the temperature, but the temperature variation does not affect BGN conductance out of the neutrality point. These results are in good agreement with what has been reported in experiments [20,23,50].

As shown in Figure 7.13, the experimental results (dark dots) [23] at $T = 55°mk$ are in good agreement with the theoretical calculations (solid line) presented in this chapter. However, the analytical solution in the degenerate approximation of the BGN conductance model can be applied to estimate BGN conductance in graphene-based devices under different gate voltages.

7.15 CONCLUSION

The characteristics of high carrier mobility, low-cost mass production, and changeable energy band structure excited by the external electric fields in BGN creates new possibilities for new electronic applications. These important properties have led to the use of BGN as a conducting ballistic channel along with the source and drain electrodes in 1D BGNFETs.

In this chapter, a mathematical model that can be analyzed using numerical methods has been derived to characterize the carrier statistics of BGN. The carrier density as a fundamental parameter incorporating the energy band structure of BGN near the Fermi level is calculated, and the carrier transport on BGN with width less than the de Broglie wavelength assumed as a 1D device is explored. The analytical model specifies that BGN carrier concentration in the degenerate limit is strongly dependent on the normalized Fermi energy and independent of temperature. This model characterizing the carrier concentration as an exponential function of normalized Fermi energy clarifies the understanding of carrier concentration in a low carrier regime. In other words, in the nondegenerate regime, the carrier concentration of BGN is a function of temperature, particularly in lower numbers of carrier concentration.

Besides considering the point that charge carriers in graphene can travel ballistically over submicron distances, the ballistic conductance of BGN is addressed in this chapter. The numerical solution of BGN ballistic conductance is introduced in addition to the analytical degenerate model of BGN conductance. Based on our model in the degenerate limit, the general conductance model is approximated by the analytical conductance model in the regions away from the neutrality point. Our method indicates that near the neutrality point, the nondegenerate approximation can be used properly showing the temperature dependence relation in conductance as well. In contrast, out of this boundary condition, it is estimated to work in the degenerate regime. Moreover, it confirms that BGN conductance is temperature dependent near the neutrality point as reported in experiments. Minimum conductance is dependent on temperature and increases by increasing the temperature near the neutrality point, but beyond the neutrality point, conductance is independent of temperature. The results obtained from the proposed model agree with the published experimental data.

REFERENCES

1. Novoselov, K.S., et al. Electric field effect in atomically thin carbon films. *Science.* 2004, 306, 666.
2. Frank, D.J., et al. Device scaling limits of Si MOSFETs and their application dependencies. *Proceedings of the IEEE.* 2001, 89(3), 259–288.
3. Meindl, J.D., Q. Chen, and J.A. Davis. Limits on silicon nanoelectronics for terascale integration. *Science.* 2001, 293(5537), 2044–2049.
4. Wenjuan Zhu, V.P., M. Freitag, and P. Avouris. Carrier scattering, mobilities, and electrostatic potential in monolayer, bilayer, and trilayer graphene. *Physical Review B.* 2009, 80(23), 235402.
5. Geim, A.K. and K.S. Novoselov. The rise of graphene. *Nature Materials.* 2007, 6(3), 183–191.
6. Castro, E.V., M.P. Lopez-Sancho, and M.A.H. Vozmediano. Pinning and switching of magnetic moments in bilayer graphene. *New Journal of Physics.* 2009, 11, 095017.
7. Nobelprize.org. The Nobel Prize in Physics 2010. 2010. Available from http://nobelprize.org/nobel_prizes/physics/laureates/2010.
8. Fiori, G. and G. Iannaccone. Performance analysis of graphene bilayer transistors through tight-binding simulations. *IWCE-13: 2009 13th International Workshop on Computational Electronics.* Beijing, 2009, pp. 301–304.
9. Zarenia, M., et al. Electrostatically confined quantum rings in bilayer graphene. *Nano Letters.* 2009, 9(12), 4088–4092.
10. Mao, Y.L., G.M. Stocks, and J.X. Zhong. First-principles study of the doping effects in bilayer graphene. *New Journal of Physics.* 2010, 12, 033046.
11. Ragheb, T. and Y. Massoud. On the modeling of resistance in graphene nanoribbon (GNR) for future interconnect applications. *Computer-Aided Design, 2008. ICCAD 2008. IEEE/ACM International Conference.* San Jose, CA, 2008.
12. Wang, X.R., et al. Room-temperature all-semiconducting sub-10-nm graphene nanoribbon field-effect transistors. *Physical Review Letters.* 2008, 100(20), 206803.
13. Liao, L., et al. High-speed graphene transistors with a self-aligned nanowire gate. *Nature.* 2010, 467, 305–308.
14. Zaharah Johari, M.T.A., D.C.Y. Chek, N. Aziziah Amin, and R. Ismail. Modelling of graphene nanoribbon Fermi energy. *Journal of Nanomaterials.* 2010, 2010, 909347.
15. Ahmadi, M.T., et al. Graphene nanoribbon conductance model in parabolic band structure. *Journal of Nanomaterials.* 2010, 2010, 753738.
16. Adam, S. and S. Das Sarma. Boltzmann transport and residual conductivity in bilayer graphene. *Physical Review B.* 2008, 77(11), 115436.
17. Hill, E.W., et al. Graphene spin valve devices. *IEEE Transactions on Magnetics.* 2006, 42(10), 2694–2696.
18. Xu, H.Y., T. Heinzel, and I.V. Zozoulenko. Edge disorder and localization regimes in bilayer graphene nanoribbons. *Physical Review B.* 2009, 80(4), 045308.
19. Rutter, G.M., et al. Structural and electronic properties of bilayer epitaxial graphene. *Journal of Vacuum Science & Technology A.* 2008, 26(4), 938–943.
20. Russo, S., et al. Double-gated graphene-based devices. *New Journal of Physics.* 2009, 11(9), 095018.
21. Bolotin, K.I., et al. Ultrahigh electron mobility in suspended graphene. *Solid State Communications.* 2008, 146(9–10), 351–355.
22. Morozov, S.V., et al. Giant intrinsic carrier mobilities in graphene and its bilayer. *Physical Review Letters.* 2008, 100(1), 016602.
23. Oostinga, J.B., et al. Gate-induced insulating state in bilayer graphene devices. *Nature Materials.* 2008, 7(2), 151–157.

24. Kargar, A. Analytical modeling of current in graphene nanoribbon field effect transistors. *Nanotechnology*. 2009, 8, 710–712.

25. Ujiie, Y., et al. Regular conductance fluctuations indicative of quasi-ballistic transport in bilayer graphene. *Journal of Physics: Condensed Matter*. 2009, 21(38), 382202.

26. Xu, N. and J.W. Ding, Conductance growth in metallic bilayer graphene nanoribbons with disorder and contact scattering. *Journal of Physics: Condensed Matter*. 2008, 20(48), 485213.

27. Wang, Y.X. and S.J. Xiong. Disorder and localization of electrons in bilayer graphene. *European Physical Journal B*. 2009, 67(1), 63–69.

28. Latil, S. and L. Henrard. Charge carriers in few-layer graphene films. *Physical Review Letters*. 2006, 97(3), 036803.

29. Koshino, M. Electronic transport in bilayer graphene. *New Journal of Physics*. 2009, 11, 095010.

30. Fal'ko, V.I. Electronic properties and the quantum Hall effect in bilayer graphene. *Philosophical Transactions of the Royal Society A: Mathematical Physical and Engineering Sciences*. 2008, 366(1863), 205–218.

31. Zhang, Y.B., et al. Direct observation of a widely tunable bandgap in bilayer graphene. *Nature*. 2009, 459(7248), 820–823.

32. Ma, R., et al. Quantum Hall effect in biased bilayer graphene. *Europhysics Letters*. 2009, 87(1), 17009.

33. Nilsson, J., et al. Electronic properties of graphene multilayers. *Physical Review Letters*. 2006, 97(26), 266801.

34. Nakada, K., et al. Edge state in graphene ribbons: Nanometer size effect and edge shape dependence. *Physical Review B*. 1996, 54(24), 17954–17961.

35. Castro, E.V., et al. Electronic properties of a biased graphene bilayer. *Journal of Physics: Condensed Matter*. 2010, 22(17), 175503.

36. Misu A, E.E. Mendez, and M.S. Dresselhaus. Near infrared reflectivity of graphite under hydrostatic pressure. I. Experiment. *Journal of Physical Society Japan*. 1979, 47, 199–207.

37. Stauber, T., et al. Fermi liquid theory of a Fermi ring. *Physical Review B*. 2007, 75, 115425.

38. Novoselov, K.S., et al. Two-dimensional gas of massless Dirac fermions in graphene. *Nature*. 2005, 438(7065), 197–200.

39. Pierret, R. *Advanced Semiconductor Fundamentals*. Englewood Cliffs, NJ: Prentice Hall, 2003.

40. Kittel, C. *Introduction to Solid State Physics*, 18th edn. New York: John Wiley & Sons, 2005.

41. Neamen, D.A. *Semiconductor Physics and Devices: Basic Principles*, 3rd edn. New York: McGraw-Hill, 2003.

42. Datta, S. *Quantum Transport: Atom to Transistor*. New York: Cambridge University Press, 2005.

43. Lai, Y.H., et al. Landau levels in a simple hexagonal bilayer graphene. *Physics Letters A*. 2008, 372(3), 292–298.

44. Masubuchi, S., et al. Fabrication of graphene nanoribbon by local anodic oxidation lithography using atomic force microscope. *Applied Physics Letters*. 2009, 94(8), 082107.

45. Horsell, D.W., et al. Mesoscopic conductance fluctuations in graphene. *Solid State Communications*. 2009, 149(27–28), 1041–1045.

46. Datta, S. *Electronic Transport in Mesoscopic Systems*. Cambridge, UK: Cambridge University Press, 2002.

47. Charlier, J.C., et al. Electron and phonon properties of graphene: Their relationship with carbon nanotubes. *Carbon Nanotubes*. 2008, 111, 673–709.

48. Castro, E.V., et al. Biased bilayer graphene: Semiconductor with a gap tunable by the electric field effect. *Physical Review Letters*. 2007, 99, 216802.
49. Williams, J.R., et al. Quantum Hall conductance of two-terminal graphene devices. *Physical Review B*. 2009, 80(4), 045408.
50. Adam, S. and M.D. Stiles. Temperature dependence of the diffusive conductivity of bilayer graphene. *Physical Review B*. 2010, 82(7), 075423.

8 Trilayer Graphene Nanoribbon Field Effect Transistor Modeling

Meisam Rahmani, Hatef Sadeghi,
Seyed Mahdi Mousavi, Mohammad
Taghi Ahmadi, and Razali Ismail

CONTENTS

8.1 INTRODUCTION TO GRAPHENE

Diamond and graphite are the materials that are extensively investigated in nanoelectronics. Since diamond's unusual bonding attributes make it hard to be naturally created, recently discovered allotropes like fullerenes and nanotubes are taken closely into consideration by scientists researching in the fields of biology, chemistry, physics, and material science [1,2]. All the structural studies on graphite, fullerenes, and nanotubes start with graphene. Due to its excellent electronic [3] and optical [4,5] characteristics, graphene nanoribbon (GNR) has recently been considered for future electron device applications. Consequently, prototype structures indicating excellent performance for transistors [6] interconnects [7], electromechanical switches [8], and infrared emitters and biosensors [9] have been demonstrated. It is illustrated that neither the velocity of the charge carriers at the upper part of the valence band nor the velocity at the lower part of the conduction band is reduced, which is a common attribute of most materials. In fact, it stays constant throughout the bands [8]. The low scattering rates and the electronic structure of graphene result in an excellent electronic transmission, which is adjustable by doping or using electrostatic field. It can serve as interconnects due to its high conductivity. Since it can be gated, it could be used as a channel in novel transistors. Because of the stable and inert properties of graphene, it could be adopted to make large regions of low defect densities and low electronic scattering rates [10]. The method of preparing multilayer GNFET is both appealing and amazing. A significant contributing reason to graphene's importance is the specific nature of its charge carriers, which is related to the fact that its charge carriers mimic relativistic particles, and are simply described using the Dirac equation rather than the Schrödinger equation. The Schrödinger equation contributes in giving an illustration of electronic properties of materials in condensed matter physics [11,12]. Graphene's high electron propagation, which is essential for achieving high-speed and high-performance transistors, is its key advantage; however, for GNR MOSFETs being used in realistic IC applications, material properties such as the band gap (EG) should be closely controlled. The GNR MOSFET can serve as a substitute device, with which the GNR chirality challenge will be dominated. A GNR MOSFET is an appliance in which GNR can be applied as the channel of a FET-like device [13]. In this study, we investigate theoretically a trilayer graphene nanoribbon (TGN) model in FETs.

8.2   GNR

The flat monolayer of carbon atoms closely arranged in a honeycomb lattice is called graphene. It is also considered to be a hexagonal planar arrangement of carbon atoms, which is the binding structure of the graphitic materials (nanotubes and fullerenes containing pentagons) [14]. Among the currently revealed carbon nanostructures, there are GNRs with distinguished properties for narrative adoption. This element has proven to be a universally recognized and reliable material for achieving nanoelectronic tools in current decades as a result of the unique electronic and electrical characteristics such as quantum electronic transport, charge carriers displaying massless Dirac fermions, tunable band gap, long spin-diffusion, length thermodynamic stability, extremely high charge-carrier mobility, and wonderful mechanical firmness [15–17]. Many properties of graphene are considered in recognizing it as a genuine 2D material. It is referred to as a 2D material due to the possession of a unique blend of semiconductor–zero density of states (DOS) and a metal gaplessness with rare electronic excitations analyzed with Dirac fermions moving in a curved space [17]. There exists a thoughtless disarray of electrons in graphene and electron–electron interactions, which has a long mean free path [18,19]. Hence, it was revealed that the characteristics are dissimilar to those of normal metals and semiconductors, because graphene is made of a robust and flexible structure with distinct phonon modes which may not be seen in common 3D solids. Owing to this fact, graphene is recognized to be the most flexible system in condensed matter research [17].

Graphene not only has condensed key properties, but it is also being used in a variety of applications, ranging from chemical sensors to transistors. Looking at a trilayer with an inhomogeneous charge, distribution may change the electrostatic potential in a variety of layers [17]. The breaking down of the equivalence between layers when they are greater than two may result in the absence of an applied electric field. Interestingly, it is observable that a gap may occur in a stack with Bernal ordering and four layers when the electronic charge of both surface layers is identifiable from the two inner ones. The presence of charge in homogeneity notwithstanding, systems that possess a greater value of layers are observed not to indicate any gap [20,21]. Another method of modeling the narrow nanoscaled pieces of graphene is the application of GNR MOSFET with its accurate boundary to the substrate, with which it is likely to subdue the CNT chirality control problems [22]. Considering the presented situations, the GNR MOSFET could be considered as a potential 2D tool, with which the short channel properties such as the drain barrier lowering (DIBL) and the subthreshold swing (S) mainly developed by electrostatics of the tool are likely to be reduced. These parameters make a significant contribution in nanoscale MOSFET performance, as long as the mentioned parameters are established appropriately [23,24].

Because of the unavailability of band gap, graphene may not be likely to be put in an off-state, hence making it relevant to create transistors with regard to its huge on–off ratios. In the case of carbon nanotubes, reducing the size of graphene to generate 1D graphene wires referred to as nanoribbon may provoke a gap. The termination of nanoribbon can be done in two ways called zigzag and armchair, with which the dimension of the band gap is identified when joined with the width of the ribbon [25]. Estimating the ribbon direction, with the aid of the perpendicular states in a nanoribbon, results in less boundary condition, which is called k-vectors. In case these allowed k-vectors cut across the Brillouin zone, we can identify the presence of a gap and also recognize the size of the alienation in energy of the valence and conduction bands [26].

A GNR transistor schematic is indicated in Figure 8.1. Graphene has an important benefit over carbon nanotubes of controlling the gap, where semiconducting

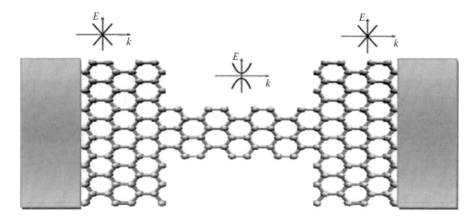

FIGURE 8.1   Schematic of GNR transistor.

and metallic tubes are created by using a top-down method (etching away 2D graphene); however, it is done in an unforeseen attitude. There is the need for specific control over the width and edge termination of the nanoribbon at the atomic level using the top-down approach. This is methodically made [27,28] with the aid of oxygen plasma etching [29,30] the particular control of the edge termination, which continues to be an experimental task.

8.3 BAND STRUCTURE OF GRAPHENE

The atomic layer of hexagonally arranged carbon atoms forming graphene is indicated in Figure 8.2. The unit cell shown by the dashed lines in Figure 8.2a consists of two atoms, tagged A and B, and possesses the lattice vectors $\vec{a}_1 = a(1,0)$ and $\vec{a}_2 = a(1/2, \sqrt{3}/2)$. As indicated in Figure 8.2b, the reciprocal is hexagonal in nature with its high symmetry points of Γ, K, and M. Two unequal points exist in the Brillouin zone, K and K' [31].

The important low-energy properties of the band structure may be captured by a tight-binding approximation. The linear combination of Bloch functions can be written with the eigenfunctions of graphene $\varphi_i(\vec{k}, \vec{r}) = \frac{1}{\sqrt{N}} \sum_{\vec{R}}^{N} e^{i\vec{k}\vec{R}} \chi_j(\vec{r} - \vec{R})$, built up from the atomic wave functions χ_j at site j, as

$$\psi_i(\vec{k}, \vec{r}) = \sum_{i'=0}^{n} c_{ii'}(\vec{k}) \varphi'_i(\vec{k}, \vec{r}) \tag{8.1}$$

The eigenenergies of this system are derived by [31]

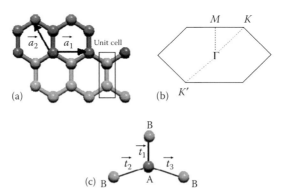

FIGURE 8.2 (**See color insert.**) (a) Lattice of graphene. Unit cell vectors $\vec{a}_1$ and $\vec{a}_2$ allocate the unit cell. (b) The reciprocal lattice and the high symmetry points Γ, K, and M. (c) Each A atom is surrounded by three nearest neighbor B atoms. Vectors $\vec{t}_1$, $\vec{t}_2$, and $\vec{t}_3$ connect the A atoms to the three B atoms.

$$E_i(\vec{k}) = \frac{\langle \psi_i | H | \psi_i \rangle}{\langle \psi_i | \psi_i \rangle} = \frac{\sum\limits_{jj'=0}^{n} c_{ij}^* c_{ij'} H_{jj'}(\vec{k})}{\sum\limits_{jj'=0}^{n} c_{ij}^* c_{ij'} S_{jj'}(\vec{k})} \tag{8.2}$$

where $H_{jj'}(k) = \langle \psi_j | H | \psi'_j \rangle$ and $S_{jj'}(\vec{k}) = \langle \psi_i | \psi_i \rangle$. The result is derived by minimizing Equation 8.2 with respect to coefficient c_{ij}^* leading to a secular equation

$$\det[H - ES] = 0 \tag{8.3}$$

The easiest clarification is derived by looking at the nearest neighbor interaction; thus, only H_{AA}, H_{BB}, and H_{AB} where A and B are the two atoms of the graphene unit cell need to be evaluated. The evaluation of energies of H_{AA} and H_{BB} gives

$$H_{AA} = H_{BB} = \frac{1}{N} \sum_{\vec{R}.\vec{R}} e^{i\vec{K}(\vec{R}-\vec{R'})} \left\langle \chi_A(\vec{r}-\vec{R'}) \middle| H \middle| \chi_A(\vec{r}-\vec{R}) \right\rangle = \varepsilon_{2p} \tag{8.4}$$

where ε_{2p} is the orbital energy of the 2p level. The off-diagonal elements $H_{AB} = H_{BA}$ are estimated for the three nearest neighbor B atoms (see Figure 8.2c). Adopting three vectors $\vec{t_1}$, $\vec{t_2}$, and $\vec{t_3}$

$$H_{AB} = \gamma_H (e^{i\vec{k}\vec{t_1}} + e^{i\vec{k}\vec{t_2}} + e^{i\vec{k}\vec{t_3}}) = \gamma_h t(k) \tag{8.5}$$

$$S_{AB} = \gamma_S (e^{i\vec{k}.\vec{t_1}} + e^{i\vec{k}.\vec{t_2}} + e^{i\vec{k}\vec{t_3}}) = \gamma_s t(k) \tag{8.6}$$

where

$$\gamma_h = \left\langle \chi_A(\vec{r}-R') \middle| H \middle| \chi_B(\vec{r}-\vec{R}+\vec{t}) \right\rangle$$

$$\gamma_s = \left\langle \chi_A(\vec{r}-R') I \chi_B(\vec{r}-\vec{R}+\vec{t}) \right\rangle$$

The determinants for H and S are then

$$H = \begin{pmatrix} \varepsilon_{2p} & \gamma_h t(k) \\ \gamma_h t(k) & \varepsilon_{2p} \end{pmatrix}$$

$$S = \begin{pmatrix} 1 & \gamma_s t(k) \\ \gamma_s t(k) & 1 \end{pmatrix}$$

From the two matrices, the result to Equation 8.3 is

$$E(\vec{k}) = \frac{\varepsilon_{2p} \pm \gamma_h u(\vec{k})}{1 \pm \gamma_s u(\vec{k})} \tag{8.7}$$

$$u(\vec{k}) = \left(1 + 4\cos\frac{\sqrt{3}k_y a}{2} \cos\frac{k_x a}{2} + 4\cos^2\frac{k_x a}{2} \right)^{\frac{1}{2}} \tag{8.8}$$

Take into consideration that the energy dispersion curves for $E > 0$ meets the curves for $E < 0$ at the k points in the Brillouin zone. The energy E is linear for small values of $\vec{k}$ around these points.

The exclusive band structure of GNR is close to the k point, which is a zero-band gap semiconductor with an energy dispersion relation,

$$E(\vec{k}) = \hbar k v_F \tag{8.9}$$

For small values of wave vector $\vec{k}$ around the k and $\vec{k}$ points, the energy E is linear in $\vec{k}$. The analysis adopts only input indices at the Fermi velocity $v_F = 3ta_{c-c}/2 = \text{m s}^{-1}$, and hopping between π orbitals situated at the closest neighbor atoms is (≈ 3 eV) [32]. In the valence band $s = -1$ and in the conductance band $s = +1$, and considering a touch at six points in the 2D Brillouin zone boundary (Dirac point k, k'), if $E = 0$, $\hbar$ is Planck's constant. To estimate the dispersion relation of GNR, a closer neighbor orthogonal tight-binding approach based on a single p_z orbital is adopted. The Hamiltonian of GNR with armchair or zigzag in 1D k-point equivalent to the direction along the GNR is the Fourier transform given by

$$H_{k_m} = \sum_l H_l \exp\left[-i \cdot k_m (z_l + z_0) \right] \tag{8.10}$$

where

$k_m = \pi m/L$ is the vector (where $m \in [0, 1]$ and the length of the unit cell is L)
z_l is the unit cell site

Solving for eigenvalues of Hamiltonian for k and k' points in the Brillouin zone, the dispersion relation can be derived. As a result of the inverse relationship between mobility and band gap, it was concluded that GNR operated as FETs and can accomplish mobility appreciably higher than those of silicon and hence may be better suited for low-power applications.

8.4 BASIC GEOMETRIES OF GRAPHENE

Two kinds of GNR consisting of armchair and zigzag edges are indicated in Figure 8.3. Depending on the width of the nanoribbon, the zigzag and armchair GNRs are metallic or semiconducting electronic features. The armchair is considered semiconducting in case $N = 3p$ or $N = 3p + 1$, and when the number of dimmer lines $N = 3p + 2$, it shows a semimetal behavior (p is integer) [33,34].

GNR including both armchair and zigzag are grouped based on the ribbon width numbers indicated by "N," which is the number of armchair or zigzag carbon chains moving in the direction of the finite length. The widths of ribbons with zigzag and armchair edges are consecutive as

$$W_z = \sqrt{3}N \frac{a_0}{2} \tag{8.11}$$

$$W_a = N \frac{a_0}{2} \tag{8.12}$$

where $a_0 = 2.49$ Å is the graphene lattice constant. Considering the past literature on GNR, it was realized that two-thirds of ribbons with armchair edges are semiconducting [25]. The bands of zigzag GNR are moderately flat around the Fermi energy. Conversely, the band of metallic armchair GNR is linear around the

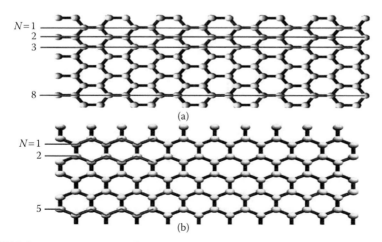

FIGURE 8.3 (a) Arrangement of armchair and (b) arrangement of zigzag GNR.

Fermi energy [34,35]. These estimations where carried out ab initio tight-binding method for GNRs with hydrogen-terminated armchair or zigzag-shaped edges that have a nonzero direct band gap [36]. There is an inversely proportional relationship between the ribbon width and energy band gaps of GNR [37].

It was revealed that the zigzag GNR is metallic in nature; while the zero conductance at the Fermi energy of armchair GNR indicates that it is semiconducting. The DOS curves also showed that electronic bands are not present for occupation near the Fermi energy armchair GNR and is not so for zigzag GNR consisting of conduction close to the Fermi energy [35]. Different structures are calculated by moving one layer over the other in the direction of the infinite edge, indicating that atoms are at the center of the hexagons in the upper layer [38]. The electronic transport in GNR of specified width, which has a wide range of conductance values due to various types of edges geometries, is presented. Discrepancies are shown by the TGN from the anticipated superimposed curves due to interlayer interactions.

Stacking arrangements and edge geometries are variables causing the GNR to behave like semiconductors or show metallic properties. The absence of an energy gap in its electronic spectra for over five layers' stacking is the main reason why graphene could not be relevant as an electronic substance [39]. The electrical features are determined by the band gap of a material being one of its fundamental features. The graphene layer involves complex engineering in order to open up a gap in graphene's electronic spectra. However, it is shown that a single-layer armchair structure has a band gap of 0.22 eV, which can be engineered to create apparatuses. This gap decreases as the sample thickness increases and eventually results in a zero gap [40]. The result of this experiment shows a promising direction for band gap engineering of graphene.

8.5   TGN

Multilayers of graphene can be piled up independently relying on the horizontal shift between consecutive graphene planes, which results in a variety of electronic properties and band structures [41]. Currently, the experiments conducted about the multilayers of graphene could be relevant in the creation of new electronic devices [32,42–44]. There are two forms of bulk graphite called ABA- (AB, hexagonal, or Bernal) and ABC- (rhombohedral) stacked TGN with different stacking manners. In TGN with ABA stacking, the electric field produces band overlap following a proper linear screening that is explained by the Thomas–Fermi approximation, whereas in ABC-stacked TGN, it opens an energy gap in the surface-state band at low energy, which results in a strong nonlinear to the field amplitude–based screening effect [45]. There is spatial inversion symmetry in the lattice of multilayer with even-numbered layers similar to that of monolayer graphene. This situation results in a valley degeneracy, in which time-reversal symmetry is not involved [46,47]. TGN exhibits a variety of electronic properties that are strongly reliant on the interlayer stacking sequence [48,49]. Since our focus is on TGN as a 1D device, quantum confinement effect will be assumed in two directions. In other words, one Cartesian direction is greater than the de Broglie wavelength. It is also notable that

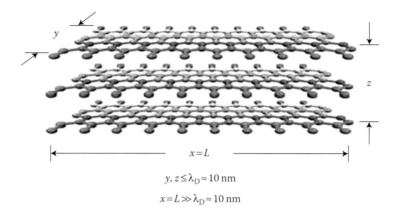

$$y, z \leq \lambda_D \approx 10 \, nm$$

$$x = L \gg \lambda_D \approx 10 \, nm$$

FIGURE 8.4 TGN as a 1D material with quantum confinement effect on two Cartesian directions.

the electrical property of TGN is a strong function of interlayer stacking [48]. As shown in Figure 8.4, because of the quantum confinement effect, a digital energy is taken for granted in y and z directions, while it is an analog type in x direction.

8.6 CONFIGURATIONS OF TGN

TGN is shown to have different electronic properties that are strongly dependent on the interlayer stacking sequence [48]. As shown in Figure 8.5a, ABA-stacked TGN with width and thickness less than the de Broglie wavelength ($\lambda_D = 10$ nm) can be assumed as a 1D material. TGN with ABA stacking is modeled in form of three honeycomb lattices with pairs of in-equivalent sites as $\{A_1, B_1\}$, $\{A_2, B_2\}$, and $\{A_3, B_3\}$ which are located in the top, center, and bottom layers, respectively. An effective-mass model utilizing the Slonczewski–Weiss–McClure parameterization has been adopted, in which every parameter can be compared by relevant parameter in the tight-binding model. As shown in Figure 8.5a, γ_0 describes nearest neighbor (A_i–B_i) pairing within every layer and γ_1 is strong nearest-layer pairing between sites (B_1–A_2 and A_2–B_3) which lie directly on top or under each other. In addition, the weak nearest-layer pairing between sites A_1–B_2 and B_2–A_3 (A_1–A_2, B_1–B_2, A_2–A_3, and B_2–B_3) are explained by γ_3 and γ_4. Considering these coupling alone, a degeneracy point would occur at each of two in-equivalent corners (K), of the hexagonal Brillouin zone, although degeneracy can be divided by next-nearest-layer coupling γ_2 (between A_1 and A_3), γ_5 (between B_1 and B_3), that is the on-site energy difference between A_1, B_2, A_3 and B_1, A_2, B_3 [50].

ABC-stacked TGN and its lattice have three layers as a coupled form, in which every coupled form has carbon atoms settling on a honeycomb lattice. A_1–B_1 pairs are located in the top, while A_2–B_2 pairs are in the heart of the layers, and finally A_3–B_3 pairs are placed at the bottom of the layers [51–53]. The arrangement of these layers is given in Figure 8.5b.

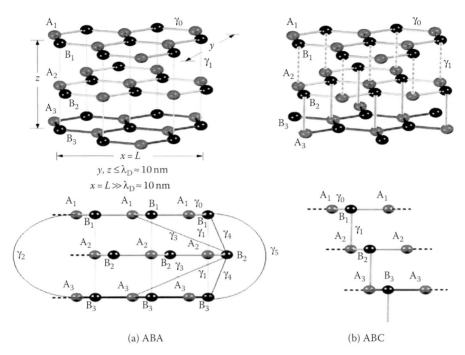

(a) ABA   (b) ABC

FIGURE 8.5 (**See color insert.**) Structures of TGN with (a) ABA (Bernal) stacking and (b) ABC (rhombohedral) stacking.

The ABA-stacked TGN is thermodynamically stable and common. For ABA-stacked TGN, the effective mass model describing the electronic property was developed for the bulk system, and also for few-layer systems. The TGN with ABC stacking has a quite different electronic structure from that of ABAs. The low energy band of ABC-stacked TGN is given by the surface states localized at outermost layers, and an energy gap in its band is opened by the interlayer potential asymmetry, while in ABA-stacked TGN, potential asymmetry causes a band overlapping [50]. The graphene plane being a 2D honeycomb lattice is the origin of the stacking order in multilayer graphene with A and B, two nonequivalent sublattices. The stacking order in bulk samples depends on having low energy for that electronic structure [24].

From Figure 8.5 we can see that the stacking types highly affect the electronic properties of the two types, ABA and ABC. A hybrid of low energy electronic band structure of monolayer and bilayer bands in the presence of interlayer asymmetry is the origin of TGN with ABA stacking. ABC-stacked TGN and its low energy band structure cannot simply be formalized similar to monolayer and bilayer. Interestingly, its low energy bands consist of a cubic formation of both monolayer and bilayer; however, they do not resemble any of them [51]. Hence, Berry's phase has a correlation in terms of cubic dispersion and chirality. In addition, interlayer asymmetry's application is likely to affect the spectrum of energy gap by increasing it slightly. The change in the shape of the Fermi circle over a degeneracy point forms trigonal warping, which is located in the hexagonal Brillouin zone, specifically at k points placed in nonequivalent corners [51–53].

ABC-stacked TGN is regarded as chirality generalization of monolayer and Bernal bilayer graphene. In ABC-stacked TGN, each layer possesses inequivalent triangular A and B sublattices. As shown in Figure 8.6a, each neighboring layer pair creates an AB-stacked BGN with the upper B sublattice located directly on top of the lower A sublattice, and the upper A placed above the center of a hexagonal plaquette of the layer below. The assortment of interlayer hopping processes is illustrated in Figure 8.6b. γ_0 and γ_1 represent the adjacent neighbor interlayer and interlayer hopping, respectively. γ_2 is used to indicate direct hopping between the trilayer low energy sites, δ is applied as the on-site energy difference of A_1 and B_3 regarding the high energy sites, γ_3 describes hopping between the low energy sites of a AB-stacked bilayer (i.e., $A_i \leftrightarrow B_{i+1}$, $i = 1, 2$), and finally γ_4 is incorporated to correspond low and high energy sites located on different layers (i.e., $A_i \leftrightarrow A_{i+1}$ and $B_i \leftrightarrow B_{i+1}$, $i = 1, 2$) [45,46].

Existence of different stacking orders corresponding to trilayer has created the energy gap, which has an effect on the electric field. This effect was investigated clearly. The energy gap of TGN is highly dependent on the selection of stacking, and also the energy gap of TGN is larger than the Bernal stacking [41]. Furthermore, the equivalent energy gap of ABC-stacked TGN can be compared with that of BGN with Bernal stacking. The circular asymmetry of the spectrum, which is a result of the trigonal warping, changes the size of the electronic gap studied in Ref. [41]. Applying an external electric field to the interlayer potential asymmetry, we have to take into consideration the screening effect as we did in the case of ABA-stacked TGN and BGN. Previously, the effect of the interlayer screening in thin graphite films has been studied [50]. The structure of ABA and ABC-stacked TGN has different values of Ns with perpendicular electric field. In TGN with ABC stacking, the surface of the energy band gives rise to a nonlinear screening effect energy gap which is now open. An overlapping in the band of ABA-stacked TGN takes place by

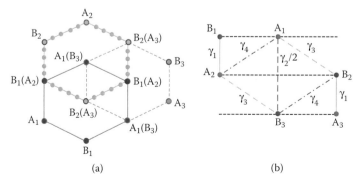

(a)   (b)

FIGURE 8.6 (See color insert.) (a) Lattice structure of ABC-stacked TGN and (b) schematic of the unit cell of ABC-stacked TGN.

applying an electric field, and this band overlapping is associated with linear screening [50]. By the Thomas–Fermi approximation, through opening an energy gap of TGN with ABC stacking, the surface state bands controlling the low energies make a high nonlinear screening effect [50].

8.7 BAND STRUCTURE OF TGN WITH ABA STACKING

As shown in Figure 8.7, the unit cell of graphene embraces two atoms used in the calculation of low energy band dispersion of ABA-stacked TGN. In order to calculate the low energy band, the tight-binding model on electrons hopping in the honeycomb lattice is adopted. As indicated by the tight-binding calculations, the band structure of TGN shows an overlap between the top of the valence band and the bottom of the conduction band, which eventually results in a finite DOS of the Fermi level. The interlayer symmetry of the trilayer can be broken through the external perpendicular electric field which means that the low energy parabolic electron and hole bands can move to lower energies [54,55].

As indicated in Figure 8.7, GNR as a 1D solid with two atoms in its unit cell is incorporated with the assumption that there is one orbital per atom leading to a Hamiltonian matrix [56]

$$\sum_m [H_{nm}]\{\varphi_m\} = E\{\varphi_n\} \tag{8.13}$$

where $\{\varphi_{m(\text{or } n)}\}$ is a column vector indicating the wave function in unit cell m (or n). Based on the theory that uses only one orbital, the $2p_z$ orbital, per carbon atom, a (2×2) matrix describing the conduction and the valence bands can be made:

$$h_0 = -t(1 + e^{i\vec{k}\cdot\vec{a_1}} + e^{i\vec{k}\cdot\vec{a_2}}) = -t(1 + 2e^{ik_x a_0}\cos k_y b_0) \tag{8.14}$$

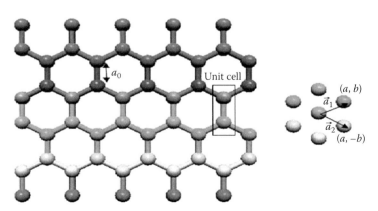

FIGURE 8.7   Schematic of the graphene unit cell with lattice vectors $\vec{a_1}$ and $\vec{a_2}$.

where a_0 is the distance between two neighbor carbon atoms. For each value of wave vector, there are a couple of eigenvalues, one positive and one negative, leading to two branches in the $E(k)$ plot, which is expected from two basic functions per unit cell; therefore, the eigenvalues are given by

$$E = \pm |h_0| = \pm t \sqrt{1 + 4\cos k_y b_0 \cos k_x a_0 + 4\cos^2 k_y b_0} \qquad (8.15)$$

where

t is the overlap energy

k_x and k_y are unit vector components

The analytical calculation of the energy levels is being complicated because the electrostatic fields split the symmetry between the three layers [57,58]. Thus, we consider a situation where the upper layer is at potential Δ, the lower layer is at potential $-\Delta$, and the layer at the center is with zero potential. At the limit $v_F |k| \leq \Delta \leq t_\perp$ perturbation theory [89] can be adopted to derive the energy [59] as

$$\varepsilon_k \approx \pm \frac{\Delta v_F |k|}{\sqrt{2} t_\perp} \left(1 - \frac{v_F^2 |k|^2}{\Delta^2} \right) \qquad (8.16)$$

where

$t_\perp$ is the hopping energy

v_F is the Fermi velocity

So Equation 8.16 can be simplified into

$$\varepsilon_k = \pm(\alpha k - \beta k^3) \qquad (8.17)$$

where

$$\alpha = \frac{v_F \Delta}{t_\perp \sqrt{2}}$$

$$\beta = \frac{v_F^3}{t_\perp \sqrt{2\Delta}}$$

k is the wave vector

Equation 8.17 approximately describes the existence of states at zero energy when $v_F |k| = \Delta$ [32]. By employing this model, we have plotted the band structure of TGN with ABA stacking in Figure 8.8.

Apparently, in the whole procedure, overlap will be seen in the band energy at least in one point.

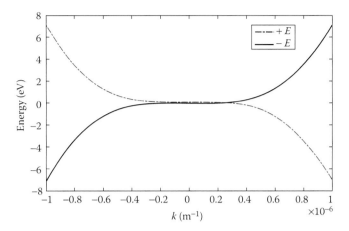

FIGURE 8.8   Band structure of ABA-stacked TGN with overlap.

8.8   DOS

The DOS and its relation with the density of electrons and holes by the Fermi distribution function can be modeled [60]. From the TGN energy equation, the DOS reveals the availability of energy states, which is defined as a function of the wave vector given by

$$\text{DOS} = \frac{1}{2\pi \left(\alpha - 3\beta k^2\right)} \tag{8.18}$$

Equation 8.17 designates that wave vector can be written as

$$k(E) \rightarrow \left\{ \frac{\left(\frac{2}{3}\right)^{\frac{1}{3}} \alpha}{\left(-9E\beta^2 + \sqrt{3}\sqrt{-4\alpha^3\beta^3 + 27E^2\beta^4}\right)^{\frac{1}{3}}} + \frac{\left(-9E\beta^2 + \sqrt{3}\sqrt{-4\alpha^3\beta^3 + 27E^2\beta^4}\right)^{\frac{1}{3}}}{2^{\frac{1}{3}}3^{\frac{2}{3}}\beta} \right\} \tag{8.19}$$

The generalization of the model of an electron in a 3D infinite potential is well utilized to derive the DOS function. By employing the wave vector relation in the DOS definition in a simplified form, we get

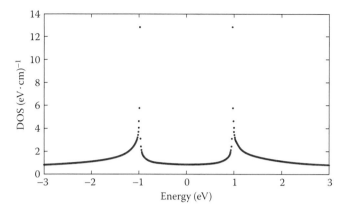

FIGURE 8.9 The DOS of TGN with ABA stacking.

$$\text{DOS} = \cfrac{1}{A - \cfrac{B}{\left(DE + \sqrt{F + E^2}\right)^{\frac{2}{3}}} - C\left(DE + \sqrt{F + E^2}\right)^{\frac{2}{3}}} \tag{8.20}$$

where
   E is the energy band structure
   A, B, C, D, and F are defined in the appendix

The infinite DOS at the Fermi level indicates that there are free carriers [61]. In addition, the low DOS spectrum exposes several prominent peaks. As shown in Figure 8.9, the DOS for TGN with ABA stacking at room temperature is plotted.

8.9 CARRIER STATISTICS OF ABA-STACKED TGN

As a main parameter, electron concentration in a band is calculated by integrating the Fermi probability distribution function over the energy as [62]

$$n = \int \text{DOS}(E) f(E) dE \tag{8.21}$$

where
   $f(E) = 1/[1 + \exp(E - E_F/k_B T)]$ is the Fermi–Dirac distribution function
   k_B is the Boltzmann constant
   E_F is the 1D Fermi energy
   T is the temperature

Biased TGN carrier concentration is modified as

$$
n = \int_{0}^{\eta} \frac{(k_{B}T)dx}{\left\{ A - \frac{B}{(k_{B}T)^{\frac{2}{3}}\left\{ D(x + E_{CO}) + \sqrt{N + (x + E_{CO})^{2}} \right\}^{\frac{2}{3}}} - C(k_{B}T)^{\frac{2}{3}}\left\{ D(x + E_{CO}) + \sqrt{N + (x + E_{CO})^{2}} \right\}^{\frac{2}{3}} \right\} (1 + e^{x - \eta})}
$$

(8.22)

where

$$
x = \frac{E - E_{c}}{k_{B}T}
$$

$$
\eta = \frac{-E_{c} + E_{F}}{k_{B}T}
$$

N, E_{CO} are defined in the Appendix

8.9.1 GENERAL MODEL OF ABA-STACKED TGN CARRIER CONCENTRATION

Based on the presented model, TGN carrier concentration is a function of normalized Fermi energy (η) as shown in Figure 8.10.

8.9.2 CARRIER CONCENTRATION OF ABA-STACKED TGN IN THE NONDEGENERATE REGIME

We note from Equation 8.23 that the Fermi level with distance more than $3k_{B}T$ from either the conduction or the valance band edge within the band gap illustrates

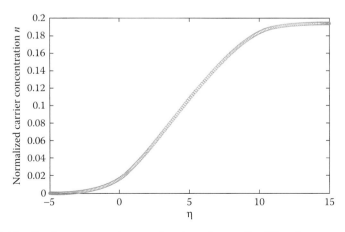

FIGURE 8.10 Carrier concentration as a function of normalized Fermi energy.

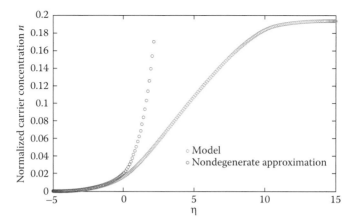

FIGURE 8.11   Comparison between presented model and nondegenerate approximation.

a nondegenerate condition. In case of a nondegenerate region, we can neglect 1 in comparison with an exponential function because of the high difference between x and η. Similar to the conventional 1D device, the carrier concentration indicates a normalized Fermi energy dependence effect for electron concentration as shown in Figure 8.11.

$$n_{ND} = \int_{0}^{\eta} \left\{ A - \frac{\dfrac{e^{-(x-\eta)}(k_B T)\,dx}{B}}{(k_B T)^{\frac{2}{3}}\left\{D(x+E_{CO})+\sqrt{N+(x+E_{CO})^2}\right\}^{\frac{2}{3}} - C(k_B T)^{\frac{2}{3}}\left\{D(x+E_{CO})+\sqrt{N+(x+E_{CO})^2}\right\}^{\frac{2}{3}}} \right\} \quad (8.23)$$

The presented carrier concentration model can be approximated by nondegenerate approximation, particularly in lower numbers of $\eta(\eta \leq -1)$. In addition, the effect of a normalized Fermi energy (η) on a nondegenerate limit as a notable parameter is reported.

8.9.3 CARRIER CONCENTRATION OF ABA-STACKED TGN IN THE DEGENERATE REGIME

In terms of a semiconductor, in case the Fermi level is situated at less than $3k_B T$ far away from the conduction and the valence bands, or located within a band, degenerate approximation will play an important role on carrier statistics study. In other words, in this regime $e^{\frac{x-\eta}{k_B T}}$ in comparison with 1 can be neglected because the

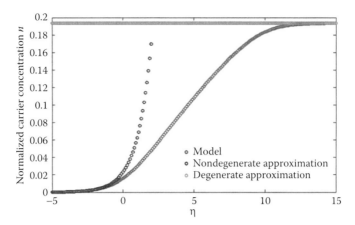

FIGURE 8.12 Comparison between presented model, nondegenerate, and degenerate approximations.

amount of $x - \eta$ is very small. Therefore, the carrier concentration for the degenerate region (n_D) can be written as

$$n_\mathrm{D} = \int_0^\eta \left\{ A - \frac{(k_\mathrm{B}T)\mathrm{d}x}{(k_\mathrm{B}T)^{\frac{2}{3}}\left\{ D(x + E_\mathrm{CO}) + \sqrt{N + (x + E_\mathrm{CO})^2} \right\}^{\frac{2}{3}}} \right.$$
$$\left. - C(k_\mathrm{B}T)^{\frac{2}{3}}\left\{ D(x + E_\mathrm{CO}) + \sqrt{N + (x + E_\mathrm{CO})^2} \right\}^{\frac{2}{3}} \right\} \quad (8.24)$$

As shown in Figure 8.12, we can approximate the presented model by degenerate approximation, and, particularly, we can incorporate it in higher numbers of η. This figure also indicates that TGN carrier concentration in a degenerate limit is unassociated with normalized Fermi energy.

In future, nanoelectronic devices will be scaled down to nanoscale size to meet Moore's law, and, therefore, will operate in a degenerate limit, which makes the degenerate approximation more dominant in future nanoscale device modeling.

8.9.4 EFFECT OF APPLIED VOLTAGE ON THE INTRINSIC FERMI LEVEL

As indicated in Figure 8.9, there are two sharp peaks at the Fermi energy. There is a direct relationship between the DOS function and the carrier effective mass. In fact, a greater effective mass indicates a higher DOS function [61]. Based on the assumed energy relation in TGN, the effective mass is a function of the applied voltage that is embedded in the wave vector given by

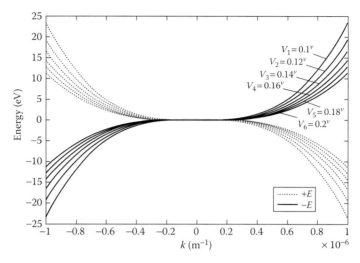

FIGURE 8.13 The effect of applied voltage on the curvature of the *E–K* graph (effective mass) in TGN with ABA stacking.

$$m^* = \hbar^2 \left[\frac{1}{6k\beta} \right]$$

(8.25)

As shown in Figure 8.13, as the applied voltage increases, the curvature of the *E–K* graph rises. In the other words, the carrier effective mass has increased. During this process, not only are the same number of electrons and holes maintained, but also the intrinsic Fermi level moves away from the band gap midpoint.

8.10 CHANNEL CONDUCTIVITY OF ABA-STACKED TGN FIELD EFFECT TRANSISTOR

The conductance model of multilayer GNRs (BGN and TGN) as a FET channel being based on the Landauer formula is presented [63,64]. Besides, the reported agreement with experimental study lends support to our model. The presented model demonstrates that minimum conductivity increases dramatically by temperature. It also draws parallels between TGN and BGN, in which similar thermal behavior is observed, while in the case of monolayer graphene, the minimum conductivity being nearly unchangeable with changing temperature shows a distinction. Maxwell–Boltzmann approximation is employed to form the conductance of TGN near the neutrality point. The analytical model in the degenerate regime in comparison with reported data proves that TGN-based transistors operate in the degenerate regime just like 1D semiconductors. Moreover, our model confirms that under similar conditions, the conductivity of TGN is less than BGN, as reported in some experiments.

Layers of graphene can be stacked differently depending on the horizontal shift of graphene planes. Every individual graphene multilayer sequence behaves like a new material in which different stacking of graphene sheets leads to different electronic properties [65,67]. Recently, unique properties of mono- and few-layer graphene have attracted great attention and have been proposed as promising candidates for future nanoelectronics. Ballistic transport phenomenon at room temperature, anomalous quantum Hall effect, and tunable band gap by applied perpendicular electric field and magnetic field have been observed experimentally [68–73]. General electronic properties of graphene-based materials are varied by increasing the number of the layers, but the low energy electronic properties in graphene layers depend on a large number of parameters. Since theorists have been unable to agree on the details of the electronic structure, it needs more research to establish the electrical properties of these materials [66]. In addition, the number and configuration of graphene layers play significant roles in realizing either metallic or semiconducting electronic behavior [69–71].

One of the important aspects of the electronic properties of graphene is its conductivity as a channel in FETs [74,75]. Recently, experimental attention has turned toward the properties of TGN [71,76], and a tunable three-layer graphene single-electron transistor has been experimentally realized [65,77]. In this section, the conductance of TGN with ABA stacking is modeled based on the 1D Landauer formula. Conductance of TGN in the degenerate and the nondegenerate perturbations are studied, as well as its temperature dependence. The analytical model in the degenerate regime out of the neutrality point is also approximated near the neutrality point as discussed in the Maxwell–Boltzmann approximation. Finally, a good agreement with experimental data is reported. Comparison between the BGN conductance model in Ref. [75] with the TGN conductance model presented in this chapter confirms that the conductivity of BGN is higher than TGN, which has already been reported in some experiments [66]. Very high carrier mobility can be achieved in graphene-based materials [78], which make them promising candidates for nanoelectronic devices.

Recently, electron and hole mobilities as high as 2×10^5 cm^2 V^{-1} s^{-1} have been reached for suspended graphene [66]. Ballistic transport has also been observed up to room temperature in these materials [69,82]. The perpendicular external applied electric or magnetic fields are expected to induce band crossing variation in Bernal-stacked TGNs [78,83–85]. Therefore, TGN is found to be a semimetal with its behavior different from that of single-layer and bilayer graphene in some aspects. The response of ABA-stacked TGN to an external electric field is different from that of mono- or bilayer graphene. In fact, rather than opening a gap in bilayer graphene, the magnitude of overlap in TGN is tuned [76]. On the other hand, overlap between the conduction and valence bands takes place in the band structure of TGNs, which can be controlled by a perpendicular external electric field. The band overlap increases with increasing the external electric field, which is independent of the electric field polarity. Moreover, it is shown that effective mass remains constant when external electric field is increased [65,66,79,80,86]. A spectrum of full tight-binding Hamiltonian of HOPG stacking (ABA) TGN was obtained in Refs. [81,86–88]. The presence of electrostatic fields breaks the symmetry between the three layers. Varying the band overlap by electric field is a unique property of ABA-stacked TGN that had not been previously found in other semimetallic systems [66].

8.11   CONDUCTANCE MODEL

For 1D TGNFET, a GNR channel is assumed to be ballistic, and as shown in Figure 8.14, the current from the source to the drain can be given by the Boltzmann transport equation, and in the Landauer formula it can be written [90,91] as

$$G = \frac{2q^2}{\hbar} \int_{-\infty}^{+\infty} M(E)T(E)\left(-\frac{df}{dE}\right) dE \qquad (8.26)$$

where

- q is the electron charge
- $\hbar$ is Planck's constant
- $T(E)$ is the transmission probability
- f is the Fermi–Dirac distribution function

High carrier mobility reported from experiments in graphene leads one to assume a completely ballistic carrier transportation in it, which implies that the average probability of injected electron at one end transmitting to the other end is approximately equal to 1 ($T(E) = 1$) [91].

The number of modes that are above the cutoff at energy E in the transmission channel by including the spin effect is

$$M(E) = \frac{dE}{ldk} = \frac{\alpha - 3\beta k^2}{l} \qquad (8.27)$$

The number of modes in corporation with the Landauer formula indicates that conductance of TGN can be written as

$$G = \frac{2\alpha q^2}{lh} \int_{-\infty}^{+\infty} \left(-\frac{d}{dE}\left(\frac{1}{1+e^{\frac{E-E_F}{k_BT}}}\right)\right) dE + \frac{-6\beta q^2}{lh} \int_{-\infty}^{+\infty} k^2 \left(-\frac{d}{dE}\left(\frac{1}{1+e^{\frac{E-E_F}{k_BT}}}\right)\right) dE \qquad (8.28)$$

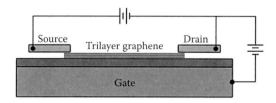

FIGURE 8.14   Simple 1D TGNFET biased by voltage V, where TGN is used as a channel between the drain and the source (The width of TGN is assumed to be less than 10 nm.).

where momentum (k) can be derived by using Cardano's solution for *cubic equations* [92]. Equation 8.28 can be assumed in the form of $G = N_1 G_1 + N_2 G_2$, where $N_1 = 2\alpha q^2/lh, N_2 = -6\beta q^2/lh$. Since G_1 is an odd function, its value is equivalent to zero and $G = N_2 G_2$.

$$
G_2 = \int_{-\infty}^{+\infty} \left(\left(-\frac{E}{2\beta} + \sqrt{\left(\frac{-\alpha}{3\beta} \right)^3 + \left(\frac{E}{2\beta} \right)^2} \right)^{\frac{1}{3}} + \left(-\frac{E}{2\beta} - \sqrt{\left(\frac{-\alpha}{3\beta} \right)^3 + \left(\frac{E}{2\beta} \right)^2} \right)^{\frac{1}{3}} \right)^2 \left(-\frac{d}{dE} \left(\frac{1}{1 + e^{\frac{E - E_F}{k_B T}}} \right) \right) dE
$$

(8.29)

This equation can be solved numerically by employing the partial integration method and using the simplification form, where $x = (E - \Delta)/k_B T$, $\eta = (E_F - \Delta)/k_B T$. Thus, the general conductance model of TGN is obtained as

$$
G_2 = -\int_{-\infty}^{+\infty} \left\{ \left(\frac{2k_B T}{1 + e^{x - \eta}} \right) \times \left(\frac{-\dfrac{k_B T}{2\beta} - \dfrac{x k_B T + \Delta}{4\beta^2 \sqrt{-\dfrac{\alpha^3}{27\beta^3} + \dfrac{E^2}{4\beta^2}}}}{3 \left(-\dfrac{x k_B T + \Delta}{2\beta} - \sqrt{-\dfrac{\alpha^3}{27\beta^3} + \dfrac{\left(x k_B T + \Delta \right)^2}{4\beta^2}}} \right)^{2/3}} + \frac{-\dfrac{k_B T}{2\beta} + \dfrac{x k_B T + \Delta}{4\beta^2 \sqrt{-\dfrac{\alpha^3}{27\beta^3} + \dfrac{E^2}{4\beta^2}}}}{3 \left(-\dfrac{x k_B T + \Delta}{2\beta} + \sqrt{-\dfrac{\alpha^3}{27\beta^3} + \dfrac{\left(x k_B T + \Delta \right)^2}{4\beta^2}}} \right)^{2/3}} \right) \right.
$$
$$
\left. \times \left(\left(-\dfrac{x k_B T + \Delta}{2\beta} - \sqrt{-\dfrac{\alpha^3}{27\beta^3} + \dfrac{\left(x k_B T + \Delta \right)^2}{4\beta^2}}} \right)^{1/3} + \left(-\dfrac{x k_B T + \Delta}{2\beta} + \sqrt{-\dfrac{\alpha^3}{27\beta^3} + \dfrac{\left(x k_B T + \Delta \right)^2}{4\beta^2}}} \right)^{1/3} \right) \right\} dx
$$

(8.30)

Figure 8.15 shows the conductance of TGN versus gate voltage. It is clearly shown that the conductivity of TGN increases by increasing the magnitude of gate voltage.

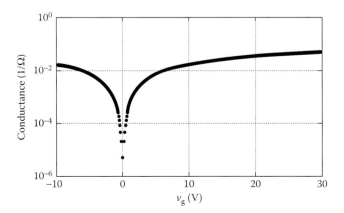

FIGURE 8.15   Numerical solution of a general TGN conductance model.

Temperature-dependent TGN conductance has been studied experimentally [66]. The measurement shows that minimum conductivity increases dramatically when temperature increases. This is comparable with bilayer graphene, in which similar thermal behavior is shown, while the minimum conductivity of monolayer graphene is nearly unchangeable with changing temperature [66,70]. The presented model confirms the reported experimental data as well.

Based on the reported model, minimal conductivity increases as the temperature increases. Figure 8.16 shows the conductance of TGN in 100 k and 300 k, with which the temperature effect on conductivity of TGN is demonstrated. The Fermi energy distribution function is estimated to be one in a degenerate limit, implying that the probability of filling the energy states is equal to one. Therefore, in degenerate condition, the general conductance model can be written as

$$
G_D = - \int_{-\infty}^{+\infty} \left(\frac{-\dfrac{1}{\beta} - \dfrac{E}{2\beta^2 \sqrt{-\dfrac{\alpha^3}{27\beta^3} + \dfrac{E^2}{4\beta^2}}}}{3\left(-\dfrac{E}{2\beta} - \sqrt{-\dfrac{\alpha^3}{27\beta^3} + \dfrac{E^2}{4\beta^2}} \right)^{2/3}} + \frac{-\dfrac{1}{\beta} + \dfrac{E}{2\beta^2 \sqrt{-\dfrac{\alpha^3}{27\beta^3} + \dfrac{E^2}{4\beta^2}}}}{3\left(-\dfrac{E}{2\beta} + \sqrt{-\dfrac{\alpha^3}{27\beta^3} + \dfrac{E^2}{4\beta^2}} \right)^{2/3}} \right)
$$

$$
\left(\left(-\dfrac{E}{2\beta} - \sqrt{-\dfrac{\alpha^3}{27\beta^3} + \dfrac{E^2}{4\beta^2}} \right)^{1/3} + \left(-\dfrac{E}{2\beta} + \sqrt{-\dfrac{\alpha^3}{27\beta^3} + \dfrac{E^2}{4\beta^2}} \right)^{1/3} \right) dE \qquad (8.31)
$$

Fortunately, in a degenerate limit, the general model can be solved analytically, which illustrates accurate results in this regime. As shown in Figure 8.17, the analytical solution of TGN conductance and the general solution of TGN conductance are in good agreement in a degenerate limit.

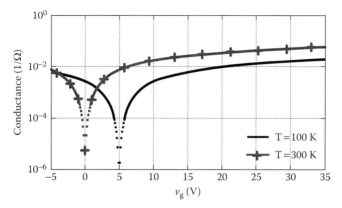

FIGURE 8.16 Temperature dependence of TGN conductance.

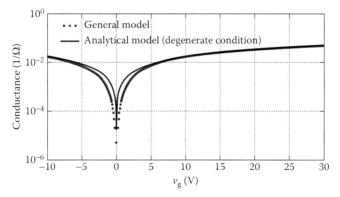

FIGURE 8.17   Analytical model of TGN conductance in degenerate condition and a general model of TGN conductance.

$$
G_{\mathrm{D}} = \frac{\left(-\dfrac{9E}{\beta} - \sqrt{\dfrac{-12\alpha^3 + 81E^2}{\beta^3}} \right)^{2/3} + \left(-\dfrac{9E}{\beta} + \sqrt{\dfrac{-12\alpha^3 + 81E^2}{\beta^3}} \right)^{2/3}}{6.87} \tag{8.32}
$$

It is apparent that the analytical solution, which is obtained in the degenerate condition, can fit in the general numerical solution out of the neutrality point. In addition, it is expected that TGN operates in the degenerate region as a channel between the drain and the source. Yet, based on the good agreement between the analytical and general models, the analytical model can be employed as a conductance model to predict TGN behavior in TGN-based devices.

As shown in Figure 8.18, the experimental results of TGN conductance [66,70] in $T = 50$ mk are in good agreement with the theoretical calculations (dark dots) presented in this chapter at the same temperature. The comparison between the presented model and published data indicates that a better performance of ABA-stacked TGN-based transistor is recommended in the degenerate regime.

Applying the Maxwell–Boltzmann approximation to the conductance model of TGN, we obtain the conductance model in the nondegenerate condition. Figure 8.19 shows the conductance model of TGN in the nondegenerate regime compared with the general model derived from Equation 8.30, which can be utilized in the neutrality point.

Considering the model of bilayer graphene, with nanoribbon as a channel in BGNFET taken from Ref. [75], and the TGN model obtained in this chapter, the comparison between bilayer and trilayer conductance in room temperature showing similar external applied electric field and channel length demonstrates that BGN conductivity is more than TGN conductivity as illustrated in Figure 8.20.

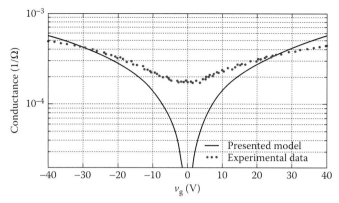

FIGURE 8.18 Conductance of TGN versus gate voltage. Comparison of the presented model and reported experimental data from Ref. [66] indicate that better performance of ABA-stacked TGN-based transistor is predicted in the degenerate regime.

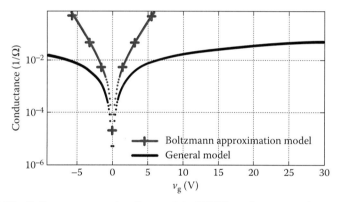

FIGURE 8.19 Boltzmann approximation model of TGN conductance and a general model of TGN conductance.

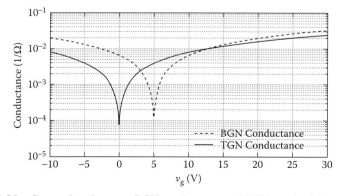

FIGURE 8.20 Comparison between BGN conductance and TGN conductivity.

8.12   CONCLUSION

Trilayer GNR with length more than the de Broglie wavelength is assumed as a 1D device since quantum confinement effect indicates digital type behavior of carriers on two Cartesian directions. In this chapter, carrier density incorporation with the energy band of TGN with ABA stacking close to the Fermi level is predicted by the numerical method. Our result shows that carrier concentration can be approximated by degenerate and nondegenerate approximation in some numbers of normalized Fermi energy. Based on the analytical model, in a nondegenerate limit, carrier concentration is a function of normalized Fermi energy. On the contrary, carrier concentration is not a correspondent of normalized Fermi energy in a degenerate limit. The numerical solution of the DOS for ABA-stacked TGN has also been introduced. We have discussed the effect of external applied voltage which is per-pendicular to the surface of TGN with ABA stacking. We have also concluded that by increasing the applied voltage, the carrier effective mass is increased, and also the position of the intrinsic Fermi level is changed. In fact, the Fermi level's position has remained far from the conduction and valance bands, and it has also been shifted from the band gap midpoint, resulting in achieving an equal number of electrons and holes. A common hexagonal structure in graphite is Bernal (ABA) stacking, in which the applied external electric field changes the amount of overlap of its conduction and valence bands, resulting in a semimetal. Finally, based on the Landauer formula, we present the conductance model of ABA-stacked TGN as a FET channel. TGN conductance can be estimated by either the analytical model in the degenerate regime out of the neutrality point or by the Maxwell–Boltzmann approximation in the nondegenerate regime in the neutrality point. The proposed model is in good agreement with the reported data from other experiments, which illustrates that minimum conductivity dramatically increases as temperature increases. Our model also indicates that better performance of an ABA-stacked TGN-based transistor can be seen in the degenerate regime, like what we expect in a conventional 1D semiconductor. Moreover, our model confirms that under similar conditions, the conductivity of TGN is less than BGN as reported in some experiments.

APPENDIX

$$A = -6.2832\alpha$$

$$B = 14.3849\alpha^2\beta$$

$$C = \frac{2.7444}{\beta}$$

$$D = -9\beta^2$$

$$F = \frac{-0.1690\alpha^3}{\beta}$$

$$E_{CO} = \frac{E_c}{k_B T}$$

$$N = \frac{-0.1690\alpha^3}{\beta(k_B T)^2} = \frac{F}{(k_B T)^2}$$

REFERENCES

1. Y. G. Gogotsi, et al. Materials: Transformation of diamond to graphite. *Nature*. 401, 663–664, 1999.
2. T. Irifune, et al. Materials: Ultrahard polycrystalline diamond from graphite. *Nature*. 421, 599–600, 2003.
3. P. L. McEuen, et al. Single-walled carbon nanotube electronics. *IEEE Transactions on Nanotechnology*. 1, 78–85, 2002.
4. J. A. Misewich, et al. Electrically induced optical emission from a carbon nanotube FET. *Science*. 300, 783–786, 2003.
5. J. Chen, et al. Bright infrared emission from electrically induced excitons in carbon nanotubes. *Science*. 310, 1171–1174, 2005.
6. A. Javey, et al. Ballistic carbon nanotube field-effect transistors. *Nature*. 424, 654–657, 2003.
7. J. Li, et al. Bottom-up approach for carbon nanotube interconnects. *Applied Physics Letters*. 82, 2491–2493, 2003.
8. J. E. Jang, et al. Nanoelectromechanical switches with vertically aligned carbon nanotubes. *Applied Physics Letters*. 87, 163114–163116, 2005.
9. J. Kong, et al. Nanotube molecular wires as chemical sensors. *Science*. 287, 622–625, 2000.
10. A. K. Geim and K. S. Novoselov. The rise of graphene. *Nature Materials*. 6, 183–191, 2007.
11. T. O. Wehling, et al. Molecular doping of graphene. *Science Nano Letters*. 8, 173–177, 2004.
12. D. J. K. S. Novoselov, et al. Two-dimensional atomic crystals. *Proceedings of the National Academy Science*. 102, 10451–10453, 2005.
13. F. Schwierz. Graphene transistors. *Nature Nanotechnology*. 5, 487–496, 2010.
14. T. S. Li, et al. Electron transport in nanotube-ribbon hybrids. *European Physical Journal B*. 70, 497–505, 2009.
15. S. Entani, et al. Interface properties of metal/graphene heterostructures studied by micro-Raman spectroscopy. *Journal of Physical Chemistry C*. 114, 20042–20048, 2010.
16. Y. T. Zhang, et al. Spin polarization and giant magnetoresistance effect induced by magnetization in zigzag graphene nanoribbons. *Physical Review B*. 81, 165404–165409, 2010.
17. A. H. Castro Neto, et al. The electronic properties of graphene. *Reviews of Modern Physics*. 81, 109–162, 2009.
18. A. V. Rozhkova, et al. Electronic properties of mesoscopic graphene structures: Charge confinement and control of spin and charge transport. *Physics Reports*. 503, 77–114, 2011.
19. Y. H. Hu, et al. Thinnest two-dimensional nanomaterial—Graphene for solar energy. *ChemSusChem*. 3, 782–796, 2010.
20. F. Guineaa, et al. Electronic properties of stacks of graphene layers. *Solid State Communications*. 143, 116–122, 2007.

21. E. V. Castro, et al. Electronic properties of a biased graphene bilayer. *Journal of Physics: Condensed Matter*. 22, 175503, 2010.
22. H. Li, et al. Carbon nanomaterials for next-generation interconnects and passives: Physics, status, and prospects. *IEEE Transactions on Electron Devices*. 56, 1799–1821, 2009.
23. M. Lundstrom and J. Guo. *Nanoscale Transistors: Device Physics, Modeling and Simulation*. Springer-Verlag, New York, pp. 217, 2005.
24. I. Saad, et al. The dependence of saturation velocity on temperature, inversion charge and electric field in a nanoscale MOSFET. *Nanoelectronics and Materials*. 3, 17–34, 2010.
25. K. Nakada, et al. Edge state in graphene ribbons: Nanometer size effect and edge shape dependence. *Physical Review B*. 54, 17954–17961, 1996.
26. R. Saito, et al. *Physical Properties of Carbon Nanotubes*. Imperial College Press, London, p. 259, 1998.
27. X. Li, et al. Chemically derived, ultrasmooth graphene nanoribbon semiconductors. *Science*. 319, 1229–1232, 2008.
28. L. Jiao, et al. Narrow graphene nanoribbons from carbon nanotubes. *Nature*. 458, 877–880, 2009.
29. Z. Chen, et al. Graphene non-ribbon electronics. *Physica E*. 40, 228–232, 2007.
30. M. Y. Han, et al. Energy band-gap engineering of graphene nanoribbons. *Physical Review Letters*. 98, 206805, 2007.
31. J. R. Williams. *Electronic Transport in Graphene: p–n Junctions, Shot Noise and Nanoribbons*. Harvard University Press, Cambridge, MA, 170 p., 2009.
32. F. Guinea, et al. Electronic states and Landau levels in graphene stacks. *Physical Review B*. 73, 1–8, 2006.
33. Y. Takane. Anomalous enhancement of the Boltzmann conductivity in disordered zigzag graphene nanoribbons. *Journal of the Physical Society of Japan*. 79, 024711–024720. 2010.
34. H. Kumazaki and D. S. Hirashima. Spin-polarized transport on zigzag graphene nanoribbon with a single defect. *Journal of the Physical Society of Japan*. 78, 6 p, 094701, 2009.
35. E. Watanabe, et al. Ballistic thermal conductance of electrons in graphene ribbons. *Physical Review B*. 80, 085404–085409, 2009.
36. H. Cheraghchi and H. Esmailzade. A gate-induced switch in zigzag graphene nanoribbons and charging effects. *Nanotechnology*. 21, 205306, 2010.
37. S. S. Yu, et al. Electronic properties of nitrogen/boron-doped graphene nanoribbons with armchair edges. *IEEE Transactions on Nanotechnology*. 9, 78–81, 2010.
38. K. S. Novoselov, et al. Two-dimensional gas of massless Dirac fermions in graphene. *Nature*. 438, 197, 2005.
39. S. Y. Zhou, et al. Substrate-induced bandgap opening in epitaxial graphene. *Nature Materials*. 6, 727–731, 2007.
40. M. F. Craciun, et al. Trilayer graphene: A semimetal with gate-tunable band overlap. *Nature Nanotechnology*. 4, 383–388, 2009.
41. A. A. Avetisyan, et al. Stacking order dependent electric field tuning of the band gap in graphene multilayers. *Physical Review B*. 81, 115432, 2010.
42. M. T. Ahmadi, et al. MOSFET-Like carbon nanotube field effect transistor model. *Nanotech Conference & Expo 2009, Technical Proceedings*. 3, 574–579, 2009.
43. M. Terronesa, et al. Graphene and graphite nanoribbons: Morphology, properties, synthesis, defects and applications. *Nano Today*. 5, 351–372, 2010.
44. M. I. Katsnelson, et al. Chiral tunnelling and the Klein paradox in graphene. *Nature Physics*. 2, 620–625, 2006.

45. F. Zhang, et al. Band structure of ABC-stacked graphene trilayers. *Physical Review B.* 82, 035409, 2010.

46. M. Koshino and E. McCann. Parity and valley degeneracy in multilayer graphene. *Physical Review B.* 81, 115315, 2010.

47. F. Guinea, et al. Interaction effects in single layer and multi-layer graphene. *European Physical Journal, Special Topics.* 148, 117–125, 2007.

48. S. Yuan, et al. Electronic transport in disordered bilayer and trilayer graphene. *Physical Review B.* 82, 235409, 2010.

49. M. Aoki and H. Amawashi. Dependence of band structures on stacking and field in layered graphene. *Solid State Communications.* 142, 123–127, 2007.

50. M. Koshino. Interlayer screening effect in graphene multilayers with ABA and ABC stacking. *Physical Review B.* 81, 125304, 2010.

51. M. Koshino and E. McCann. Trigonal warping and Berry's phase Nπ in ABC-stacked multilayer graphene. *Physical Review B.* 80, 165409, 2009.

52. M. Koshino and E. McCann. Gate-induced interlayer asymmetry in ABA-stacked trilayer graphene. *Physical Review B.* 79, 125443, 2009.

53. Y. Xu and S.-H. Ke. The infrared spectra of ABC-stacking tri- and tetra-layer graphenes. Mesoscale and nanoscale physics. *Physical Review B.* 84, 245433, 2010.

54. M. F. Craciun, et al. Tuneable electronic properties in graphene. *NanoToday Press.* 6, 42–60, 2011.

55. M. Bresciani, et al. Simple and efficient modeling of the *E–k* relationship and low-field mobility in graphene nano-ribbons. *Solid-State Electronics.* 54, 1015–1021, 2010.

56. S. Datta. *Quantum Transport: Atom to Transistor.* Cambridge University Press, New York, pp. 113–114, 2005.

57. M. T. Ahmadi, et al. The high-field drift velocity in degenerately-doped silicon nanowires. *International Journal of Nanotechnology.* 6, 601–617, 2009.

58. M. T. Ahmadi, et al. The ultimate ballistic drift velocity in carbon nanotubes. *Journal of Nanomaterials.* 2008, 8 p, 769250, 2008.

59. F. Guinea, et al. Electronic states and Landau levels in graphene stacks. *Physical Review B.* 73, 245426, 2006.

60. M. T. Ahmadi, et al. Carrier velocity in carbon nano tube field effect transistor. *IEEE International Conference on Semiconductor Electronics.* Johor Bahru, Malaysia, November 25–27, pp. 519–523, 2008.

61. D. A. Neamen. *Semiconductor Physics and Devices*, 3rd Edition. McGraw Hill, New York, NY, 2003.

62. V. K. Arora. Failure of Ohm's Law: Its implications on the design of nanoelectronic devices and circuits. *Proceedings of the IEEE International Conference on Microelectronics.* 14, 15–22, 2006.

63. Hatef Sadeghi, et al. Channel conductance of ABA stacking trilayer graphene nanoribbon field effect transistor. *Modern Physics Letters B*, 26, 1250047, 2012. doi:10.1142/S0217984912500479 (IF=0.6).

64. S. M. Mousavi, et al. Bilayer graphene nanoribbon conductance model in parabolic structure. *Presented in ESciNano2010 AIP Conference Proceedings.* 1341(1), pp. 388–390. DOI:10.1063/1.3587025.

65. A. A. Avetisyan, et al. Stacking order dependent electric field tuning of the band gap in graphene multilayers. *Physical Review B.* 81, 115432, 2010.

66. M. F. Craciun, et al. Trilayer graphene is a semimetal with a gate-tunable band overlap. *Nature Nanotechnology.* 4, 383–388, 2009.

67. J. H. Warner. The influence of the number of graphene layers on the atomic resolution images obtained from aberration-corrected high resolution transmission electron microscopy. *Nanotechnology.* 21, 255707, 2010.

68. J. A. Yan, et al. Phonon dispersions and vibrational properties of monolayer, bilayer, and trilayer graphene: Density-functional perturbation theory. *Physical Review B*. 77, 125401, 2008.

69. C. Berger, et al. Ultrathin epitaxial graphite: 2D electron gas properties and a route toward graphene-based nanoelectronics. *Journal of Physical Chemistry B*. 108, 19912–19916, 2004.

70. W. J. Zhu, et al. Carrier scattering, mobilities, and electrostatic potential in monolayer, bilayer, and trilayer graphene. *Physical Review B*. 80, 235402, 2009.

71. P. Sutter, et al. Electronic structure of few-layer epitaxial graphene on Ru(0001). *Nano Letters*. 9, 2654–2660, 2009.

72. K. H. Ding, et al. Localized magnetic states in biased bilayer and trilayer graphene. *Journal of Physics: Condensed Matter*. 21, 182002, 2009.

73. M. Ezawa. Supersymmetry and unconventional quantum Hall effect in monolayer, bilayer and trilayer graphene. *Physica E: Low-Dimensional Systems & Nanostructures*. 40, 269–272, 2007.

74. M. T. Ahmadi, et al. Graphene nanoribbon conductance model in parabolic band structure. *Journal of Nanomaterials*. 2010, 4, 2010.

75. H. Sadeghi, et al. Ballistic conductance model of bilayer graphene nanoribbon (BGN). *Journal of Computational and Theoretical Nanoscience*. 8, 1993–1998, 2011.

76. M. Koshino and E. McCann. Gate-induced interlayer asymmetry in ABA-stacked trilayer graphene. *Physical Review B*. 79, 125443, 2009.

77. J. Guttinger, et al. Coulomb oscillations in three-layer graphene nanostructures. *New Journal of Physics*. 10, 125029, 2008.

78. K. F. Mak, et al. Electronic structure of few-layer graphene: Experimental demonstration of strong dependence on stacking sequence. *Physical Review Letters*. 104, 176404, 2010.

79. F. Zhang, et al. Band structure of ABC-stacked graphene trilayers. *Physical Review B*. 82, 035409, 2010.

80. C. L. Lu, et al. Absorption spectra of trilayer rhombohedral graphite. *Applied Physics Letters*. 89, 221910, 2006.

81. E. McCann and M. Koshino. Spin-orbit coupling and broken spin degeneracy in multilayer graphene. *Physical Review B*. 81, 241409, 2010.

82. P. N. Nirmalraj, et al. Nanoscale mapping of electrical resistivity and connectivity in graphene strips and networks. *Nano Letters*. 11, 16–22, 2011.

83. W. Zhu, V. Perebeinos, M. Freitag, and P. Avouris. Carrier scattering, mobilities, and electrostatic potential in monolayer, bilayer, and trilayer graphene. *Physical Review B*. 80, 235402, 2009.

84. G. M. Rutter, et al. Structural and electronic properties of bilayer epitaxial graphene. *Journal of Vacuum Science & Technology A*. 26, 938–943, 2008.

85. S. Russo, et al. Double-gated graphene-based devices. *New Journal of Physics*. 11, 095018, 2009.

86. A. A. Avetisyan, et al. Electric-field control of the band gap and Fermi energy in graphene multilayers by top and back gates. *Physical Review B*. 80, 195401, 2009.

87. F. Guinea, et al. Electronic states and Landau levels in graphene stacks. *Physical Review B*. 73, 245426, 2006.

88. S. Latil, et al. Massless fermions in multilayer graphitic systems with misoriented layers: Ab initio calculations and experimental fingerprints. *Physical Review B*. 76, 201402(R), 2007.

89. T. Kat. *Perturbation Theory for Linear Operators*, vol. 132. Springer Verlag, Berlin, 1995.

90. S. Datta. *Quantum Transport: Atom to Transistor*. Cambridge University Press, New York, 2005.
91. S. Datta. *Electronic Transport in Mesoscopic Systems*. Cambridge University Press, Cambridge, UK, 2002.
92. A. D. Polyanin. *Cubic Equation*. 2004. Available at http://eqworld.ipmnet.ru/en/solutions/ae/ae0103.pdf.

9 Graphene Nanoribbon Transistor Model
Additional Concepts

*Mahdiar Ghadiry, Mahdieh Nadi,
and Asrulnizam Abd Manaf*

CONTENTS

9.1 INTRODUCTION

Moore's law has been predicted as the trend for silicon technology over the last four decades [1]. Along the way, it has provided the fundamental complementary metal–oxide–semiconductor (CMOS) technology for today's global information society. It was thought that silicon will be the dominant technology for at least 10 more years [1]. However, some uncertainties in the behavior of transistors under extreme low dimensions make silicon questionable for future circuit technology [2]. Therefore, the end of the silicon era has been predicted for quite some time now due to technical reasons.

Short channel effect is one of the most concerning issues in nanoscale devices. When the junction depth is comparable to the channel length, which is the case in nanoscale metal–oxide–semiconductor field-effect transistors (MOSFETs), the barrier height decreases causing the threshold voltage to reduce. Further, when a high drain voltage is applied to a short channel device, the barrier is lowered even more, resulting in further decrease of the threshold voltage. This phenomenon is called drain-induced barrier lowering (DIBL). Eventually, the device reaches the punch-through condition, when the gate is completely unable to control the current flow.

In addition, as the transistor lengths are entering the sub-100 nm regime, silicon technology suffers from the impact of several secondary effects, which have been ignored in the past. These secondary effects include subthreshold leakage, hot-electron effects, oxide-tunneling, gate leakage, gate-induced drain leakage (GIDL), band-to-band tunneling, device parameter variations, reverse and normal short/narrow channel effects, source/drain series resistance, quantum effects on threshold voltage, velocity saturation and overshoot effects, gate resistance, interconnect parasitics, etc.

On the contrary, by using graphene it seems possible to make devices with channels that are extremely thin and that will allow graphene field-effect transistors (FETs) to be scaled to shorter channel lengths and higher speeds without encountering the adverse short-channel effects that restrict the performance of existing devices. Consequently, one of the most promising future options to enhance silicon is the introduction of carbon-based electronics [3].

Therefore, new material-based device concepts such as nanowire FETs, carbon nanotube (CNT) FETs, etc. have been suggested. Among these, CNT FETs, because of their excellent electronic properties, have been comprehensively explored and notable research concentration especially on their high carrier mobility and conductance has been reported [3,4]. However, there are still some unresolved issues concerning control on the chirality of CNTs, making them questionable for use in realistic applications [3–5]. Recent experimental studies [1,6] show the possibility of fabricating graphene nanoribbon (GNR) transistors, and the potential of GNR as an alternative approach to overcome CNT's chirality issues has been investigated. As a result, many researchers have been attracted to this field and provided several models for GNR's properties [1,4,7–12]. Nevertheless, there is lack of work in modeling its behavior near the drain junction and its breakdown voltage.

The length of velocity saturation region (LVSR) of FETs or the width of the drain region where impact ionization and carrier velocity saturation occur is one of the most important parameters in nanoscale devices. It controls the lateral drain breakdown voltage [13,14], substrate current, hot-electron generation [15,16], and drain current at the drain region [17–19]. In an FET, if the applied drain voltage is higher than the drain saturation voltage, the electric field near the drain junction will be higher than the critical field strength, which results in carrier velocity saturation. In addition, high electric field near the drain junction causes impact ionization and substrate current generation [20].

Although several models are available for saturation region of silicon-based MOSFETs [14–16,18–24], there is still plenty of room for research in modeling of this region for GNR transistors. In order to gain insight into reliability issues of these devices, close analysis of this region is necessary. In addition, these kinds of models open the way to design power transistors based on graphene. Since there is no experimental data yet for the length of saturation region in graphene-based transistors, analytical modeling seems to be a powerful tool in this case to provide some estimation.

In this chapter, analytical models for surface potential, lateral electric field, and length of saturation region are presented and the behavior of GNR transistors is studied in the saturation region.

9.2 SATURATION REGION MODELING OF GRAPHENE NANORIBBON TRANSISTORS

A schematic cross section of top-gated graphene nanoribbon field-effect transistor (GNRFET) is shown in Figure 9.1, where t_{OX} is the oxide thickness of top gate with a dielectric constant of ε_{OX}; t_G, W, and L are the GNR's thickness, width, and the channel length, respectively.

Generally, to model the potential distribution, Poisson's equation is solved [10,22,26,27]. Applying Gauss' law for the GNRFET presented in Figure 9.1 and using the similar approach presented in [10,27], the one-dimensional (1D) Poisson's equation for a GNR could be written as

$$\frac{d^2V(x)}{dx^2} - \frac{V(x)}{\lambda^2} = -\frac{qN}{\varepsilon_G} - \frac{V_G + V_{BI}}{\lambda^2} \tag{9.1}$$

where

$\quad$ $V(x)$ is the surface potential of GNR at any point along the x-direction inside the saturation region $(0 < x < \Delta L)$

$\quad$ $\varepsilon_G = 3 \times 10^{-11}$ F/m is GNR dielectric constant

$\quad$ $V_G = 0.1$ V and q are gate voltage and charge magnitude, respectively [28]

The parameter $\lambda = (\varepsilon_G t_G t_{OX}/\varepsilon_{OX})^{-2}$ is the relevant length scale for potential variation [29]. Built-in voltage V_{BI} in GNR with a bandgap $E_G = h v_F /3 w_G$ can be calculated using the following equation [10]:

$$V_{BI} = \frac{h v_F}{6 q w_G} - V_T \ln\left(\frac{N}{n_i}\right) \tag{9.2}$$

where

$\quad$ $v_F \sim 10^6$ m/s is the Fermi velocity

$\quad$ $V_T = k_B T/q$ is thermal voltage

$\quad$ n_i and N are the intrinsic and doping carrier concentrations

Boundary conditions for Equation 9.1 can be defined as $V(0) = V_S$, $V(\Delta L) = V_D$, $E(0) = E_S$, where V_S, V_D, ΔL, and E_S are saturation voltage at onset of saturation region, drain voltage, LVSR, and saturation surface electric field [27], respectively. Solving Equation 9.1 with the defined boundary conditions and taking $A = -qN/\varepsilon_G - (V_G + V_{BI})/\lambda^2$ yield

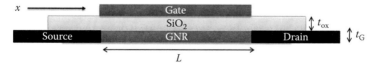

FIGURE 9.1 (See color insert.) Schematic cross section of a top-gated GNRFET [9]. Typical device parameters are doping concentration $N = 5 \times 10^{16}$ m^{-2}, $t_{OX} = 1$ nm, $L = 20$ nm, and $w_G = 5$ nm.

$$V(x) = -\lambda^2 A + (\lambda^2 A + V_S)\cosh(x/\lambda) + \lambda E_S \sinh(x/\lambda) \qquad (9.3)$$

Since $E(x) = -dV(x)/dx$, surface electric field distribution $E(x)$ can simply be obtained using Equation 9.3.

$$E(x) = \left(\lambda A + \frac{V_S}{\lambda}\right)\sinh(x/\lambda) + E_S \cosh(x/\lambda) \qquad (9.4)$$

For the cases where $x/\lambda > 3$ (near the drain region), the $\sinh(x/\lambda)$ and $\cosh(x/\lambda)$ can be approximated to $0.5 \exp(x/\lambda)$. Therefore, Equations 9.3 and 9.4 are reduced to

$$V(x) \approx 0.5(\lambda^2 A + V_S + \lambda E_S)\exp(x/\lambda) - \lambda^2 A \qquad (9.5)$$

$$E(x) \approx 0.5\left(\lambda A + \frac{V_S}{\lambda} + E_S\right)\exp\left(\frac{x}{\lambda}\right) \qquad (9.6)$$

In order to calculate ΔL, either Equation 9.3 or 9.5 can numerically be solved at $x = \Delta L$. Using Equation 9.5, we have

$$\Delta L \approx L - \frac{\lambda V_S \dfrac{1}{2}\exp\left(\dfrac{\Delta L}{\lambda}\right)}{V_D + \lambda^2 A - \dfrac{1}{2}\exp\left(\dfrac{\Delta L}{\lambda}\right)(\lambda^2 A + V_S)} \qquad (9.7)$$

To calculate E_S and V_S, the electric field in the region between source and the saturation region can be assumed to be linear [26,27,29]. As a result, it can be concluded that $d^2V(L_E)/dx^2 = -E_S/L_E$, where L_E is the effective channel length defined as $L_E = L - \Delta L$. Using Equation 9.1 again in linear region with the mentioned boundary condition, we have

$$E_S = \frac{L_E}{\lambda^2 - L_E^2}\left(\frac{t_G t_{OX} qN}{\varepsilon_{OX}} + V_G + V_{BI}\right) \qquad (9.8)$$

Since λ is much less than L_E ($0.5 < \lambda < 5$ nm and $8 < L_E < 17$ nm) for most cases of this study, the term λ^2 in Equation 9.8 can be neglected with respect to L_E^2. Thus, Equation 9.8, can be further reduced to

$$E_S \approx \frac{-1}{L_E}\left(\frac{t_G t_{OX} qN}{\varepsilon_{OX}} + V_G + V_{BI}\right) \qquad (9.9)$$

In addition, V_S can be computed as Equation 9.10 considering $V(0) = 0$.

$$V_S \approx \frac{t_G t_{OX} qN}{\varepsilon_{OX}} + V_G + V_{BI} \qquad (9.10)$$

9.2.1 Simulation Results and Discussion

Figure 9.2 shows the electric field distribution in the lateral direction with the different drain voltages. The values for V_D have been chosen based on saturation voltage V_S, which has been almost 0.23 V in this simulation. Figure 9.2 indicates that the potential distribution along the nanoribbon surface is similar to the profile of an abrupt junction at the edges of p-base/drift region and drain/drift region junctions [22]. In other words, it shows that the electric field profile follows an exponential form depending on the distance from the source. Equation 9.6 also highlighted this fact.

Figure 9.3 shows the saturation electric field at different drain–source voltages, channel thicknesses, and channel lengths. As can be seen, increasing V_D results in slight increase in E_S, when t_{OX} is in the range of ~1–5 nm. In addition, this figure shows that by increasing t_{OX} and L, the saturation electric field increases and decreases, respectively.

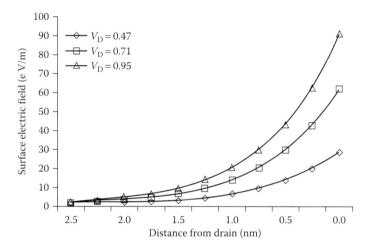

FIGURE 9.2 Surface electric field distribution at different drain voltages.

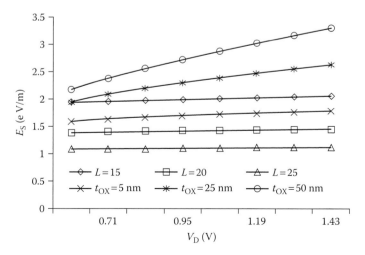

FIGURE 9.3 Saturation electric field at different drain voltages.

Figure 9.4 depicts the effect of the drain voltage and channel length on the length of the saturation region. The higher the drain voltage and the longer the channel, the longer the ΔL. In addition, the figure shows that the ratio of $\Delta L/L$ increases as L decreases. In Figure 9.5, it is shown that increasing oxide thickness causes increase in ΔL. In addition, the term $d(\Delta L)/d(V_D)$ increases as t_{OX} increases.

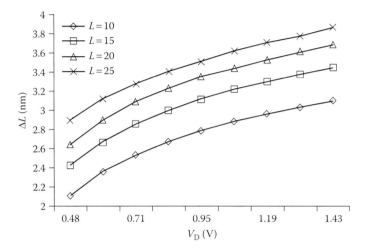

FIGURE 9.4   LVSR with different channel lengths and drain voltages.

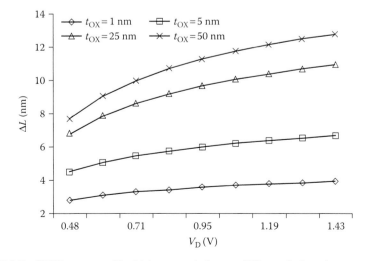

FIGURE 9.5   LVSR versus oxide thickness variations at different drain voltages.

9.3 ONE-DIMENSIONAL MODEL FOR SATURATION REGION OF DOUBLE-GATE GNR TRANSISTOR

A schematic cross section of double-gate (DG)-GNRFET is shown in Figure 9.6, where t_{OX} is the oxide thickness of the top gate with a dielectric constant of ε_{OX}; t_G, W, and L are the thickness, width, and the channel length of the GNR, respectively.

Applying Gauss law on the device presented in Figure 9.6 yields

$$-\int_0^x \varepsilon_{ox}E_{ox1}dx - \int_0^x \varepsilon_{ox}E_{ox2}dx - \int_0^{t_G} \varepsilon_G E_s dt + \int_0^{t_G} \varepsilon_G E(x)dt = -q\int_0^x \int_0^{t_G} (n + N_d)dxdt \quad (9.11)$$

where

E_S is the surface electric at onset of saturation region
ε_G is GNR's dielectric constants
ε_{OX} is oxide dielectric constants
E_{OX} is the electric field of the oxide/channel interface
N_d is the doping concentration

By solving Equation 9.12 and taking $E_{OX}(x) = V_G - V_{BI} - V(x)/t_{OX}$, the following equation is obtained:

$$\frac{d^2V(x)}{dx^2} - \frac{V(x)}{2\lambda^2} + \frac{(V_G - V_{BI})}{2\lambda^2} = \frac{q(n + N_d)}{\varepsilon_G} \quad (9.12)$$

where

$V(x)$ is the surface potential of GNR at any point along the x-direction inside the saturation region ($0 < x < \Delta L$)
$V_G = 0.5$ V is top-gate and back-gate voltage
q is the charge magnitude

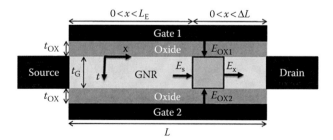

FIGURE 9.6 (**See color insert.**) Schematic cross section of a DG-GNRFET [9]. Typical device parameters are doping concentration $N_d = 5 \times 10^{16}$ m^{-2}, $t_{OX} = 1$ nm, $L = 20$ nm, and $w_G = 5$ nm.

The parameter $\lambda = (\varepsilon_G t_G t_{OX}/\varepsilon_{OX})^{-2}$ is the relevant length scale for potential variation [29].

Boundary conditions for Equation 9.12 can be defined as $V(0) = V_S$, $V(\Delta L) = V_D$, and $E(0) = E_S$, where E_S, V_S, V_D, and ΔL are saturation electric field, drain saturation voltage, drain voltage, and LVSR, respectively [27]. Solving Equation 9.12 with the defined boundary conditions and taking $A = -q(N_d + n)/\varepsilon_G - (V_G - V_{BI})/2\lambda^2$ yield

$$V(x) = -2\lambda^2 A + (2\lambda^2 A + V_S)\cosh\left(\frac{x}{\sqrt{2}\lambda}\right) + \sqrt{2}\lambda E_S \sinh\left(\frac{x}{\sqrt{2}\lambda}\right) \quad (9.13)$$

Since $E(x) = -dV(x)/dx$, the surface electric field distribution $E(x)$ can be obtained using Equation 9.13.

$$|E(x)| = \left(\sqrt{2}\lambda A + \frac{V_S}{\sqrt{2}\lambda}\right)\sinh\left(\frac{x}{\sqrt{2}\lambda}\right) + E_S \cosh\left(\frac{x}{\sqrt{2}\lambda}\right) \quad (9.14)$$

For the cases where $x/\lambda > 3$ (near the drain region), sinh (x/λ) and cosh (x/λ) can be approximated to $0.5 \exp(x/\lambda)$. Therefore, Equations 9.13 and 9.14 are reduced to

$$V(x) \approx 0.5(2\lambda^2 A + V_S + \sqrt{2}\lambda E_S)\exp\left(\frac{x}{\sqrt{2}\lambda}\right) - 2\lambda^2 A \quad (9.15)$$

$$|E(x)| \approx 0.5\left(\sqrt{2}\lambda A + \frac{V_S}{\sqrt{2}\lambda} + E_S\right)\exp\left(\frac{x}{\sqrt{2}\lambda}\right) \quad (9.16)$$

In order to calculate ΔL, either Equation 9.4 or 9.6 can numerically be solved at $x = \Delta L$. For example, using Equation 9.6 we have

$$\Delta L \approx L - \frac{\lambda V_S \dfrac{1}{\sqrt{2}}\exp\left(\dfrac{\Delta L}{\sqrt{2}\lambda}\right)}{V_D + 2\lambda^2 A - \dfrac{1}{2}\exp\left(\dfrac{\Delta L}{\sqrt{2}\lambda}\right)(2\lambda^2 A + V_S)} \quad (9.17)$$

Since the focus of this chapter is on modeling of the saturation region, the electric field between the source and the saturation region $(0 < x < \Delta L)$ can be assumed to be linear for simplicity [26,27,29,31,32]. As a result, it can be simply concluded that

$$\frac{d^2 V(L_E)}{dx^2} = \frac{E_S}{L_E} \quad (9.18)$$

where L_E is the effective channel length defined as $L_E = L - \Delta L$. Using Gauss law again inside the linear region results in a same differential equation as Equation 9.12. Solving Equation 9.12 with the new boundary condition of (9.18) yields

$$E_S = \frac{L_E}{2\lambda^2 - L_E^2}\left(\frac{t_G t_{OX} q(N_d + n)}{\varepsilon_{OX}} + V_G - V_{BI}\right) \tag{9.19}$$

Since λ is much less than L_E ($0.5 < \lambda < 5$ nm and $8 < L_E < 17$ nm) for most cases of this study, the term λ^2 in Equation 9.19 can be neglected with respect to L_E^2. Thus, Equation 9.19 can be reduced to

$$|E_S| \approx \frac{1}{L_E}\left(\frac{t_G t_{OX} qN}{\varepsilon_{OX}} + V_G + V_{BI}\right) \tag{9.20}$$

In addition, V_S can be computed as Equation 9.21 by integrating Equation 9.20 assuming the source voltage $V(0)$ equal to 0 V.

$$V_S \approx \frac{t_G t_{OX} qN}{\varepsilon_{OX}} + V_G + V_{BI} \tag{9.21}$$

9.3.1 SIMULATION RESULTS AND DISCUSSION

Figure 9.7 shows the electric field distribution in the lateral direction with the different drain voltages. The values for V_D have been chosen based on the saturation voltage V_S, which has been almost 0.36 V in this simulation. This figure shows that the electric field profile follows an exponential form depending on the distance from the drain. Equation 9.7 also highlighted this fact.

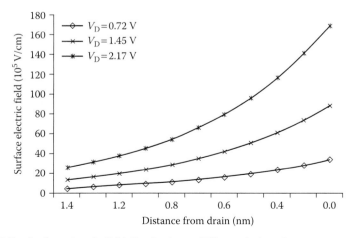

FIGURE 9.7 Surface electric field distribution at different drain voltages.

Figure 9.8 shows the saturation electric field at different drain–source voltages, channel thicknesses, and channel lengths. As can be seen, increasing V_D results in slight increase in E_S. In addition, this figure shows that by increasing t_{OX} and L, the saturation electric field increases and decreases, respectively.

Figure 9.9 depicts the effect of the drain voltage and channel length on the length of saturation region. The higher the drain voltage and longer the channel, the longer the ΔL. In addition, the figure shows that the ratio of $\Delta L/L$ increases as L decreases.

In Figure 9.10, it can be seen that increasing the oxide thickness causes increase in ΔL. In addition, the term $d(\Delta L)/d(V_D)$ increases as t_{OX} increases.

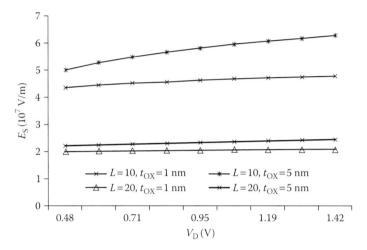

FIGURE 9.8 Saturation electric field at different drain voltages, oxide thicknesses, and channel lengths.

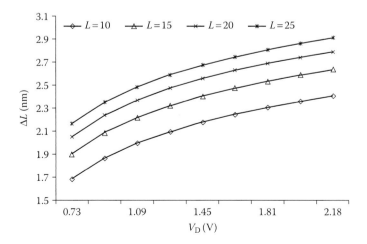

FIGURE 9.9 LVSR with different channel lengths and drain voltages.

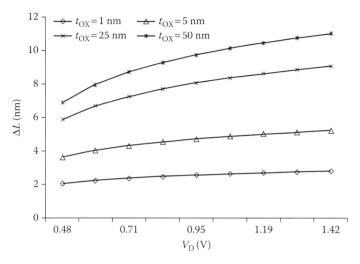

FIGURE 9.10 LVSR versus oxide thickness variations at different drain voltages.

9.4 TWO-DIMENSIONAL MODEL FOR SATURATION REGION OF DOUBLE-GATE GNR TRANSISTOR

A schematic cross section of DG-GNRFET is shown in Figure 9.11, where t_{OX} is the oxide thickness of front and back gates with a dielectric constant of ε_{OX}. t_G, ε_G, W, and L are the thickness, dielectric constant, width, and length of the GNR, respectively.

Generally, to model the potential distribution, Poisson's equation is solved [21].

$$\nabla^2 \Psi(x, y) = \frac{-qN_d}{\varepsilon_G}, \quad 0 \leq x \leq t_G, \quad 0 \leq y \leq L \quad (9.22)$$

where
$\psi(x, y)$ is the potential at any point (x, y) in the GNR
q is the electric charge magnitude
N_d is the doping concentration of the GNR

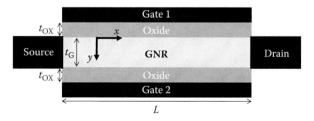

FIGURE 9.11 (**See color insert.**) Schematic cross section of a DG-GNRFET.

Ignoring the built-in potential of the source/drain channel junction [21], the boundary conditions of Equation 9.1 is defined as $\psi(0, 0) = 0$ and $\psi(0, L) = V_{DS}$. In addition, as the electric flux along the front and back GNR/oxide interface is continuous, the potential function must satisfy

$$\frac{\partial \Psi(x, y)}{\partial x}\bigg|_{x=0} = \frac{\varepsilon_{OX}}{\varepsilon_G} \times \frac{\Psi(0, y) - V_{g1}}{t_{OX}} \tag{9.23}$$

and

$$\frac{\partial \Psi(x, y)}{\partial x}\bigg|_{x=t_G} = \frac{\varepsilon_{OX}}{\varepsilon_G} \times \frac{V_{g2} - \Psi(t_G, y)}{t_{OX}} \tag{9.24}$$

where
$V_{g1} = V_{GS1} - \Phi_{FB1}$
$V_{g2} = V_{GS2} - \Phi_{FB2}$
V_{GS1} and V_{GS2} are gate–source voltages for front and back gates, respectively
Φ_{FB1} and Φ_{FB2} are front and back flat band voltages, respectively

According to Refs. [21,33], $\psi(x, y)$ can be decomposed into two parts:

$$\Psi(x, y) = V(x) + U(x, y) \tag{9.25}$$

where
$V(x)$ is the 1D solution of the Poisson equation

and

$$\frac{\partial^2 V(x)}{\partial x^2} = \frac{-qN_d}{\varepsilon_G} \tag{9.26}$$

which accounts for long-channel effects and $U(x, y)$ is the solution of Poisson equation, which deals with 2D short channel effects. Using Equations 9.25 and 9.26 in Equation 9.1, $U(x, y)$ satisfies the Laplace equation

$$\frac{\partial^2 U(x, y)}{\partial x^2} + \frac{\partial^2 U(x, y)}{\partial y^2} = 0 \tag{9.27}$$

As Equation 9.25 shows, the boundary conditions of $\psi(x, y)$ can be also written in two parts. Therefore, by separating Equations 9.21 and 9.22 the boundary conditions can be expressed as

$$\frac{\partial V(x)}{\partial x}\bigg|_{x=0} = \frac{\varepsilon_{OX}}{\varepsilon_G}\left(\frac{V(0) - V_{g1}}{t_{OX}}\right) \tag{9.28}$$

and

$$\frac{\partial V(x)}{\partial x}\bigg|_{x=t_G} = \frac{\varepsilon_{OX}}{\varepsilon_G}\left(\frac{V_{g2} - V(t_G)}{t_{OX}}\right) \tag{9.29}$$

Solving Equation 9.26 with the boundary conditions of Equations 9.28 and 9.29 yields

$$V(0) = V_0 = \frac{C_G}{2C_G + C_{OX}} \left[V_{g2} + V_{g1} \left(\frac{C_{OX}}{C_G} + 1 \right) + qN_d t_G \left(\frac{1}{C_G} + \frac{1}{C_{OX}} \right) \right] \quad (9.30)$$

In this chapter, we study the device at zero gate bias condition, $V_{GS1} = V_{GS2} = 0$; therefore, Equation 9.30 can be reduced to

$$V_0 = \frac{qN_d(t_G C_{OX} + \varepsilon_G)}{C_{OX}(2C_G + C_{OX})} \quad (9.31)$$

The boundary conditions for Equation 9.27 are expressed as

$$U(0,0) = -V_0 \quad (9.32)$$

$$U(0,L) = V_{DS} - V_0 \quad (9.33)$$

$$\left. \frac{\partial U(x,y)}{\partial x} \right|_{x=0} = \frac{\varepsilon_{OX}}{\varepsilon_G} \times \frac{U(0,y)}{t_{OX}} \quad (9.34)$$

$$\left. \frac{\partial U(x,y)}{\partial x} \right|_{x=t_G} = -\frac{\varepsilon_{OX}}{\varepsilon_G} \times \frac{U(t_G,y)}{t_{OX}} \quad (9.35)$$

The solution of Equation 9.27 can be obtained by the separation of variables method [33]. The solution at the surface is given by the following exponential series:

$$U(0,y) = \sum_{n=1}^{\infty} A_n \exp(\lambda_n y) + B_n \exp(-\lambda_n y) \quad (9.36)$$

where

$$A_n = \frac{(V_{DS} - V_0)\exp(-\lambda_n L) + V_0 \exp(-2\lambda_n L)}{1 - \exp(-2\lambda_n L)} \quad (9.37)$$

$$B_n = -V_s - A_n \quad (9.38)$$

λ_n is potential variations parameter defined as

$$t_G \lambda_n = \left(\frac{C_G}{2C_{OX}} \right) \left[(t_G \lambda_n)^2 - \left(\frac{C_{OX}}{C_G} \right)^2 \right] \tan(t_G \lambda_n) \quad (9.39)$$

As t_G is a small value (in order of 10^{-9}), $\tan(t_G\lambda_n)$ can be approximated to $t_G\lambda_n$. Thus, λ_n is given by

$$\lambda_n = \frac{1}{t_G}\sqrt{1 + \frac{2C_{OX}}{C_G}} \quad (9.40)$$

In order to get a simple solution of $U(0, y)$, it can be approximated to only the first term ($n = 1$) of the series in Equation 9.36 according to [21]. Thus, Equation 9.36 reduces to Equation 9.37 taking into account our approximation that $\exp(-x) \approx 0$ for $x > 3$. This approximation is justified if y and $L > 3\lambda$, which is relevant in this study.

$$U(0, y) = (V_{DS} - V_0)e^{\lambda(y-L)} - V_0 e^{-\lambda y} \tag{9.41}$$

Thus, using Equation 9.25, the surface potential along y can be written as

$$\Psi(0, y) = V_0 + (V_{DS} - V_0)e^{\lambda(y-L)} - V_0 e^{-\lambda y} \tag{9.42}$$

In addition, the lateral electric field along the channel can be expressed as derivation of Equation 9.42 over y.

$$E(0, y) = -\lambda \left[(V_{DS} - V_S)e^{\lambda(y-L)} + V_0 e^{-\lambda y} \right] \tag{9.43}$$

By taking $y = L - L_D$ and $\psi(0, y) = V_{sat}$, and solving Equation 9.42 for L_D, we have

$$L_D = L - \frac{1}{\lambda} \ln \left(\frac{\dfrac{(V_{DS} - V_S)}{e^{\lambda L}} e^{2\lambda(L - L_d)} - V_0}{V_{sat} - V_0} \right) \tag{9.44}$$

which can be solved numerically. In Equation 9.23, V_{sat} and L_D are drain saturation voltage and LSVR, respectively.

The proposed equations simply explain the relations of surface potential, electric field, and length of saturation region with t_{OX}, t_G, V_{DS}, and L.

9.4.1 SIMULATION RESULTS AND DISCUSSION

In this section, the profile of surface electric field and potential variation is shown. In addition, the effect of several parameters such as drain–source voltage, oxide thickness, channel length, and doping concentration on the length of saturation region is studied.

For the purpose of model verification, we compared the calculated values using the proposed model with the simulated results by MEDICI simulation tool for a Si-based device. As shown in Figure 9.12, there is a good agreement between the simulation results and model at different doping concentrations, oxide thicknesses, and distances from the drain. Once the surface potential model is verified, the LVSR model is also proved, because it is the solution of the surface potential for $\psi(0, y) = V_{sat}$.

Figure 9.13a and b shows that the potential and electric field distribution along the nanoribbon surface are similar to the profile of surface potential in an abrupt

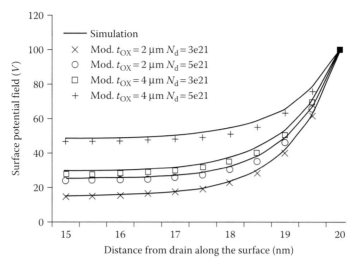

FIGURE 9.12 Comparison of the results extracted from 2D numerical simulator and model.

junction [22]. Figure 9.13c indicates that increasing the doping concentration in the channel region results in a significant increase in the electric field near the drain junction. Doping concentration has been set to be in order of $10^{25}\,\mathrm{m}^{-3}$ to be an influential factor in the electric field and length of saturation region.

Figure 9.13d shows that expanding the oxide thickness causes decrease in LVSR. This is because, expanding the oxide thickness results in higher saturation voltage and lateral electric field and thus a decrease in LVSR. It is worth mentioning that the calculation of LVSR is done for $V_{DS} > V_{sat}$. Therefore, wherever $V_{DS} < V_{sat}$, there is a missing point in the charts. For example, in Figure 9.13b and f, LVSR cannot be calculated for a few V_{DS}es.

Figure 9.13e shows the dependence of L_D on L. As can be seen in this figure and Equation 9.23, there is a direct relation between L and L_D. Finally, Figure 9.13f shows that by applying higher doping concentration, the saturation voltage increases and L_D decreases.

9.5 SUMMARY

Three analytical models for surface potential, electric field, and LVSR of double-gate- and single-gate GNR transistors were proposed. In addition, the behavior of the GNR FETs in the saturation region was investigated using the proposed models. Furthermore, using the presented models, the effects of device parameters such as ribbon thickness, doping concentration, and channel length were examined. As expected from the small geometry of the device, high lateral electric field was seen near the drain junction, being an issue for reliability of these devices. In addition, LVSR was found to be more than half of the given channel length in some cases.

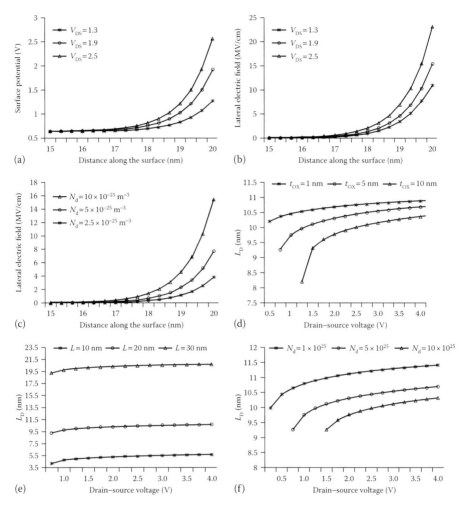

FIGURE 9.13 The effect of several parameters on the profile of surface potential, electric field distribution, and LSVR. Typical parameters are $N_d = 5 \times 10^{25}$ m^{-3}, $L = 20$ nm, $t_{OX} = 1$ nm, and $t_G = 0.4$ nm.

REFERENCES

1. Lemme, M.C., et al. A graphene field-effect device. *IEEE Electron Device Letters*. 2007. 28(4): 282–284.
2. Ghadiry, M.H., et al. Modelling and simulation of saturation region in double gate graphene nanoribbon transistors. *Semiconductors Journal*. 2011. 46(1): 126–129.
3. Ghadiry, M.H., A.K. A'Ain, and M. Nadisenejani. Design and analysis of a novel low PDP full adder cell. *Journal of Circuits, Systems, and Computers*. 2011. 20(3): 439–445.
4. Liang, G., et al. Computational study of double-gate graphene nano-ribbon transistors. *Journal of Computational Electronics*. 2008. 7(3): 394–397.
5. Ghadiry, M.H., et al. Design and analysis of a new carbon nanotube full adder cell. *Journal of Nanomaterials*. 2011(36): 1–6.

6. Berger, C., et al. Electronic confinement and coherence in patterned epitaxial graphene. *Science*. 2006. 312(5777): 1191–1196.

7. Giovannetti, G., et al. Doping graphene with metal contacts. *Physical Review Letters*. 2008. 101: 026803.

8. Fang, T., A. Konar, and D.J.H. Xing. Carrier statistics and quantum capacitance of graphene sheets and ribbons. *Applied Physics Letters*. 2007. 91(9): 092109.

9. Meric, I., et al. Current saturation in zero-bandgap, topgated graphene field-effect transistors. *Nature Nanotechnology*. 2008. 3: 654–659.

10. Zhang, Q., et al. Graphene nanoribbon tunnel transistors. *IEEE Electron Device Letters*. 2008. 29(12): 1344–1346.

11. Cheli, M., G. Fiori, and G. Iannaccone. A semianalytical model of bilayer-graphene field-effect transistor. *IEEE Transactions on Electron Devices*. 2009. 56(12): 2979–2986.

12. Fiori, G. and G. Iannaccone. Ultralow-voltage bilayer graphene tunnel FET. *IEEE Electron Device Letters*. 2009. 30(10): 1096–1098.

13. Schütz, A., S. Selberherr, and H.W. Pötzl. A two-dimensional model of the avalanche effects in MOS transistors. *Solid-State Electronics*. 1982. 25(3): 177–183.

14. Wong, H. A physically-based MOS transistor avalanche breakdown model. *IEEE Transactions on Electron Devices*. 1995. 42(12): 2197–2202.

15. Arora, N.D. and M.S. Sharma. MOSFET substrate current model for circuit simulation. *IEEE Transactions on Electron Devices*. 1991. 38(6): 1392–1398.

16. Fang, F.F. and A.B. Fowler. Hot electron effects and saturation velocities in silicon inversion layers. *Journal of Applied Physics*. 1970. 41(4): 1825–1831.

17. Frohman-Bentchkowsky, D. and A.S. Grove. Conductance of MOS transistors in saturation. *IEEE Transactions on Electron Devices*. 1969. 16(1): 108–113.

18. Baum, G. and H. Beneking. Drift velocity saturation in MOS transistors. *IEEE Transactions on Electron Devices*. 1970. 17(6): 481–482.

19. Gildenblat, G., et al. PSP: An advanced surface-potential-based MOSFET model for circuit simulation. *IEEE Transactions on Electron Devices*. 2006. 53(9): 1979–1993.

20. Wong, H. and M.C. Poon. Approximation of the length of velocity saturation region in MOSFET's. *IEEE Transactions on Electron Devices*. 1997. 44(11): 2033–2036.

21. Imam, M.A., M.A. Osman, and A.A. Osman. Threshold voltage model for deep-submicron fully depleted SOI MOSFETs with back gate substrate induced surface potential effects. *Microelectronics Reliability*. 1999. 39(4): 487–495.

22. Yang, W., et al. A novel analytical model for the breakdown voltage of thin-film SOI power MOSFETs. *Solid-State Electronics*. 2005. 49(1): 43–48.

23. Guo, Y., W. Zhigong, and S. Gene. A three-dimensional breakdown model of SOI lateral power transistors with a circular layout. *Journal of Semiconductors*. 2009. 30(11): 114006.

24. El-Mansy, Y.A. and A.R. Boothroyd. A simple two-dimensional model for IGFET operation in the saturation region. *IEEE Transactions on Electron Devices*. 1977. 24(3): 254–262.

25. Cheli, M., P. Michetti, and G. Iannaccone. Model and performance evaluation of field-effect transistors based on epitaxial graphene on SiC. *IEEE Transactions on Electron Devices*. 2010. 57(8): 1936–1941.

26. Krizaj, D., G. Charitat, and S. Amon. A new analytical model for determination of breakdown voltage of resurf structures. *Solid-State Electronics*. 1996. 39(9): 1353–1358.

27. El Banna, M. and M. El Nokali. A pseudo-two-dimensional analysis of short channel MOSFETs. *Solid-State Electronics*. 1988. 31(2): 269–274.

28. Stauber, T., N.M.R. Peres, and F. Guinea. Electronic transport in graphene: A semiclassical approach including midgap states. *Physical Review B*. 2007. 76: 205423.

29. Wong, H. Drain breakdown in submicron MOSFETs: A review. *Microelectronics Reliability*. 2000. 40(1): 3–15.

30. Cheli, M., P. Michetti, and G. Iannaccone. Model and performance evaluation of field-effect transistors based on epitaxial graphene on SiC. *IEEE Transactions on Electron Devices*. 2010. 57(8): 1939–1941.

31. Kolhatkar, J.S. and A.K. Dutta. A new substrate current model for submicron MOSFETs. *IEEE Transactions on Electron Devices*. 2000. 47(4): 861–863.

32. Singh, A.K. Study of avalanche breakdown (MI) mode in sub micron MOSFET device. *Microelectronics International*. 2005. 22(1): 16–20.

33. Harper, P.G. and D.L. Weaire. *Introduction to Physical Mathematics*. 1985, Cambridge, Cambridge University Press.

10 Silicon Nanowire Field Effect Transistor Modeling

Amir Hossein Fallahpour and
Mohammad Taghi Ahmadi

CONTENTS

10.1   INTRODUCTION

Silicon is a wonderful material for current electronic devices and there has been deep research on various aspects, both fundamental as well as applications. Nowadays, it has become possible to produce silicon nanowires with a control on diameter. Metal–oxide–semiconductor field effect transistor (MOSFET) scaling over the past years has enabled us to pack millions of MOS transistors on a single chip. Gordon Moore predicted that the number of transistors capable of being integrated would double every 2 years. According to the International Technology Roadmap for Semiconductors (ITRS), the scaling of complementary metal–oxide–semiconductor (CMOS) technology will continue for at least another decade. After 40 years of advances in integrated circuit (IC) technology, the scaling of silicon MOSFET has entered the nanometer dimension with the introduction of 22 and 45 nm process technology that offers low power, high-density, and high-speed generation of processor with the latest technological advancement. Currently, chips are being made with gate lengths ranging from 22 to 45 nm, and still more scaling is expected. In spite of the 15 nm gate lengths having been scheduled for production by the end of this decade, the 6 nm gate length p-channel FET has been recently applied in devices demonstrated by Intel, AMD, and IBM. Such one-dimensional (1D) structures have the potential to be used for the development of very small silicon-based devices such as sensors, logical gates, memories, and systems for optoelectronic and biological applications. Molecular level detection of gases and viruses, single electron transistors, memories, light sources, and logical circuits are among the applications that have been already achieved. Nevertheless, the scaling of the Si MOSFET below 20 nm may soon meet its fundamental physical limitations. This threshold makes possible the application of novel devices and structures such as nanowires in new physics as the future nanoelectronics due to their unique electronic properties. IC technology has been regarded as one of the most important inventions in engineering history. The tremendous progress in IC technology in the past four decades has become the driving power of the information technology (IT) revolution, which has marvelously changed our lives and the whole world. The secret of the miracle in IC technology originates from certain factors including scaling down the dimension of each transistor, taking advantage of the basic element of ICs, and increasing the total number of transistors in one IC chip. To further reduce the size of active devices such as field-effect transistors (FETs), the emergence of new structures and materials is highly required. Nanowire FETs illustrate this miniaturization trend. Silicon nanowire (SiNW) FETs as extremely downscaled MOSFETs are promising candidates to replace traditional bulk-devices.

According to ITRS, MOSFETs will reach the sub-10 nm dimensions by 2018. There is, however, a consensus that new device architectures and materials are needed to continue the downscaling after the 45 nm technology node. The ultimate goal of device modeling and characterization is to predict the properties and performance of the device through modeling and simulations.

10.1.1 SCALING OF CMOS DEVICES AND MODELING

In IC technology, the downscaling of CMOS device dimensions has been constantly pursued over the last 40 years to increase the packing density and the performance of ICs. According to the ITRS, the MPU physical gate length will be scaled down to as small as 7 nm by the year 2018 [1]. As the device characteristic enters into the nanoscale, we are likely to deal with various challenges in the modeling and simulation of carrier transport in MOSFET devices. First, a three-dimensional (3D) simulation capability is strongly necessary since the conventional planar, single-gate structure is slowly being replaced by some kinds of nonplanar, multiple-gate structures to cover up various short-channel effects such as the threshold voltage roll-off and the drain-induced barrier lowering (DIBL) [2]. In addition, we must include quantum mechanical effects in the calculation of carrier transport because of the characteristic of MOSFET devices being comparable to the size of the wave packet of electrons. Since most of the currently available transport models for device simulation are derived from the semi-classical Boltzmann transport equation [3,4], they have to be accurate to either take in some important features of the quantum effects or be completely replaced by more exact quantum transport models. In addition, since the channel length of nanoscale MOSFET devices becomes close to the mean free path, a significant part of electrons injected from the source to the channel does not experience scattering before they exit the channel. Then, the carrier transport is quasi-ballistic, and the carrier distribution function in the channel in the operation condition is far from that of the equilibrium condition. Accordingly, the information contained in the macroscopic variables is not enough to reproduce the highly nonequilibrium distribution function observed in the quasi-ballistic transport condition, and it is very complicated to find reasonable results from the macroscopic transport models [5–7].

10.1.2 MOS TRANSISTOR

The basic concept of the FET was invented in 1930 [8]. Since the 1960s, the development has increased. The Si MOSFET was used in ICs [9]. Device dimensions have been progressively shrinking, and circuit complexity and industry progress have also been growing exponentially. This quick scaling has brought Si technology to the point where various fundamental physical phenomena are beginning to obstruct the path to further progress. These effects include tunneling through the gate insulator, quantum confinement issues, tunneling through the bandgap, discrete atomistic effects in the doping and at interfaces, interface scattering, and thermal problems associated with very high power densities. Because of these complexities, many new changes to the basic MOSFET technology are being explored, including the discovery of new gate geometries and multiple gates, the use of extremely thin channel, the use of strain to increase mobility, and application of fully depleted silicon layers for the FET channel. However, power constraints should be taken into consideration with respect to the new devices. In fact, for these new devices, power dissipation and the temperature increase due to very high power densities creates problems, showing that each technology design point must be optimized to gain maximum performance [10].

The MOSFET is the basic structure for very large scale integration (VLSI) circuits and microprocessors. The MOSFET is typically referred to as unipolar or majority carrier device since current in a MOSFET is primarily transported by carriers of one polarity (e.g., electrons in an n-channel device). An n-channel (p-channel) MOSFET consists of a p-type (n-type) substrate with two n+ (p+) areas (the source and the drain). The gate electrode is generally made of doped polysilicon and is disconnected from the semiconductor channel by a thin gate oxide layer. Application of a positive (negative) voltage at the gate terminal for an n-channel (p-channel) MOSFET leads to an increase of negative (positive) charge at the oxide–silicon interface, making a channel for current conduction. The transistor effect is reached by adjusted gate electric field using gate-source bias to induce a MOS transistor to be commonly known as the "field-effect transistor."

10.1.3 CMOS Technology Overview

In bulk CMOS technology, the gates are n-type and p-type polysilicon and are topped with a metal silicide lowering the gate series resistance. The gates are patterned down to minimum dimensions that are 50%, or even smaller, of the general lithographic feature size gained via special lithographic and lateral etching techniques. Shallow trench isolation (STI) is used to separate the FETs resulting in a very high circuit density. A combination of deep and shallow implants is used for the source and drains, and these must be carefully engineered to be decreased. Shallow angled ion implants are used to create the so-called halo doping profiles being higher near the source and drain, and lower in the middle of the channel. Considering the fact that the halos are defined in relation to the edges of the gates, the average doping in the halo overlap region increases when the gate length shrinks. This rise in doping tends to compensate for the natural decrease of threshold voltage (V_T) that occurs in very short MOSFETs. This makes the use of FETs with shorter gate lengths possible, while in cases with no doping rise, shorter gate lengths would not likely happen [11]. To keep wire delay under control, a hierarchy of wiring sizes is usually used [12,13].

The floating body removes the dependency on source-to-substrate voltage of bulk MOSFET enhancing several types of circuitry.

10.1.4 Scaling of MOSFETs

It has been 45 years since the invention of the MOSFET. Through these years we have seen fast and stable improvement in the development of IC technology. The main driving force behind this progress has been the downscaling of the MOSFET's dimensions, which was first suggested in 1974 by Dennard et al. Starting with gate lengths of 10 μm in 1970, we have achieved about 32 nm in 2007 (presently known as 65 nm technology node) corresponding to roughly a 15% decrease each year [14]. Two kinds of scaling methods are usually used, constant voltage scaling and constant electric field scaling.

Constant voltage scaling is wholly a geometric approach where the power supply is kept constant, while the transistor's dimensions are scaled down by a factor α. However, decreasing the channel length and gate dielectric thickness raises the electric field in the channel. This primarily improves the mobility in the channel, but as the field starts to rise further than 1 MV/cm, the current gain decreases starts falling due to saturation of the carrier velocity.

The constant field scaling approach involves reducing the transistor's dimensions along with power voltage supply in order to maintain the electric field strength in the channel and to certify the same transistor physics and operation. Therefore, the power per transistor reduces quadratically; as a result, the power density (P/LW) stays constant instead of exploding as in a constant voltage scaling approach. However, the transistor's speed (fT) only rises linearly. Current density also increases linearly (I/W^2) rather than quadratically. Due to its profitability, the computer industry fundamentally pursued constant voltage scaling for the period 1973–1993. During 1993–2003, although the power supply voltage was subject to reduction due to the novel technology generation, this pace of technological advances did not fulfill the requirements of the constant electric field scaling approach [15]. It is clear that the drain voltage (V) has not been reducing as rapidly as channel length (L), which means that the transistor's electric fields have been rising rather than staying constant for a constant field scaling approach.

10.1.5 Scaling Issues and Approaches

Over the last three decades, by decreasing the transistor lengths and the gate oxide thickness along with the reduction in supply voltage, there has been a stable improvement in transistor performance, a reduction in transistor size, and a reduction in cost per function. As CMOS dimensions start approaching the nanometer regime (<100 nm), we take into consideration new effects in the device performance from new physical phenomenon. To keep up the rate of improvement in the device performance with constant downscaling, modifications to device fabrication and design are necessary. In particular, a collection of unwanted phenomenon problems occur, collectively called "short channel effects," which obstruct further improvement in transistor downscaling.

10.1.6 Short Channel Effect

One of the main challenges in transistor scaling is the "short channel effects" that become more noticeable with gate lengths smaller than 100 nm. As MOSFETs get smaller in dimensions, the source and drain regions become closer to each other. In a long channel device, the source and drain regions are far away from each other and so separated that their reduction regions have no effect on the potentiality in the central region under the gate. In a short channel device, however, the source–drain distance is comparable to the MOS reduction width in the vertical direction, and the source–drain potential has a powerful effect on the surface band bending under the gate [16]. Since some of the charge is shared among the source and the drain in the channel, the net effective charge being controlled via the gate is reduced, which results in lowering the threshold voltage. It can be observed that in short channel lengths, the electric field under the gate has a 2D nature. Regarding

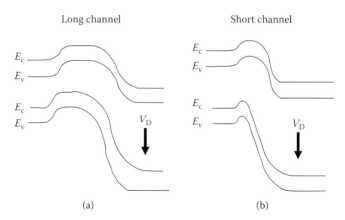

FIGURE 10.1   Band diagram for (a) a long channel and (b) a short channel device.

the points in the lower parts of the scale, new techniques are necessary to obtain sufficient control and reproducibility in the threshold voltage for subsequent technology nodes. This is related to the fact that a short channel effect occurs when a high drain bias is applied to a short channel device making the barrier height at the source end of the channel affected by the drain bias, which decreases the threshold voltage as seen in Figure 10.1. This effect is recognized as DIBL. For the short channel device, the drain voltage has an effect on the electron barrier height, which appears as a variation in the threshold voltage with drain bias. This effect gets worse as the channel length decreases, which leads to the punch through condition when an extreme drain current flows and the gate loses all control. Since the power supply voltage is also scaled down following the constant field scaling approach, there is a reduced gate voltage swing available so that the net effect of the DIBL phenomenon is to cause increased off-current leakage for the transistor.

A third short channel effect is related with the transistor's turnoff characteristics. The subthreshold slope (SS) is identified as the state in which the changes in gate voltage require producing a decade change in drain current (units of mV/dec). As we scale down the transistor with reducing channel lengths, the SS degrades (enlarge in magnitude from its theoretical minimum of 60 mV/dec [16]). Ideally, the SS should be as small as possible to rapidly turn on or off the transistor, but the device's physics limits it to about 60 mV/dec. SS degradation causes enlarged off-state leakage current and induces an extra process variability in device and circuit fabrication.

10.1.6.1 Gate Oxide Thickness

Following the constant field scaling approach, the gate oxide SiO_2 thickness is scaled down in conjunction with gate length. This downscaling aids to keep short channel effects under control and to maintain electrostatic integrity. However, an oxide thickness of 1.2 nm, which is used in the 90 nm logic technology, contains only a five-atom thick oxide layer, indicating the approach of a physical boundary beyond which the carrier tunneling current through the gate rises dramatically. The gate oxide tunneling current rises exponentially as the gate oxide thickness decreases, so that it approaches the drain on-current (I_{on}) at an oxide thickness of 1 nm. This is clearly an undesirable circumstance; the gate oxide leakage current is an unwanted parasitic current. Another issue related with a very thin gate oxide is the loss of inversion charge, which leads to smaller gate capacitance and hence smaller transconductance [17]. Quantum mechanics says that the peak of the inversion charge density lies at a short distance from the Si–SiO$_2$ surface. This reduces the depletion capacitance and effectively decreases the total gate capacitance. A third effect known as the polysilicon gate depletion effect also occurs with a thinner gate oxide. A thin space charge forms in the heavily doped polysilicon gate close to the gate oxide surface, which acts to reduce the overall gate capacitance [17]. For a polysilicon doping of 10^{20} cm^{-3} and a 2 nm oxide, roughly 20% of the inversion charge is lost at the 1.5 V gate due to the combined effects of polysilicon gate depletion and inversion-layer quantization [17].

10.1.6.2   New Materials

Scaling of MOSFET transistors has led to enhanced performance due to gate length reduction. However, intrinsic semiconductor properties such as electron and hole mobility for the silicon lattice cannot be scaled. Beyond the 90 nm technology node, novel innovations in transport have been thought of to enhance the channel carrier mobility and the MOSFET performance. One approach has been to utilize strain using the silicon grown on an underlying SiGe layer [18].

In strained silicon, a layer of Si$_x$Ge$_{1-x}$ is primarily grown during epitaxial growth by adding a few Ge atoms near the wafer's crystalline surface. Since Ge has a bigger lattice constant (5.65 Å) than Si (5.4 Å), the produced crystal structure is larger as a subsequent silicon layer grown on the SiGe is strained. The top Si layer grown over the Si$_x$Ge$_{1-x}$ surface is in tensile strain as the Si atoms try to align according to the expanded lattice. This strain in the Si causes a decrease in the effective electron (hole) mass and a rise in electron (hole) mobility. This increased mobility enhances the transistor's switching delay and leads to a larger transconductance and higher drain current. An increase of 10–20% in the MOSFET's drive current with strained silicon technology for a 50 nm length channel has also been reported [18]. Problems related to the thinning of the oxide layer presents one of the major challenges for continued MOSFET downscaling, since the physical restrictions of oxide scaling are being approached with only few atomic layers for the current technology node. Therefore, it is essential to use an alternate gate dielectric to solve the rising gate leakage current problem associated with the thinning gate oxide (SiO$_2$). Using a high-k dielectric allows the application of a thicker dielectric, while providing similar gate capacitance and so equivalent transistor performance. The gate capacitance for a parallel plate capacitor is $C = (k\varepsilon_0 A)/d$, where k is the native dielectric constant (=3.9 for SiO$_2$), ε_0 is permittivity of free space (=0.0885 pF/cm), A is the area of the capacitor, and d is the dielectric thickness. To keep the same gate capacitance, the same k/d ratio is needed. Thus, using a high-k dielectric allows a thicker dielectric given by $d_{high-k} = d_{ox} \times (k_{high-k}/3.9)$. For a high-$k$ dielectric with $k_{high-k} = 16$ and $d_{ox} = 1$ nm, the thickness for the high-k of $d_{high-k} \sim 4$ nm. After about a decade of research, hafnium-oxide-based materials such as HfO$_2$, HfSi$_x$O$_y$, HfO$_x$N$_y$, and HfSi$_x$O$_y$N$_z$ ($k_{HfO_2} \sim 25$) have appeared as leading candidates to replace SiO$_2$ gate dielectrics in advanced CMOS applications [19]. Downscaling the gate dielectric also requires replacing polysilicon as the gate electrode material. Metal gate technology involves no poly-depletion effects, and offers much better threshold voltage control [19]. Over the last few years, research in metal gate technology has recognized several promising candidates such as W, Ti, Mo, Nb, and their derivatives such as WN, TiN, and TaN [20]. An option to metal gates is to fabricate fully silicided gates. It involves exchanging the poly-Si into silicides, which are in direct contact with the gate dielectric after their fabrication, e.g., MoSi, WSi, TiSi, HfSi, PtSi, CoSi, and NiSi [19]. Nowadays, some companies are planning to introduce *high-k dielectric* and *metal gates* for their 45 nm technology node [21].

10.1.7 ISSUES AT NANOSCALE LEVEL

In decreasing the length of MOSFETs beyond 50 nm, technology requires additional innovations to deal with barriers forced by fundamental physics. The classical approach used to scale the conventional MOSFET starts to fail at such a small scale, and new issues emerge such as short channel effects, which are important to overcome in order to continue the scaling trend. The issues most often mentioned are

1. Current tunneling due to thin gate oxide
2. Quantum mechanical tunneling of carriers from the source to the drain [22,23]
3. Threshold voltage increase due to quantum confinement [24]
4. Random dopant-induced fluctuations

As the gate length is decreased to around 10 nm, gate control over the channel region is reduced, and there is an increased source–drain tunneling of electrons. This leads to an increased off-current and degradation in the SS. It is still an arguable topic to find out whether source–drain tunneling or the device's electrostatics degradation is the limiting factor for scaling. A simulation study by Wang and Lundstrom [23] showed that source–drain tunneling might set the scaling limit well below 10 nm. MOSFETs with gate lengths around 10 nm need to have a thinner channel layer to ensure adequate device turnoff. With new device designs such as ultrathin body (UTB) FETs, where the MOSFET is fabricated on an extremely thin silicon layer on an oxide substrate (SOI), it is very important to have a body thickness below 10 nm to maintain electrostatic integrity. Due to quantum confinement effects in UTB FETs, the threshold voltage rises with reducing channel width. The quantum mechanical narrow channel effect occurs because electrons in the inversion layer are not only located away from the surface, but also occupy separate energy levels in the channel. A larger surface potential is needed to populate the inversion layer, which increases the threshold voltage [17]. The classical threshold condition $\psi_s = 2 \times \psi_b$ (ψ_s = surface potential) can be modified to $\psi_s = 2 \times \psi_b + \Delta\psi_{QM}$ to consist of the shift of inversion charge density away from the surface, where $\Delta\psi_{QM} = (kT/q)\ln(Q_i^{CL}/Q_i^{QM})$ [17]. Thus, the threshold voltage shift because of the quantum effect can be calculated as $\Delta V_t = dV_g / d\psi_s \times \Delta\psi_{QM} = m \times \Delta\psi_{QM}$, where m is the body-effect coefficient normally between 1.2 and 1.5 [25]. For a thin SOI silicon film thickness, the quantum effect becomes significant as the silicon thickness is decreased below ~5 nm. Random fluctuation of the number of dopant atoms in the channel was predicted as a limiting factor in transistor scaling back in the 1970s [26]. Recall that constant field scaling requires the substrate doping to increase at the rate of scaling factor α. However, in a deviation from constant field scaling, undoped channels have been adopted for FETs below 100 nm gate lengths due to a statistical variation in the doping level, known as random dopant induced fluctuation. Random dopant fluctuation compared with the conventionally doped for 24 MOSFETs with various random "atom" distributions for $L = 100$ nm, $W = 50$ nm, $t_{ox} = 30$ Å, and an average uniform substrate doping of 8.6×10^{17} cm^{-3} [25]. Variation in threshold voltage can be analytically shown to be [40]

$$\sigma_{Vt} = \frac{q}{C_{OX}} \sqrt{\frac{N_a \times W_{dm}}{3 \times L \times W}} \times \left(1 - \frac{X_s}{W_{dm}} \right)^{\frac{3}{2}}$$

where
W_{dm} is maximum gate depletion width
X_s is the width of low impurity region
N_a is substrate doping
L is the channel length
W is the channel width
C_{OX} is oxide capacitance

While channel doping previously has been used to adjust the MOSFET's threshold voltage, metal gates with suitable work function are now used to adjust the threshold voltage.

10.2 NONCLASSICAL DEVICE STRUCTURES

As the MOSFET scaling process continues, ITRS expects that the semiconductor industry would require channel lengths in the range of around 10 nm by 2015 [14]. In addition to the introduction of new materials and increasing bulk MOSFET performance, new device concepts are required to continue scaling into the 10 nm gate length [27,28]. Advanced MOSFET structures such as UTB FET, FinFET, dual-gate FET (DGFET), gate-all-around (GAA) FET, and tri-gate FET suggest the opportunity to continue scaling beyond the bulk because they provide a sharper SS, reduced short channel effects, and better carrier transport.

10.2.1 Ultra Thin-Body Single Gate MOSFET

In this section, the UTB single gate MOSFET device structure is illustrated. The fundamental concept of this device is implied as having a thin silicon channel with an underlying insulation oxide to eliminate the leakage current paths through the substrate and to decrease parasitic capacitances resulting in an increase in the device's speed. Since almost all the off-state current flows through the bottom of the body, it is desired to change the semiconductor substrate with an insulating dielectric. However, thicker source and drain structures are necessary to minimize parasitic source/drain series resistance [29]. Increased poly-Si S/D contact areas were employed to decrease the parasitic resistance of the device [30]. Drain currents of 400 µA/µm are achieved at a drain voltage of 1 V and $V_gV_t = 2$ V.

10.2.2 DUAL-GATE FET/FINFET

A dual-gate device FET structure allows for more aggressive device scaling, since short channel effects are further suppressed by doubling the effective gate control. There have been several variations suggested for DGFET structure, but the majority of them suffer from process complexity, between these, the FinFET appears as the most practical design. The channel consists of a thin vertical fin which is covered by the gate on three sides. The FinFET is a double gate FET as the gate oxide is thin on the vertical sidewall, but is wide on the top. The fin width is an important parameter for the device as it identifies the body thickness and the short channel effects are dependent on it. For efficient gate control, it is required that the fin width is half of the gate length or less [31]. Due to the vertical nature of a FinFET channel, it has (110) oriented surfaces when fabricated on a standard (100) wafer. This crystal orientation leads to increased hole mobility, but decreased electron mobility [31]. The main advantage of the FinFET over the planar MOSFET is that it offers reduction of short channel effects. It is clear from the figures that a DGFET gives a more ideal SS and better DIBL characteristics.

10.2.3 OMEGA-GATE FET/TRI-GATE FET

Omega-gate and tri-gate FETs are multi-gate transistors having three-sided gate structures. Omega-gate FET has the gate increasing into the substrate on the sides, making an effective fourth gate, which gives better gate control than a tri-gate FET. For future device designs, companies are working with tri-gate FET design as well as the omega-gate FET [32,33].

10.3 SILICON NANOWIRE TECHNOLOGY

Semiconductor nanowires are single crystal structures with a diameter of nanometers showing interesting and new properties because of their 1D feature and confinement. The nanowire approach to nanoscale MOSFET fabrication offers the possibility for ultimate scaling of the MOSFET using GAA-FET device structures. Since the addition of more gates (two or more) improves MOSFET performance, as well as reducing short channel effects, the GAA FET is attractive for extremely short channel MOSFET. A number of attempts have been made at fabricating distinct silicon nanowire transistors using both GAA device and back gate. ICs using nanowires have successfully been fabricated [34]. Since 2000, much research work on the fabrication of nanowire devices has been carried out. In addition, the diameter of silicon nanowire can be controlled during the synthesis of nanowires, multishell Si–Ge nanowire heterostructures, and also

ICs. A nanowire device was grown via vapor–liquid–solid (VLS) mechanism using gold nanoclusters as the catalysts. This fabrication process using Au nanoclusters led to high-quality single-crystal nanowire growth with a fine controlled diameter. Silane (SiH_4) was used as the vapor phase reactant [35]. The fabricated nanowires were deposited onto oxidized silicon substrates with electrodes separated about 800–2000 nm. Thermal annealing was done to enhance the contact and passivate the Si-SiO_x traps. A significant observation from the outcome is that there is an improved mobility in nanowire devices. Hole mobility is estimated to be 1000 cm^2/Vs, which is noticeably bigger than bulk hole mobility (450 cm^2/Vs). Silicon nanowire devices also show higher transconductance and more ideal SS behavior. The relationship between converted nanowire MOSFET data and a bulk silicon device shows improvement on current, decrease off-current, smaller SS, and enhanced transconductance. There have been reports of the fabrication of GAA nanowire MOSFET devices [36–38]. The GAA structure is reported to lead to enhanced gate control and finer short channel performance. To form a silicon fin structure, active areas were etched out down to the buried oxide. The patterned silicon was then oxidized in dry O_2, which produced two nanowire cores, one at the top and another at the bottom of fin. The top nanowire etches out and the bottom one releases from the underlying oxide with a wet etching process. The release was followed by a 130 nm α-Si deposition and 9 nm gate oxide to form the gate dielectric and polysilicon gate electrode. After three decades of downscaling, CMOS technology continues to be the leading technology in today's era of VLSI. The limits of silicon scaling are the major challenges being faced today by quadruple-gate MOSFET fabrication technologies. ITRS predicts that MOS transistors in the year 2020 will have gate lengths of 9 nm [39]. Further nanoelectronic studies intended to understand device physics and review their ultimate scaling limits by simulation work will be useful. Taking advantage of semiclassical models with quantum corrections, triangular wire transistors and triple gate FETs with large cross sections have been recognized (i.e., the Si body thicknesses are more than 30 nm). However, full quantum effect simulations are needed to simulate SiNWTs with gate lengths below 10 nm. Currently, processor chips are approaching 100 million transistors with waste rating of over 100 W. This power is already near the battery life limit. The new technology development decreases logic transition energy limits in order of magnitude of k_BT, where k_B is Boltzmann's constant, 1.38×10^{-23} J/K, and T is absolute temperature in 0 K.

10.3.1 STRUCTURAL CHARACTERIZATION OF SiNW

Silicon nanowires are single crystalline nanostructures with the same diameters. High-resolution transmission electron microscopy (HRTEM) is used to give a detailed illustration of the structures of these nanowires [40,41]. For diameters between 3 and 10 nm, 95% of the grown SiNWs are located along the <110> direction; for diameters between 10 and 20 nm, 61% of the grown SiNWs are along the <112> direction; and for diameters between 20 and 30 nm, 64% of the grown SiNWs are along the <111> direction. These outcomes show a clear preference for growth of the smallest SiNWs and largest SiNWs along the <110> direction and the <111>

direction, respectively [42]. The nanowire surface energy has a key role in determining the growth direction for smaller diameter SiNWs [43].

10.3.2 SiNWs for Nanoelectronics

One of the influential strategies for creation and integration of nanodevices is crossed nanowire (cNW) architecture. The nanowires within a crossed array can behave as active devices or interconnects. Based on the cNW configuration, diodes and transistors, two important elements for logic, can be fabricated [44,45].

10.4 SILICON NANOWIRE MODELING

Semiconductors are the most technologically important devices these days. With the exception of simple electronic devices, most of the electronic devices that we use involve semiconductors. In order to understand how semiconductors can be used to create devices, it is essential to understand the fundamental electronic properties of semiconductors [46,47].

10.4.1 Fermi–Dirac Distribution

The probability $f(E)$ that a quantum state of energy E is occupied by an electron is given by

$$f(E) = \frac{1}{e^{\frac{E-E_F}{k_B T}} + 1} \tag{10.1}$$

where
  $f(E)$ is the occupation probability of a state of energy, E
  E_F is the Fermi energy
  k_B is Boltzmann's constant
  T is the temperature in kelvin

The distribution describes the occupation probability for a quantum state of energy E at a temperature T. If the degeneracy of the states, the number of electron energy states with the same energy, and the energies of the available electron states are known, the distribution can be used to calculate the thermodynamic properties of systems of electrons. As shown in Figure 10.2 at $E = E_F$, the probability of occupation is ½. In Maxwell–Boltzmann (nondegenerate) approximation, the "1" in the denominator is neglected as the Fermi energy E_F is below the conduction band. Since no quantum state exists in the forbidden bandgap, the minimum energy of occupation is E_c. If $\eta < -3$ ($\eta = (E - E_F)/(k_B T)$), then, 1 in the denominator of Equation 10.1 can be neglected.

$$f(E) \approx \frac{1}{e^{\frac{E-E_F}{k_B T}}} = e^{-\frac{E-E_F}{k_B T}} \tag{10.2}$$

Probability of occupation

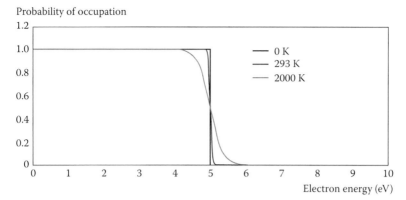

FIGURE 10.2 Fermi–Dirac distribution for some temperatures.

10.4.2 Quasi-1D Nanowire

In 1D nanowire, only one of the three Cartesian directions is much larger than the de Broglie wavelength (Figure 10.3).

The energy spectrum is analog in the x-direction and quantum-confined in the other two directions (for p-type):

$$E = E_{v1} - \frac{\hbar^2 k_x^2}{2m_1^*} \tag{10.3}$$

where

$$E_{v1} = E_{v0} - \frac{\pi^2 \hbar^2}{2m_2^* L_y^2} - \frac{\pi^2 \hbar^2}{2m_3^* L_z^2} \tag{10.4}$$

where E_v is the band edge in the quantum limit of a nanowire that is lifted from the bulk valance band edge E_{v0} by the zero-point energy in the y- and z-directions, where the length $L_{y,z} \ll \lambda_D$, λ_D is the de Broglie wavelength with a typical value of 10 nm. Equation 10.4 is applicable in the quantum limit, where only the lowest quantized state is occupied. k_x is the analog-type continuous momentum wave vector in the x-direction that can have values between $-\infty$ and $+\infty$. For a cylindrical wire with circular cross section, the energy spectrum is the same as in Equation 10.3, except the band edge which is given by

FIGURE 10.3 A nanowire with $L_{y,z} \ll \lambda_D$ and $L_x \gg \lambda_D$ for rectangular and $R \ll \lambda_D$ and $L_x \gg \lambda_D$ for circular cross section.

$$E_v = E_{v0} - \alpha_{01}^2 \frac{\pi^2 \hbar^2}{2m_1^* R^2} \tag{10.5}$$

where
$\alpha_{01} = 2.405$ is the first zero of the Bessel function of order zero, i.e., $J_0(\alpha_{01}) = 0$
$R \ll \lambda_D$ is the radius of the wire that is smaller than the de Broglie wavelength

The quantum wave in the x-direction is a propagating wave, and in the y- (z)-direction it is the standing wave. The normalized wave function is given by

$$\psi_k(x, y, z) = \frac{2}{\sqrt{\Omega}} e^{j(k_x \cdot x)} \sin\left(\frac{\pi}{L_y} y\right) \sin\left(\frac{\pi}{L_z} z\right) \tag{10.6}$$

where $\Omega = L_x L_y L_z$ is the volume of the nanowire. In Equation 10.6, the standing wave component of the wave function is replaced by the Bessel function with appropriate normalization constant for the cylindrical wires.

10.4.3 DISTRIBUTION FUNCTION

The distribution function of the kinetic energy $E_k = \hbar^2 k_x^2 / 2m_1^*$ is given by the Fermi–Dirac distribution function

$$f(E_k) = \frac{1}{e^{\frac{E_k - E_{F1}}{k_B T}} + 1} \tag{10.7}$$

where E_{F1} is considered as the 1D Fermi energy with the probability of occupation of half and the ambient temperature of T, which is related to a hole coming from the absence of an electron. On the other hand, the distribution function for holes is related to the electron distribution function; therefore, the probability of a state at energy E is empty, which is given by

$$f(p) = 1 - f(E_k) = 1 - \frac{1}{e^{\frac{E_k - E_F}{kT}} + 1} = \frac{1}{e^{\frac{E_F - E_k}{kT}} + 1} \tag{10.8}$$

10.4.4 INTRINSIC VELOCITY

The intrinsic velocity v_{i1} is a weighted average of the magnitude of the random velocity with weight equal to the distribution function and 1D density of states. An important unresolved problem is regarding the ultimate drift velocity in a nanostructure and its interdependence with the mobility. The carriers are moving in random directions, so the net velocity in equilibrium is zero as the vector sum of all velocities $\sum \vec{v}_{n(p)} = 0$. The question arisen addresses the point that what we do not care about the direction and what the average magnitude of the velocity $|v| = \sqrt{2(E_v - E)/m_h^*}$ as a function of temperature and carrier concentration is? What is named as the intrinsic velocity of any dimensional nanostructure is obtained by the following equations:

$$n_3 = \frac{1}{2\pi^2}\left(\frac{2m_n^*}{\hbar^2}\right)^{\frac{3}{2}}\int_{E_{co}}^{\infty}\frac{1}{e^{\frac{E-E_{co}+E_{co}-E_{F3}}{k_B T}}+1}\cdot(E-E_{co})^{\frac{1}{2}}dE \qquad (10.9)$$

$$n_3 = \left[\frac{1}{2\pi^2}\left(\frac{2m_n^*}{\hbar^2}\right)^{\frac{3}{2}}(k_B T)^{\frac{3}{2}}\Gamma(3/2)\right]\frac{1}{\Gamma(3/2)}\int_0^{\infty}\frac{1}{e^{x-\eta_{c3}}+1}x^{\frac{1}{2}}dx = N_{c3}\Im_{\frac{1}{2}}(\eta_{c3}) \quad (10.10)$$

With an added feature of introducing $|v| = \sqrt{2(E_v - E)/m_h^*}$ in the integrand, and dividing the total velocity by the carrier concentration p_d for a given dimensionality, the integration leads to the intrinsic velocity given by

$$v_{id} = v_{thd}\frac{\Im_{\frac{d-1}{2}}(\eta_d)}{(p/N_v)_d} = v_{thd}\frac{\Im_{\frac{d-1}{2}}(\eta_d)}{\Im_{\frac{d-2}{2}}(\eta_d)} \qquad (10.11)$$

As expected, the intirinsic velocity is constant for low carrier concentrations and is a function of carrier concentration in the degenerate domain rising with the carrier concentration, which is dramatic in the case of nanostructures. The asymmetry in the distribution of holes and electrons has made the holes (electrons) drift in the same (opposite) direction of the applied electric field. In case of a very large electric field, the majority of the holes are traveling in the negative x-direction, which is implied as the conversion of otherwise completely random motion into a streamlined one, in which the ultimate velocity per holes is equal to the intrinsic velocity v_i. Therefore, the ultimate velocity is ballistic and independent of scattering interactions. The intrinsic velocity is given by [48,53]

$$v_{i1} = v_{th}\frac{1}{\sqrt{\pi}}\frac{\Im_0(\eta_{F1})}{\Im_{-\frac{1}{2}}(\eta_{F1})} \qquad (10.12)$$

The carrier concentration per unit length p_1 for holes in a p-type device is given by

$$p_1 = N_{v1}\Im_{-\frac{1}{2}}(\eta_{F1}) \qquad (10.13)$$

For quasi-1D p-type nanowires, the ultimate average velocity per holes is a function of temperature and doping concentration.

$$v_{i1} = \sqrt{\frac{2k_B T}{\pi m_{hh}^*}}\times\frac{N_{v1}}{p_1}\Im_0(\eta_{F1}) \qquad (10.14)$$

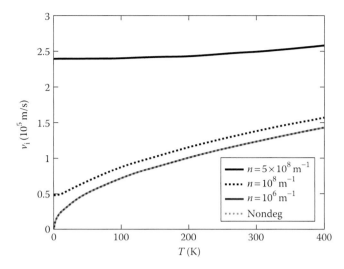

FIGURE 10.4 Velocity versus temperature for nanowire for various concentrations.

with

$$N_{v1} = \left(\frac{2m_{hh}^* k_B T}{\pi \hbar^2} \right)^{\frac{1}{2}} \tag{10.15}$$

where N_{v1} is the effective density of states for the valence band. Figure 10.4 indicates the ultimate velocity which is considered as a function of temperature. The graph for nondegenerate approximation is also shown, in which the ultimate saturation velocity is the thermal velocity that is suitable for 1D carrier motion:

$$v_{i1ND} = v_{th1} = \frac{1}{\sqrt{\pi}} v_{th} = \sqrt{\frac{2k_B T}{\pi m_{hh}^*}} \tag{10.16}$$

However, it should be considered that in case of high concentration, the velocity will be associated with concentration, making it independent of the temperature. Figure 10.5 gives an account of the graph of ultimate intrinsic velocity that is a function of carrier concentration for various temperatures. In case the temperature is low, the carriers follow the degenerate statistics, causing their velocity to be limited by an appropriate average of the Fermi velocity that is a function of carrier concentration. The ultimate saturation velocity for degenerate expression is given by

$$v_{i1D} = \frac{\hbar}{4m_{hh}^*}(p_1 \pi) \tag{10.17}$$

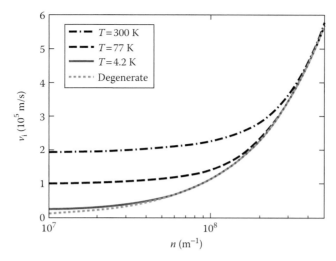

FIGURE 10.5 Velocity of the charge carriers versus doping concentration for various temperatures.

10.4.5 Nanowire Transistor

Silicon is the most common material used for nanowire FETs (SiNWFETs). The advantage of using silicon is not only due to the familiarity of this material with the semiconductor industry, but is also related to the fact that it is the material with the best processes to grow nanowires with well-controlled size and dopant levels. Figure 10.6 shows a possibility of a nanowire transistor made of a cylindrical nanowire, around which a cylindrical insulator with a metal or a polysilicon gate exists.

$$v_d = v_{sat} \frac{1}{\left[1 + \left(\frac{\varepsilon}{\varepsilon_c}\right)^\gamma\right]^{\frac{1}{\gamma}}} \tag{10.18}$$

The velocity saturation effect can be implemented in the modeling of a nanowire transistor of Figure 10.6 by using the empirical Equation 10.18 that has been tested in a number of experiments, where ε is applied electric filed and γ is a parameter [48].

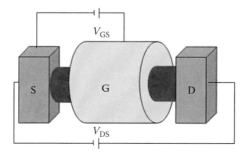

FIGURE 10.6 **(See color insert.)** Schematic of a nanowire transistor with gate dielectric.

The drain current I_D as a function of the gate voltage V_{GS} and drain voltage V_D is then obtained as

$$I_D = \frac{\mu_{ff} C_G}{2L} \frac{\left[2(V_{GS} - V_T)V_D - V_D^2\right]}{1 + \dfrac{V_D}{V_c}} \quad \text{for } 0 < V < V_{Dsat} \tag{10.19}$$

and

$$I_{Dsat} = \alpha C_G (V_{GT} - V_{Dsat})v_{sat} \quad \text{for } V_D \geq V_{Dsat} \tag{10.20}$$

with

$$V_{Dsat} = \frac{1}{(2\alpha - 1)}\left[(s - \alpha)V_c - (1 - \alpha)V_{GT}\right] \tag{10.21}$$

$$I_{Dsat} = \frac{\alpha}{(2\alpha - 1)} \frac{\mu_{ff} C_G}{2L} V_c \left[\alpha V_{GT} - (s - \alpha)V_c\right] \tag{10.22}$$

$$s = \sqrt{\left(\alpha + (1 - \alpha)\frac{V_{GT}}{V_c}\right)^2 + 2\alpha(2\alpha - 1)\frac{V_{GT}}{V_c}}$$

where
C_G is the gate capacitance per unit length
L is the effective channel length
V_T (around 0.08 V) is the threshold voltage ($v_D / v_{Dsat} = \alpha$)
μ_{ff} is the low-field ohmic mobility that is related to the mean free path ℓ_o (value of $\mu_{ff} = 500\,\text{cm}^2/\text{Vs}$ is taken in the calculations)
$v_{sat} = v_{i1}$ is the intrinsic velocity in the absence of quantum emission taken to be the thermal velocity

With the simple geometry of the nanowire transistor, the gate capacitance C_G is given by [48]

$$C_G = \frac{2\pi\varepsilon_{ins}}{\ln\left(\dfrac{R_{ins}}{R_{ewire}}\right)} \tag{10.23}$$

where
$R_{ins} = t_{ox} + R_{wire}$ is the radius of the SiO_2 insulator
t_{ox} is thickness of the gate insulator
$R_{ewire} = R_{wire} - (\varepsilon_{Si} / \varepsilon_{ins})Z_Q$ is the effective radius of the wire of Figure 10.7 with quantum distance Z_Q due to peaking of the wave function at a distance z_Q away from the interface
ε_{ins} is the permittivity of the gate insulator taken to be 3.9

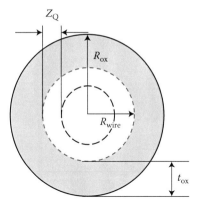

FIGURE 10.7 **(See color insert.)** Schematic of a nanowire cross section with gate dielectric.

At the onset of current saturation $\alpha = 1$, all carriers leave the channel with the saturation velocity where electric field is extremely high. Therefore, the saturation current is given by

$$I_{Dsat} = C_G(V_{GT} - V_{Dsat})v_{sat} \tag{10.24}$$

When Equations 10.19 and 10.24 at the onset of current saturation are consolidated, the drain voltage at which the current saturates is given by

$$V_{Dsat} = V_c\left[\sqrt{1 + \frac{2V_{GT}}{V_c}} - 1\right] \tag{10.25}$$

Substituting V_{Dsat} in Equation 10.24, the saturation current is given by

$$I_{Dsat} = \frac{1}{2}\frac{C_G\mu_{ff}W}{L}V_{Dsat}^2 \tag{10.26}$$

For validating the developed model, the I–V characteristics of the model can be compared with published fabrication data in Ref. [49]. Figure 10.8 shows the current–voltage characteristics of a transistor with $L = 300$ nm. The comparison with the published data is not that perfect due to the experimental geometry that is not either fully circular or fully rectangular. This prescribed model is very useful in predicting the ballistic nature of an SiNWFET. It is found that in a nanowire channel, as seen in the comparison of I–V characteristics, the channel conductance is nonzero. Hence, channel length modulation (CLM) due to pinch-off point moving in the channel is a mistaken identity that does not exist either for the long or the short channel. Thus, it can be concluded that the infinite electric field is not attainable in the drain end of an SiNWFET device as many researchers have assumed.

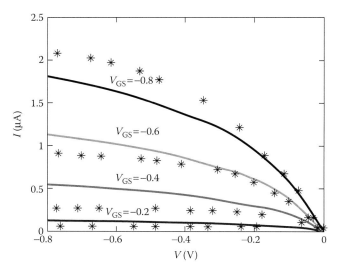

FIGURE 10.8   (See color insert.) Current–voltage characteristics in a nanowire compared with the published data taken from Ref. [49].

10.5   SIMULATION WITH SOFTWARE

10.5.1   Introduction

Normally, simulation of electronic devices involves self-consistent solution of the electrostatic potentiality and carrier distribution inside the device. Engineers have enhanced knowledge of carrier transport and semiconductor physics. Earlier treatment of electrons and holes as semiclassical particles with an effective mass was sufficient to predict the behavior of the semiconductor devices, and the drift diffusion equation was enough to describe the carrier transport. Currently, as we begin to enter 45 nm technology node, MOSFETs have reduced in size up to nanoscale dimensions, which need a reexamination of our approach to device modeling. In particular, the properties of materials can be altered using strain, hetero junction and computational grading, by which quantum effects start to show up in the nanometer regime. As a result, the conventional drift–diffusion and Boltzmann equations do not capture the role of quantum mechanics in modeling transistors in the 10 nm regime. Hence, a more sophisticated analysis of device physics is required, such as the nonequilibrium Green's function (NEGF) approach, to model devices all the way to a ballistic level (<10 nm) [50].

10.5.2   Transport Models Classification

Table 10.1 shows a classification of the existing transport models according to the channel length for FET modeling. In general, when the channel length is greater than 0.1 μm, macroscopic variables can be applied such as electron density, velocity,

TABLE 10.1
Classification of Transport Models According to Channel Length for FET

	$L > 0.1$ μm	10 nm $< L <$ 100 nm	$L < 10$ nm
Semiclassical	Drift–diffusion	Monte Carlo simulation	Analytic solution
	Hydrodynamics	Direct calculation	(from ballistic BTE)
	Six moments equation		
Quantum mechanical	Density gradient	NEGF	NEGF (no scattering)
	Effective potential	Wigner function	Schrödinger equation
		Pauli master equation	

energy, and electron temperature. Among semiclassical transport models appropriate for this area, the drift–diffusion model, the hydrodynamic model, and the six moment equation can be used for modeling FETs with long channel lengths ($L > 0.1$ μm) [51]. For short channel lengths, a quantum mechanical description is required and the density gradient and effective potential approach need to be used. When the channel length is less than 0.1 μm, the Boltzmann transport equation (BTE) must be solved by the Monte Carlo method or by a direct method. At short channel lengths (<0.1 μm) the NEGF, Wigner function, and Pauli master equation can be used to describe quantum effects [49]. For channel lengths for the silicon MOSFET in the range of 10 nm, it becomes imperative to use the NEGF transport model.

The conventional transport theories based on BTE focus on scattering-dominated transport; however, for nanoscale MOSFETs, these may operate in a quasi-ballistic transport regime, where the scattering effects become less important. To simulate nanoscale devices, the NEGF provides the best framework available.

10.5.3 NANOELECTRONICS SIMULATOR

Development in nanotechnology has led to the emergence of atomistic-level modeling. The OMEN simulator has the capability to solve the challenging 3D quantum transport problem, which can be run efficiently on large systems. OMEN, a valid tool for the Network for Computational Nanotechnology (NCN) community, is one of the simulation tools for the design and analysis of nano-electronic devices.

10.5.4 MODELING NANOELECTRONIC DEVICES

OMEN is a 3D, atomistic, and full-band quantum transport simulator designed for multi-gate UTB FETs and NWFETs. The Schrödinger equation is solved with open boundary conditions in the wave function or in the NEGF formalism [52].

10.5.4.1 Description of Simulation Software

The simulation software used here for nanoscale nanowire MOSFET device modeling is based on the work done by colleagues at Purdue University. Figure 10.6 shows the schematic diagram of the cylindrical silicon nanowire device being simulated, in which OMEN is a 2D and 3D Schrödinger–Poisson solver. The cross section in simulations has been taken to be circular and the x-direction in simulations along the nanowire length is assumed to be <100>.

The multilevel parallel implementation of OMEN and the optimization of its numerical algorithm make the simulation of nanowire with a cross section up to 22 nm^2 and UTB with a body thickness up to 10 nm and a gate length of $L_g = 40$ nm possible.

10.5.5 Simulation Tool Preview

The OMEN simulation tool discussed previously is a sophisticated yet simple-to-use simulation tool which promises real-world simulation capability. Figure 10.9 shows a snapshot of the tool's data input page. The various input parameters for the tool are

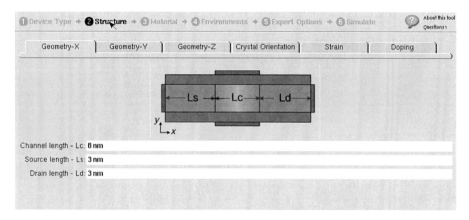

FIGURE 10.9   Snapshot of the structure page of the OMEN simulation tool.

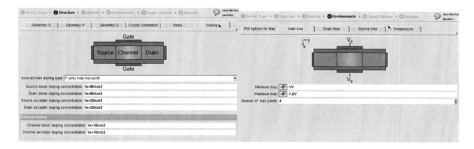

FIGURE 10.10   Snapshots of the environment and structure for set inputs of the OMEN simulation tool. (From Buddharaju, K.D., et al. Gate-all around Si-nanowire CMOS inverter logic fabricated using top-down approach, *Proceedings of the IEEE 37th European Solid-State Device Research Conference (ESSDERC 2007)*, Munich, Germany, pp. 303–306, September 11–13, 2007.)

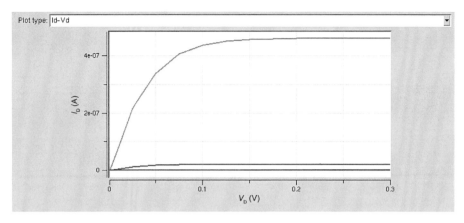

FIGURE 10.11   Snapshot of I_D–V_D curve for the default device.

the nanowire diameter (in nm), oxide thickness (in nm), gate length (in nm), source and drain extension length (in nm), source and drain doping (in/cm³), and channel doping (in/cm³), which are logged in on the structure page. The tool also allows for adjustment of the gate work function.

The drain and gate voltage can be varied in increments of as small as 0.001 V and set to a maximum achievable value of 10 V. The tool provides an option of selecting the number of valleys to be simulated for the device.

The simulation run time turns out to be about 4 h. We present a sample simulation run for a device in the following. The selected parameters seen in the input boxes as shown in Figures 10.9–10.17, namely 2.5 nm nanowire diameter, 1 nm oxide thickness, 8 nm gate length, 3 nm source/drain extension lengths, and select p-type doping also current voltage characteristic is indicated in Figure 10.11.

Following the shown snapshots of the simulation tool inputs, we start with one example to compare the result with our model step by step.

1. The type of device that we are going to simulate is defined.

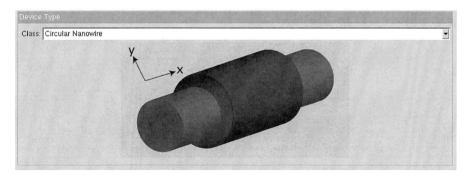

FIGURE 10.12   Snapshot of the device type page of the OMEN simulation tool.

2. The lengths of the channel, source, drain, and doping in the structure page are defined.

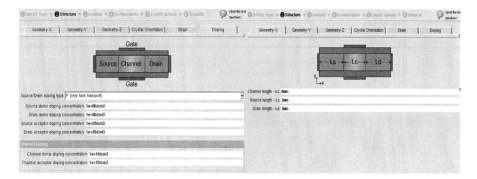

FIGURE 10.13 Snapshots of the structure page of the OMEN simulation tool.

3. The values of the gate contact work function, channel, insulator material, and so on in the material page are settled.

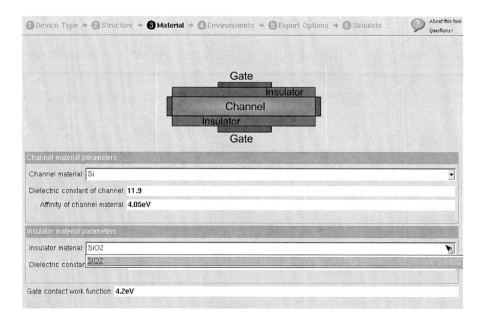

FIGURE 10.14 Snapshot of the material page of the OMEN simulation tool.

4. Plot option, drain bias, source bias, and gate bias in the environment page are defined.

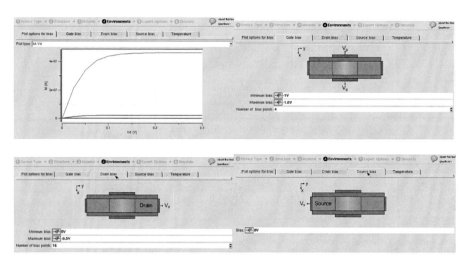

FIGURE 10.15   Snapshots of the environment page of the OMEN simulation tool.

5. Band structure calculation option and Poisson iteration in expert option are defined.

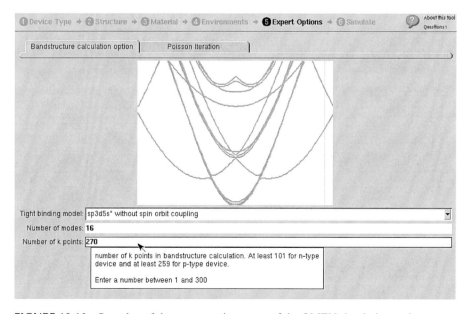

FIGURE 10.16   Snapshot of the expert options page of the OMEN simulation tool.

6. Finally start simulation. Time of process depends on the number of bias point from 1 to 4 h.

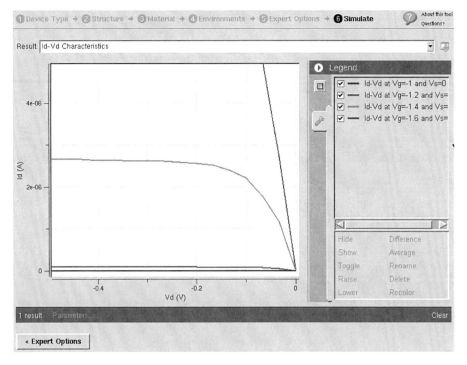

FIGURE 10.17   Snapshots of result of our example.

10.5.6   SIMULATION RESULT

The *I–V* characteristic of the developed model is then compared with the result of this simulation tool. As seen in Figure 10.18, the simulation result from the OMEN simulator and the shifted numerical result of the developed model are comparable and there is no noticeable difference between the *I–V* curves. Figure 10.19 shows the result for $V_G = -1, -1.2, -1.4$, and -1.6 V. (Each curve indicates the corresponding input.)

10.6   SUMMARY

In order to achieve downscaling of devices such as FETs, the emergence of new structures and materials is required. Nanowire is one of the candidates with which FETs can be scaled down. There are a number of high-field transport theories used to answer interdependence, including Monte Carlo simulations, energy-balance theories, path integral methods, Green's function, and many others. No clear consensus has emerged on the true nature of saturation velocity and its dependence on band structure parameters, doping profiles, or ambient temperature. The presented model based on quantum confinement and high electric field effect illustrates the velocity approach to the modeling of a p-type silicon nanowire transistor. With the development of devices in

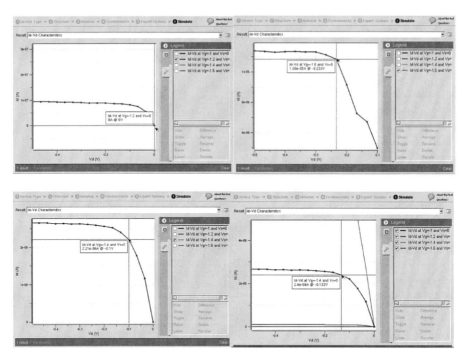

FIGURE 10.18 Snapshots of the result of GAA p-type silicon nanowire FET using OMEN simulator.

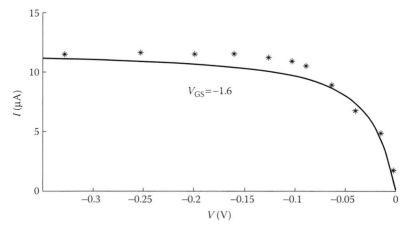

FIGURE 10.19 Comparison between simulation and developed model for $V_{GS} = -1.6$.

nanoscale dimensions, the significance of saturation velocity is highlighted. In this work, we extended the work of Arora to embrace the degenerate domain in a p-type nanowire, where in one direction, carriers have analog-typed classical spectrum, while in the other two directions, there is quantum confined or digital spectrum. This work intends to develop a physical model based on SiNWFETs device physics, which can be applied to device behavior in order to explore device matters. The results obtained are applied to the modeling of the current–voltage characteristics of a p-type SiNWFET and finally compared with published data for validation of presented model. Besides this, in this work, we use OMEN for simulation of p-type SiNWFET to compare the *I–V* characteristics of the simulation result with those of the presented model. The limitations on carrier (holes) drift due to high-field streamlining are also reported. Asymmetrical distribution function that adapts randomness in a zero-field to a streamlined one in a high electric field is employed, where the saturation velocity is always ballistic and the ultimate drift velocity is found to be dependent on thermal velocity for nondegenerately doped nanostructure. However, the ultimate drift velocity is the Fermi velocity for degenerately doped nanostructures. The transport mechanism is presented and applied to the modeling of a p-type SiNWFET. The comparison with the experimental data [49] is not perfect due to the experimental geometry that is not either fully circular or fully rectangular. This prescribed model is very useful in predicting the ballistic nature of an SiNWFET. It is found that in a nanowire channel, the channel conductance is nonzero. Hence, CLM due to the pinch-off point moving in the channel is a mistaken identity that does not exist either for the long or the short channel. Thus, it can be concluded that an infinite electric field is not attainable in the drain end of an SiNWFET device, which concurs with what many researchers have assumed.

REFERENCES

1. Semiconductor Industry Association. *International Technology Roadmap for Semiconductors 2004 Update*. 2004 [Online].
2. Y. Taur, C.H. Wann, and D.J. Frank. 25 nm CMOS design considerations. In *IEDM Technical Digest*, San Francisco, p. 789. December 1998.
3. W. van Roosbroeck. The transport of added current carriers in a homogeneous semiconductor. *Phys. Rev.* 91(2), 282–289. 1953.
4. M. Lundstrom. *Fundamentals of Carrier Transport*, 2nd ed. Cambridge University Press, Cambridge. 2000.
5. M. Lundstrom, Z. Ren, and S. Datta. Essential physics of carrier transport in nanoscale MOSFETs. In *International Conference on Simulation of Semiconductor Processes and Devices*. Washington DC, p. 1. September 2000.
6. J.-H. Rhew and M.S. Lundstrom. Drift-diffusion equation for ballistic transport in nanoscale metal-oxide-semiconductor field effect transistors. *J. Appl. Phys.* 92(9), 5196–5202, 2002.
7. T.-W. Tang, S. Ramaswamy, and J. Nam. An improved hydrodynamic transport model for silicon. *IEEE Trans. Electron Device*. 40(8), 1469–1477, 1993.
8. J.E. Lilienfeld. Method and apparatus for controlling electric currents, U.S. Patent 1,745,175, 1930.
9. P.K. Bondy. Moore's law governs the silicon revolution. *Proc. IEEE*. 86, 78–81, 1998.
10. D.J. Frank, R.H. Dennard, E. Nowak, P.M. Solomon, Y. Taur, and H.-S.P. Wong. Device scaling limits of Si MOSFETs and their application dependencies. *Proc. IEEE*. 89, 259–288, 2001.
11. Semiconductor Industry Association (SIA). *International Technology Roadmap for Semiconductors, 2003 Edition*. SEMATECH, Austin, TX. 2003. Retrieved from http://public.itrs.net.
12. G. Sai-Halasz. Performance trends in high-end processors. *Proc. IEEE*. 83, 20, 1995.

13. J.W. Sleight, P.R. Varekamp, N. Lustig, J. Adkisson, A. Allen, O. Bula, X. Chen, et al. A high performance 0.13 μm SOI CMOS technology with a 70 nm silicon film and with a second generation low-k Cu BEOL. *IEDM Technical Digest.* San Francisco, pp. 245–248, 2001.

14. International Technology Roadmap for Semiconductors. ITRS 2011 edition (http://www.itrs.net/Links/2011ITRS/Home2011.htm).

15. D.A. Muller. A sound barrier for silicon? *Nat. Mater.* 4(9), 645–647, 2005.

16. Y. Taur and T.H. Ning. *Fundamentals of Modern VLSI Devices.* Cambridge Press, New York. pp. 140–142, 1998.

17. Y. Taur. CMOS design near the limit of scaling. *IBM J. Res. Dev.* 46(2–3), 213– 222, 2002.

18. T. Ghani, M. Armstrong, C. Auth, M. Bost, P. Charvat, G. Glass, T. Hoffmann, et al. A 90 nm high volume manufacturing logic technology featuring novel 45 nm gate length strained silicon CMOS transistors. *IEDM Technical Digest.* San Francisco, pp. 978–980, 2003.

19. E.P. Gusev, V. Narayanan, and M.M. Frank. Advanced high-k dielectric stacks with poly Si and metal gates: Recent progress and current challenges. *IBM J. Res. Dev.* 50(4/5), 387–410.

20. L. Chang, Y.-K. Choi, D. Ha, P. Ranade, S. Xiong, J. Bokor, C. Hu, and T.-J. King. Extremely scaled silicon nano-CMOS device. *Proc. IEEE.* 91(11), 1860–1873, 2003.

21. http://www.intel.com/pressroom/archive/releases/20060125comp.htm.

22. M. Stadele. Influence of source–drain tunneling on the subthreshold behavior of sub-10-nm double-gate MOSFETs. In *Proceedings of European Solid-State Device Research Conference.* Firenze, Italy, pp. 135–138, September 2002.

23. J. Wang and M. Lundstrom. Does source-to-drain tunneling limit the ultimate scaling of MOSFETs? *IEDM Technical Digest.* San Francisco, pp. 707–710, December 2002.

24. H. Majima, H. Ishikuro, and T. Hiramoto. Experimental evidence for quantum mechanical narrow channel effect in ultra-narrow MOSFETs. *IEEE Electron Device Lett.* 21, 396–398, 2000.

25. Y. Taur, D.A. Buchanan, W. Chen, D.J. Frank, K.E. Ismail, S.-H. Lo, G.A. Sai-Halasz, S.G. Viswanathan, H.-J.C. Wann, S.J. Wind, and H.-S. Wong. CMOS scaling into the nanometer regime. *Proc. IEEE.* 85, 486–504, 1997.

26. B. Hoeneisen and C.A. Mead. Fundamental limitations in microelectronics-I. MOS technology. *Solid-State Electron.* 15, 819, 1972.

27. H.-S.P. Wong. Beyond the conventional transistor. *IBM J. Res. Dev.* 46, 133–168, 2002.

28. L. Chang, Y.-K. Choi, D. Ha, P. Ranade, S. Xiong, J. Bokor, C. Hu, and T.-J. King. Extremely scaled silicon nano-CMOS device. *Proc. IEEE.* 91(11), 1860–1873, 2003.

29. L. Chang. Moore's law lives on [CMOS transistors]. *IEEE Circuits Devices Mag.* 19, 35, 2003.

30. Y.-K. Choi, K. Asano, N. Lindert, V. Subramanian, T.J. King, J. Bokor, and C. Hu. Ultrathin-body SOI MOSFET for deep-sub-tenth micron era. *IEEE Electron Device Lett.* 21, 254, 2000.

31. L. Chang, Y.-K. Choi, D. Ha, P. Ranade, S. Xiong, J. Bokor, C. Hu, and T.-J. King. Extremely scaled silicon nano-CMOS device. *Proc. IEEE.* 91(11), 1860–1873, 2003.

32. F.-L. Yang, H.-Y. Chen, F.-C. Cheng, C.-C. Huang, C.-Y. Chang, H.-K. Chiu, C.C. Lee, C.-C. Chen, H.-T. Huang, C.-J. Chen, H.-J. Tao, Y.-C. Yeo, M.-S. Liang, and C. Hu. 25 nm CMOS omega FET's. *IEDM Technical Digest.* San Francisco, pp. 255–258, 2002.

33. R. Chau, B. Boyanov, B. Doyle, M. Doczy, S. Datta, S. Hareland, B. Jin, J. Kavalieros, M. Metz. Silicon nano-transistors for logic applications. *Physica E.* 19, 1–5, 2003.

34. Y. Cui. Functional nanoscale electronic devices assembled using silicon nanowire building blocks. *Science.* 291, 851, 2001.

35. Y. Cui. Diameter-controlled synthesis of single-crystal silicon nanowires. *Appl. Phys. Lett.* 78, 2214, 2001.

36. N. Singh. High-performance fully depleted silicon nanowire (diameter ≤5 nm) gate all-around CMOS devices. *IEEE Electron Device Lett.* 27(5), 383, 2006.

37. Park, J-T., et al. Pi-gate SOI MOSFET. *IEEE Electron Device Lett.* 22(8), 405–406, 2001.

38. T. Saito, T. Saraya, T. Inukai, H. Majima, T. Nagumo and T. Hiramoto. Suppression of short channel effect in triangular parallel wire channel MOSFETs. *IEICE Trans. Electron.* E85-C(5), 1073–1078, 2002.

39. H. Majima, Y. Saito and T. Hiramoto. Impact of quantum mechanical effects on design of nano-scale narrow channel n- and p-type MOSFETs. *IEDM Technical Digest.* San Francisco, p. 733, 2001.

40. Y. Cui, L.J. Lauhon, M.S. Gudiksen, J. Wang and C.M. Lieber. Diameter-controlled synthesis of single-crystal silicon nanowires. *Appl. Phys. Lett.* 78, 2214–2216, 2001.

41. Y. Wu, Y. Cui, L. Huynh, C.J. Barrelet, D.C. Bell and C.M. Lieber. Controlled growth and structures of molecular-scale silicon nanowires. *Nano Lett.* 4, 433–436, 2004.

42. R.S. Wagner. *Whisker Technology*, Ed. A.P. Levitt Wiley, New York, 1970.

43. S.L. Hong. *J. Korean Phys. Soc.* 37, 93, 2000.

44. X. Duan, C. Niu, V. Sahi, J. Chen, J.W. Parce, S. Empedocles, and J.L. Goldman. High-performance thin-film transistors using semiconductor nanowires and nanoribbons. *Nature.* 425, 274, 2003.

45. M.C. McAlpine, R.S. Friedman, S. Jin, K.-h. Lin, W.U. Wang, C.M. Lieber. High-performance nanowire electronics and photonics on glass and plastic substrates. *Nano Lett.* 3, 1531–1535, 2003.

46. M.T. Ahmadi, Z. Johari, A.H. Fallahpour, and R. Ismail. Graphenee nanoribon conductance model in parabolic band structure. *Journal of nanomaterial.* Article ID 753738, 2010.

47. A.H. Fallahpour, M.T. Ahmadi, R. Ismail. Analytical study of dift velocity in N-type silicon nanowires. *ASQED*, 252–254, 2009.

48. M.T. Ahmadi, A.H. Fallahpour, and R. Ismail. P-type silicon nanowire transistor modeling. *Int. Joint J. Conf. Comput. Electron.* 2(6), 23–25, 2009.

49. K.D. Buddharaju, N. Singh, S.C. Rustagi, S.H.G. Teo, L.Y. Wong, L.J. Tang, C.H. Tung, G.Q. Lo, N. Balasubramanian, and D.L. Kwong. Gate-all around Si-nanowire CMOS inverter logic fabricated using top-down approach. *Proceedings of the IEEE 37th European Solid-State Device Research Conference (ESSDERC 2007).* Munich, Germany. September 11–13, pp. 303–306, 2007.

50. M.P. Anantram, Mark S. Lundstrom, Dmitri E. Nikonov. Modeling of nanoscale devices. *Proceedings of the IEEE.* 96(9), September 2008.

51. S. Jin. *Modeling of Quantum Transport in Nano-Scale MOSFET Devices.* Ph.D. dissertation, Seoul National University. 2006.

52. H. Bae, S. Clark, B. Haley, and R. Hoon. *A Nano-electronics Simulator for Petascale Computing: From NEMO to OMEN.* Birck Nanotechnology Center, Purdue University, West Lafayette, IN. 2008.

53. M.T. Ahmadi, A.H. Fallahpour, and R. Ismail. Analytical study of carriers in silicon nanowires. *MASAUM J. Basic Appl. Sci.* 1(2), 233–239, 2009.

11 Silicon Nanowires/ Nanoneedles
Advanced Fabrication Methods

Habib Hamidinezhad and Yussof Wahab

CONTENTS

11.1 INTRODUCTION

In the past few years, there has been great interest in the fabrication of nanostructure materials. Nanostructures are the ones with nanoscale dimensions. The unique properties of nanomaterials is one of the main reasons causing interest in nanomaterials, which make them distinct from the bulk materials. These properties, such as large surface area, have applications for various uses. Silicon-based nanostructures, especially, are attracting interest since the techniques used to fabricate them are largely correlated with available semiconductor production processes. Silicon nanostructures' properties are different from those of bulk silicon. For example, a bandgap, which is tunable by the size of the structures, is appealing when fabricating semiconductor devices. Silicon nanostructures are nanocrystalline silicon (nc-Si) consisting of a series of nanometer-sized crystallites embedded in a matrix of amorphous silicon (a-Si) material. These materials can be used in certain electronic and optical devices such as thin film solar cells with properties different from those of crystalline and amorphous silicon based solar cells. Development of nanosize devices is a new and rapidly growing area of research. The use of silicon nanostructures potentially allows for cost-effective mass production associated with thin film photovoltaics while maintaining higher efficiency and stability, which are considered as the characteristics of crystalline solar cells (Yue et al., 2006). In recent years, investigation of the synthesis and characterization of nanowires (NWs) and nanotubes has increased due to their fundamental importance in nanotechnology. Some applications of one-dimensional (1D) nanostructures have been studied for devices using flexible substrates, nanoscale devices, and sensor applications. Investigation of the properties of NWs and their growth mechanisms have demonstrated that they have a great potential for nanoelectronic and nanophotonic applications. A silicon nanowire (SiNW) is an elongated crystalline or a-Si with the diameter ranging from 10 to 100 nm and the length of several micrometers. Researchers have focused on SiNWs because of their unique properties being significantly different from those of bulk silicon (Ma et al., 2003). The electronic bandgap of SiNWs is adjustable with the regulation NW diameter. The SiNWs' bandgap increases with the decreasing diameter of the NWs (Ma et al., 2003). It has been shown that SiNWs combust when in contact with a strong light. This is due to the enhanced photothermal effect. It indicates that a high degree of light absorbance is shown by bulk SiNWs. Growth of SiNWs below the 10 nm regime opens new pathways for novel applications. The surface effects of SiNWs can be used for application of ultrasensitive biosensors because of the increasing ratio of the surface-to-volume and the decreasing diameter of SiNWs. Furthermore, the novel materials' properties are accessible considering their confinement effects (Albuschies et al., 2006). Due to the potential applications of SiNWs in photonics and microelectronics fields, they are attracting more attention (Griffiths et al., 2007). Different methods have been used to synthesize SiNWs: chemical vapor deposition (CVD), plasma enhanced CVD (PECVD), laser ablation, and evaporation. These methods are often based on the vapor–liquid–solid (VLS) idea, in which various metals such as gold, titanium, and gallium catalytically enhance the growth of silicon wires. Initially, Wagner and Ellis (1965) described the vapor–liquid–solid (VLS) mechanism using gold as a catalyst material in 1964, and then Givargizov developed it in 1975. A basic aspect of the NWs growth based on the VLS mechanism is the necessity of the existence of a metal particle acting as a catalyst for anisotropic crystal growth (Albuschies et al., 2006). In the PECVD method, plasma treatment of catalysts has been shown to assist in forming the islands or etching the deposited thin films. One of the important steps in nanostructure growth is particle formation occurring in both the thermal and the plasma assisted methods (Griffiths et al., 2007). For low temperature growth of SiNWs, the presence of pulse silane plasma improves the density and sample coverage of SiNWs (Parlevliet and Cornish, 2006).

11.2   INTRODUCTION TO SiNWs

SiNWs have recently attracted considerable attention, because of their applications in integrated nanoscale electronics (Cui and Lieber, 2001), as well as their potentiality for exploiting the unique fundamental properties in submicron dimensions. Most of the methods in SiNWs synthesis are categorized as the "bottom-up" approaches, among which the VLS mechanism is the most commonly used method (Morales and Lieber, 1997, 1998; Wang et al., 1998; Hu and Lieber, 1999). In the VLS process of SiNW growth introduced by Wagner et al. (Wagner and Ellis, 1964; Cui et al. 2001), a metal such as gold (Au) is usually used to catalyze the decomposition of the source gas such as silane (SiH_4) or silicon tetrachloride ($SiCl_4$). In summary, the VLS growth process involves three stages. First, a liquid alloy is produced at temperatures higher than the alloy's eutectic temperature (i.e., 363°C for Au–Si). Second, the nucleate semiconductor nanostructure is produced as the alloy liquid is supersaturated. Last, when the nanostructures nucleate at the liquid/solid interface, more condensation/dissolution of nanostructure evaporates into the system and increases the amount of the semiconductor nanostructure's precipitation. Besides the VLS growth process, many other "bottom-up" mechanisms have also been investigated to grow SiNWs. Many research groups have demonstrated that SiNWs with much smaller diameters than those of Wagner's can be synthesized in bulk quantities with the VLS mechanism (Lu and Korgel, 2003). Using the CVD method, Yi Cui in Lieber's group (Cui et al., 2001) synthesized the single-crystal SiNWs incorporating the gold nanocluster-catalyzed growth with the molecular dimensioned diameter. They also showed that with this method, the smallest diameter NWs grow primarily along the ⟨100⟩ direction, whereas larger NWs grow along the ⟨111⟩ direction. Xianmao Lu and Korgel (2003) showed that their single-crystal SiNWs, achieved using gold nanocluster catalysts, were dispersed in a supercritical hexane. In Fonash's group, Yinghui Shan et al. combined the VLS technique with the lithographically fabricated permanent nanochannel growth templates to control the size, shape, orientation, and positioning of the grown NWs and nanoribbons (Shan et al., 2004). Table 11.1 shows the NW materials which have been investigated extensively using different mechanisms from the 1960s until 2010.

TABLE 11.1
Research on Nanowires Using Different Mechanisms

Type of Nanowire (Method)	Studies (Mechanism and Catalyst)	Authors (Year)
SiNWs (hydrogen reduction of SiCl$_4$)	First introduced the VLS mechanism (VLS)	Wagner and Ellis et al. (1964)
Si, Ge, GaAs, and InAs nanowires	Growth rate and whisker diameter (VLS)	Givargizov (1975)
SiNWs	Structure, TEM	Zhang et al. (1998)
SiNWs (physical evaporation)	The quantum confinement effect by photoluminescence (PL) (VLS)	Feng et al. (2000)
SiNWs	The reasons of forming amorphous SiNWs (VLS, Au/Pd)	Liu et al. (2000)
Amorphous SiNWs	Synthesis of SiNWs without supplying any gaseous or liquid Si sources, structural investigation (SLS, Ni)	Yan et al. (2000)
SiNWs (thermal evaporation)	Effects of ambient pressure on SiNW growth	Fan et al. (2001)
Amorphous SiNWs (thermal CVD)	Effect of H$_2$ gas and etching on the catalyst size and the effect of catalyst size on the formation of vertical growth NWs (VLS, Au–Pd)	Liu et al. (2001)
SiNWs (PECVD)	A detailed growth study using electron microscopy, focused ion beam preparation, and Raman spectroscopy (Au, VLS)	Hofmann et al. (2003)
SiNWs (CVD)	Structure (Au, VLS)	Niu et al. (2003a)
SiO$_2$ nanowires (thermal processing)	Description of main phases found in the NWs (Pd/Au, VLS)	Elechiguerra et al. (2004)
SiNWs (CVD)	Controlling the NW diameter along its length (Ti)	Sharma et al. (2004)
Silicon oxide nanowires (CVD)	The influence of hydrogen on the growth of gallium-catalyzed silicon oxide nanowires and the reasons and mechanisms for the growth of these nanowires (Ga, VLS)	Yan et al. (2005)
Silicon nanowires (CVD)	Growth and structure (Au, OAG [oxide-assisted growth], VLS)	Yao et al. (2005)
SiNWs (oxide-assisted growth)	Hydrothermal growth, elucidation of growth mechanism (OAG)	Pei et al. (2006)
SiNWs (PPECVD)	Influence of frequencies on density and sample coverage of SiNWs (Au)	Parlevliet and Cornish (2006)
SiNWs (LPCVD)	High-density growth of SiNWs, self-assembled Au nanoparticle formation (Au)	Albuschies et al. (2006)
SiNWs (PECVD)	Plasma-assisted growth (Au, In, Ga, VLS)	Griffiths et al. (2007)

(continued)

TABLE 11.1
Continued

Type of Nanowire (Method)	Studies (Mechanism and Catalyst)	Authors (Year)
SiNWs (CVD)	Effect of high-pressure water vapor annealing (HWA) on NW's optical properties (Au, VLS)	Salhi et al. (2007)
SiNWs (under ambient atmospheric pressure)	Morphology and microstructure analysis (SLS)	Wang et al. (2008)
SiNWs (PECVD)	Effects of substrate temperature and H_2 dilution of SiH_4 on the morphology and compositional evolution of the SiNWs (Sn, VLS)	Yu et al. (2008)
SiNWs (PECVD)	The morphology and crystalline structure (Ga, VLS)	Zardo et al. (2009)
Silicon oxide nanowires (thermal growth)	Effect of catalyst (Fe/Au)	Liu et al. (2009)
Silica nanowires (thermal evaporation of silicon monoxide)	Synthesis, characterization, and photoluminescence properties (noncatalytic)	Srivastava et al. (2009)
SiNWs (nanosphere lithography and CVD)	Ordered arrays of epitaxial SiNWs (polystyrene spheres and Au film) VLS	Lerose et al. (2010)
SiNWs (UHV CVD)	Spreading of liquid AuSi from the catalytic seed (VLS) Au	Dailey et al. (2010)
SiNWs (PECVD)	Structure and growth mechanism of Si nanoneedles (VLS) Au	Červenka et al. (2010)

11.3   SILICON

Si is the second-most abundant element on the earth (25.7%), which is usually found in the form of oxides and silicates such as sand and quartz. It has been categorized as a member of the semiconductor's family placed in the fourth-column element where there is electrical conductivity between metals and insulators. The basic information about elemental Si is listed in Table 11.2. The most common crystal structure of Si is the covalent-bonded diamond structure, in which each Si atom is bonded to four surrounding Si atoms.

The bandgap, also known as the optical gap, energy gap, or mobility gap, is an important property of semiconductors determining the optoelectronic properties of devices created from such semiconductors. There is an energy gap between the completely filled valence band (VB) and the completely empty conduction band (CB) in the intrinsic material at 0 K, which is an important feature of a semiconductor. The atomic levels of the valence shell in the elements are the origin of these states. The bandgap is the minimum amount of energy needed for an electron to jump from the VB to the CB as shown in Figure 11.1. For an intrinsic semiconductor, the Fermi energy (E_f) is located in the mid-bandgap state. At 0 K, all the electrons are confined

TABLE 11.2

Some Properties of Si

Atomic number	14
Atomic mass	28.0855 a.m.u.
Melting point	1410°C
Density at 293 K	2.329 g cm^{-3}
Dielectric constant at 293 K	11.9
Energy gap, E_g, at 300 K	1.12 eV
Color	Dark gray with bluish tinge
Crystal structure	Cubic-diamond
Space group	Fd-3m
Lattice parameters	0.54309 nm

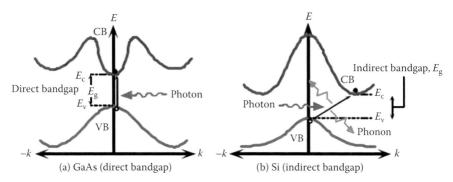

FIGURE 11.1 Energy band structures of (a) direct bandgap semiconductor, GaAs, and (b) indirect bandgap semiconductor, Si.

to the VB, and the material is a complete insulator. Above 0 K, some electrons have enough thermal energy to transit from the VB to the CB where they are free to move and conduct current through crystals (Kayali et al., 1996).

Semiconductor materials are classified by type of bandgap, including the direct bandgap semiconductors and the indirect bandgap semiconductors. GaAs, InP, etc. categorized as III–IV or II–IV semiconductors are mostly considered as direct bandgap semiconductors; however, Si, Ge, and group IV semiconductors are considered as indirect bandgap semiconductors. The classification of semiconductor materials is dependent on whether the excited electron can transit directly from the VB to the CB. Selection rules such as the conservation of spin and momentum govern electronic transitions. For every electronic transition taking place between the CB and VB, the momentum or wave vector (k) must be conserved in the whole process. There are minimum and maximum amounts in the CB and the VB for the same momentum vector of the center of the zone ($k = 0$). For the semiconductor with a direct bandgap, the conservation of momentum selection rule is satisfied. Thus, an electron with the

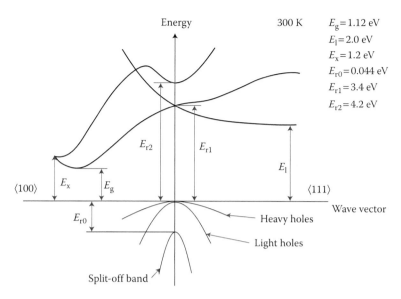

FIGURE 11.2 Band structure of bulk Si.

smallest energy will make a transition from the CB to the VB without a change in the k value, resulting in a photon of energy equal to the bandgap. Furthermore, a direct bandgap semiconductor is an efficient light emitter. In the case of Si, there is an indirect bandgap where the minimum of the CB and the maximum of the VB are located at different points of k (Figure 11.2). The top of the VB is placed at the center of the Brillouin zone (at the Γ_{25} point). The CB has six minima at the equivalent [100] directions, centered at the $\Delta_{1,C} = (0.86,0,0)\ \pi/a$ points. According to the conservation of momentum selection rule, an electron cannot transit directly from the minimum point in the Si CB to the maximum point of the VB, but must involve a change in k. The lattice vibration (phonons) provides the difference in momentum between the CB and the VB. The energy is generally given up as heat to the phonon rather than as an emitted photon. This may explain the inability of indirect gap semiconductors to emit light efficiently (Kovalev et al., 1999).

A vertical virtual transition at $k = 0$ or $0.86\ \pi/a$ with a sequence electron–phonon scattering process is caused by a photon. This is the only possible reason for the optical transitions in bulk Si that is due to the conservation rules. The transversal optical phonons (TO, $E_{TO} \approx 56$ meV), longitudinal optical phonons (LO, $E_{LO} \approx 53.5$ meV), and transversal acoustic phonons (TA, $E_{TA} \approx 18.7$ meV) are relevant phonon modes. A typical PL spectrum from pure bulk Si is shown in Figure 11.3. This spectrum consists of three lines assigned to the free exciton recombination with the assistance of TA, TO/LO, and TO + $O(\Gamma)$ phonons, respectively. Since the $O(\Gamma)$ is a center Brillouin zone phonon, the combination of TO and $O(\Gamma)$ phonons is momentum-allowed. According to the inset of Figure 11.3, the 2.5 meV splitting of the main free exciton line in TO and LO

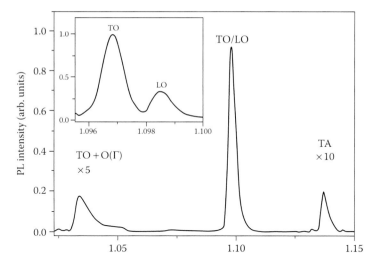

FIGURE 11.3 PL spectra of bulk, undoped Si with free exciton transitions, measured at 20 K, E_{ex} = 1.4 eV. Inset: fine splitting of the free exciton line in TO and LO momentum-conserving phonon satellites. (Adapted from Kovalev, D., Heckler, H., Polisski, G., and Koch, F., *Physica Status Solidi B*, 215, 1999.)

double can be observed at very low temperatures. The largest contribution to the PL is due to a TO phonon-assisted recombination process. The room temperature bandgap of bulk Si is 1.12 eV, where it resides in the infrared region of the light spectrum.

11.3.1 CRYSTALLINE SILICON

Nowadays, crystalline silicon (c-Si) is the basis for most integrated circuits (ICs) as it is a readily available semiconductor. c-Si is easily processed and doped to form semiconductor devices. There are three types of c-Si. The face-centered cubic (FCC) is a dark colored type of c-Si structure with a slightly bluish tinge. This type of c-Si is highly reflective when it is cleaned and polished. A sample of the c-Si is shown in Figure 11.4. c-Si is used in many optic and electronic devices such as ICs, diodes, and solar cells. The earliest photovoltaic devices or solar cells were made from the c-Si. c-Si is still commonly used for mentioned devices because of its high efficiency. Surface orientations of silicon crystals are various as (111), (100), and (110). The orientation of the crystal plane is shown by the nomenclature (x, y, z), which is defined in terms of the axes (a_1, a_2, a_3) as illustrated in Figure 2.5. It shows the unit cell of an FCC crystal structure with different crystal planes. Directions are shown by [x, y, z] described by a vector where x, y, and z are the components of the vector in terms of the axes (a_1, a_2, a_3). An example of [111] vector is shown in Figure 11.4. The arrangement of atoms in an FCC unit cell, known as the basic unit of a crystal lattice, is also shown in this figure. The structure of Si is an FCC diamond cubic crystal structure with a space group of Fd3m.

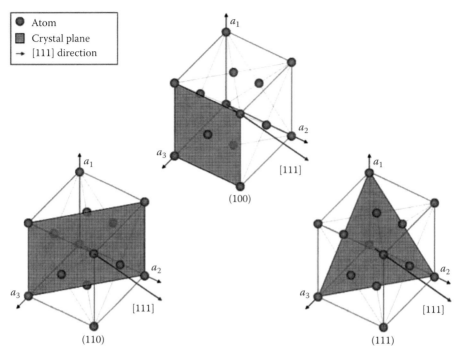

FIGURE 11.4 Face-centered cubic (FCC) unit cell with crystal planes and the [111] direction.

11.3.2 A-SI

a-Si is a glass-like material. This type of Si does not have a structurally long ranged order; in fact, it has a random orientation. The light absorption coefficient in a-Si is higher than that of c-Si due to the orientation of the material. Because of this, the thin film device fabricated from a-Si can absorb more quantities of light than the thicker c-Si slices. This enables a-Si to be used as one of the main materials in fabricating thin film devices. Moreover, making doped layers of a-Si by introducing dopant gases during the deposition of a-Si is done by a relatively direct process. a-Si can be produced using various methods including the CVD techniques such as plasma-enhanced chemical vapor deposition (PECVD), hot wire chemical vapor deposition (HWCVD), and physical deposition methods such as sputtering. One of the main uses of a-Si is the fabrication of cheap thin film solar cells. Amorphous Si are randomly structured glass-like materials. Amorphous solar cells made of such materials have fairly unique properties regardless of their disadvantages. Figure 11.5 is an illustration of this glass-like structure named as hydrogenated a-Si. There are many dangling bonds in a-Si. These bonds trap the charge carriers and accordingly reduce the efficiency of the cell. Many of the traps can be eliminated by creating a hydrogenated amorphous silicon (a-Si:H), a hydrogen and silicon alloy (Wilson, 1980). This makes the a-Si appropriate to be used in high efficiency photovoltaic applications. As seen in Figure 11.5, the hydrogen atoms attached to the dangling bonds make them passive, and thus result in elimination of the traps. The color of a-Si films is dependent on their

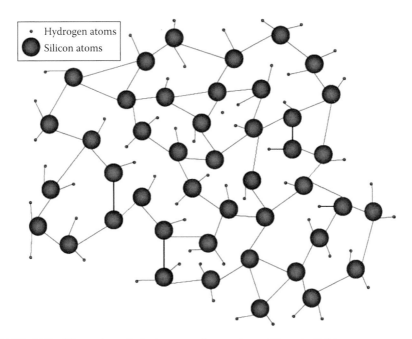

FIGURE 11.5   Illustration of a hydrogenated amorphous silicon (a-Si:H) structure.

thicknesses. A thin film of a-Si with approximately 500 nm thickness on glass has a distinctive red-orange color. However, a thinner a-Si film with thickness of approximately 200 nm film tends to turn out a yellowish color.

11.3.3   POLYCRYSTALLINE SILICON

Another type of silicon is the polycrystalline silicon. This type of silicon consists of a series of small crystallites grown together with an extremely high crystallinity or a proportion of crystalline materials. The polycrystalline silicon typically has a grain size (crystallite diameter) ranging from 10 to 30 μm and a crystalline fraction close to 100% (Cabarrocas, 2004). Usually, the polycrystalline materials are synthesized by CVD methods (Cabarrocas, 2004) or HWCVD methods (Lee et al., 2002). They can be doped and used to fabricate solar cells in some devices.

11.3.4   MICROCRYSTALLINE SILICON

Structurally, with embedding a series of micron-sized crystallinity in a-Si matrix, microcrystalline silicon (μc-Si) is formed. The μc-Si consists of crystalline, amorphous phases, and grain boundaries (Droz et al., 2003). Generally, this structure of Si has a grain size between 10 and 200 nm and a crystalline fraction ranging from 10 to 100% (Cabarrocos, 2004). Due to its higher carrier mobility and hence high electrical conductivity, μc-Si is an attractive material for electronic device applications (Das and Jana, 2004). Usually, μc-Si can be synthesized by PECVD (Fontcuberta i Morral and Roca i Cabarrocas, 2001), hot HWCVD (Carius et al., 2003), and

VHF-PECVD. The crystallinity of the material depends on the depositioning conditions (Droz et al., 2003). This type of silicon is also commonly used to produce single junctions and micromorph tandem solar cells in the photovoltaic industry (Meier et al., 2004).

11.3.5 NC-SI

Similar to microcrystalline, there is another type of silicon known as nc-Si. It consists of crystallines embedded in an a-Si matrix. Commonly, nc-Si is known as having a grain size. This grain has a size between 2 and 5 nm, and crystalline fraction between 10 and 80% (Cabarrocas, 2004). Besides the amorphous and crystalline phases, there is a layer of amorphous buffer in nc-Si that can be crystallized by post-growth annealing (Gourbilleau et al., 1999). Some researchers produced nc-Si by using radio frequency (RF) sputtering (Gourbilleau et al., 1999), CVD (Zhang and Han, 2002), and PECVD (Wang et al., 1992). This form of silicon is interesting because of its good optical and electronic properties. It can be used to fabricate electroluminescent devices (Sato et al., 2003) like solar cells (Yue et al., 2006).

11.3.6 NANOSTRUCTURED SILICON

Nanostructured silicon is the material consisting of highly textured structures with nanometer scales. These structures can stand free rather than be embedded in a substrate. Nanostructured silicon is of much interest and has novel applications because the properties of this material being different from that of bulk silicon.

11.4 NWS

NWs are nanostructures with the diameter of the order of a nanometer (10^{-9} m). Studies in nanoscience have attracted several research groups as well as candidates for nanotechnology applications. NWs can be defined as structures having a diameter of 10 nm or less and an unconstrained length. As quantum mechanical effects are important at this scale, these wires are also known by the term "quantum wires." In comparison with other low dimensional systems, it is revealed that there are two confined quantum directions for the NWs, while in the case of the electrical conductions, one unconfined direction is left. This configuration allows them to be used in electrical applications. Small diameter NWs have exhibited significantly different electrical, optical, and magnetic properties compared to their bulk (3D) crystalline counterparts, which is due to their unique electronic state density. The size and form of the NWs affect their overall properties causing their manner to be different from that of their corresponding bulk materials. There are several reasons that contribute in making the NWs different from their counterpart bulk materials. These include the increased surface area, the diameter-dependent bandgap, the very high density of electronic states, the joint density of states near the energies of their Van Hove singularities, the enhanced exciton binding energy, and the increased surface scattering for electrons and phonons. Moreover, NWs have utilities from an application standpoint. For instance, some of the critical parameters of NWs for certain properties can be controlled independently, while in their bulk counterparts, it does not make sense. In the NWs with small diameters, some specialized properties can be raised nonlinearly. This can be raised by exploiting the singular aspects of the 1D electronic density of states. Furthermore, NWs have provided a framework for the "bottom-up" approach to be applied in designing the nanostructures for nanoscience investigations and potential nanotechnology applications. These frameworks include the nanoscience development opportunities, the application of the smaller length scales in semiconductors, optoelectronic and magnetic industries, and the dramatic development of the biotechnology industry where the functions are taken at nanoscale. With a review of the current situation of NWs, it can be understood that the research in this field is of considerable broad interest. In contrast to crystalline bulk materials, the aim of this review is to focus on the properties of NWs with regard to the possible applications resulting from their unique properties and the future findings in this area. There are three types of NWs—metallic NWs (Ni, Pt, Au, Al), semiconductor NWs (Si, InP, GaN, etc.), and insulator NWs (SiO_2, TiO_2). Metallic NWs can be applied as building blocks in metal semiconductor heterostructures and metallic interconnectors. Insulator NWs are considered as candidates for future applications in silicon-on-insulator (SOI). Semiconductor NWs with band structure and a broad selection of composition are useful in semiconductor electronics. Semiconductor NWs are the critical components in nanoscale devices.

11.4.1 MORPHOLOGY

The morphology of SiNWs is strongly dependent on the method and mechanism being used for their growth. For example, in NWs grown via the VLS mechanism, a catalyst particle emerges at the tip of the NW after being grown (Wagner and Ellis, 1964). However, in other growth mechanisms, this is not observed. The growth defects such as kinking are exhibited in the crystalline NWs. Using the CVD growth method, Westwater and coworkers (1998) produced SiNWs under higher pressures (>0.3 torr) and/or lower temperatures (<440°C) where kinking tended to be exhibited. The growth direction of these SiNWs was also changed spontaneously during the growth (Westwater et al., 1998). Moreover, SiNWs are often made of a c-Si core surrounded by ana-Si oxide sheath. The thickness of the sheath is different from that of silicon oxide. Niu and coworkers (2004a) synthesized an example of this NW using low vacuum CVD. Using molten gallium as a catalyst, Sharma and Sunkara (2004) also produced NWs with a crystalline core surrounded by an amorphous oxide layer. The growth technique and growth mechanism can often affect the growth direction of SiNWs, known as the characteristic of morphology. For example, after synthesizing gallium-catalyzed NWs, Sharma and Sunkara (2004) revealed that SiNWs tend to grow along the [100] direction. In contrast, the NWs fabricated with SiH_4 and gold as a vapor phase source of silicon and a catalyst, which were treated by a laser ablation via the VLS mechanism, tended to be primarily oriented in the [112] direction, and seldom were observed in the [100] or [111] direction (Tan et al., 2002). Using the CVD method, Westwater and coworkers (1998) have produced a high-density growth of thin NWs with a high SiH_4 partial pressure resulting from a lowered defective chemical potential, which is due to the augmented

SiH_4 partial pressure. This allows the growth of thinner NWs with a higher density (Westwater et al., 1998). The catalyst droplet size (Cui et al., 2001; McIlroy et al., 2004) and the SiH_4 partial pressure (Westwater et al., 1997) affect the diameter of SiNWs grown via the VLS mechanism. In particular, larger diameter NWs were produced from a larger diameter catalyst droplet, and small diameter NWs were produced due to a higher SiH_4 partial pressure (Westwater et al., 1998). SiNWs can grow under a range of growth conditions and growth mechanisms. This range of conditions produces SiNWs with various morphologies. According to the literature, there are many ranges of diameters and lengths for SiNW. SiNWs have been synthesized using a thermal evaporation process with lengths up to 4 mm and diameter of 100 nm (Shi et al., 2005). These NWs have been grown at atmospheric pressure by using the CVD method with an average diameter of 100 nm and an areal density of 3×10^{17} NW·cm^{-2} (Kulkarni et al., 2005). NWs grown from atomic vapor in UHV can exhibit diameters from 3 to 7 nm and lengths of up to 100 nm. These NWs tend to grow in bundles with a diameter of 20–30 nm (Marsen and Sattler, 1999). By reviewing the literature, it can be seen that there are a greater number of publications focusing on the synthesis of small diameter SiNWs. There is particular interest in producing SiNWs with small diameters, as this distinctive property distinguishes them from bulk silicon. A crystalline and an amorphous phase of SiNWs with the diameter ranging from 10 to 20 nm and length of several tens of microns have also been grown using the SLS mechanism (Xing et al., 2003). Hofmann and coworkers (2003) have also grown worm-like amorphous SiNWs using PECVD via the VLS mechanism. It has been possible to observe the growth of SiNWs in UHV-CVD using *in situ* transmission electron microscopy (TEM). With the observation of the surface of silicon, Ross et al. (2005) understood that the surface of NW is a sawtooth-like facet shaped at an angle to the direction of growth. A thermodynamic origin has been proposed for the sawtooth faceting (Ross et al., 2005).

11.4.2 PROPERTIES OF SINWS

SiNWs have unique properties with many of them being changed by the morphology of the NW. The morphology of the NW includes the diameter, length, crystallinity, growth orientation, and the presence of any features such as a catalyst's tip, the number of kinks (growth defects) present in the NWs, and the density of NW growth. These parameters of NWs can also affect properties such as the bandgap and electrical characteristics. SiNWs and other nanostructures tend to differ from bulk silicon due to these properties. How the properties of nanostructured and nanoscale silicon differ from that of bulk silicon is one of the very interesting points of investigation.

11.4.2.1 Structural Properties

Using various measurements and microscopy techniques, we can conduct the study and research of nanostructures. These techniques enable us to characterize the materials with tiny size scales, and even have the objects scaled down to individual atoms. The study and characterization of the structural properties of NWs are essentially important for different types of applications. With this process, we can establish a reproducible relationship between the NWs' desired functionality and their structural and geometrical specifications. The properties of NWs are sensitively dependent on geometrical configuration and surface conditions. This dependency is due to the enhanced surface-to-volume ratio in NWs. Besides, in NWs made of the same type of materials, heterologous properties might emerge. This is due to differences in surface conditions, crystalline size, crystal phase, and aspect ratio, and also dependence on the synthesis methods and conditions of preparation. The properties of NWs are strongly dependent on size, shape, and structure. In special cases, in which the NW diameter is the same as that of the wavelength of the charge carrier, the quantum confinement effect shifts the energy states, and accordingly induces visible photoluminescence in Si. Measuring the electrical or mechanical properties of these NWs is a difficult task, due to the small size of the objects. The TEM–STM technique has increasingly been used during the past years, for example, for the characterization of gold NWs, carbon nanotubes, and gold point contacts. Measurement of Raman spectroscopy (RS) on SiNWs and a comparison of its results with that of bulk silicon have been conducted previously. The diameter-dependent downshift of the c-Si peak is one of the main items of interest that occur in the Raman spectra for SiNWs. In comparison to bulk c-Si, the Raman spectrum of SiNWs has shown a downshift in wavenumber of the TO c-Si peak (Yu et al., 1998a). This shift is dependent on the diameter of the NWs (Wang et al., 2000; Fukata et al., 2005a) as the smaller diameter NWs have a larger shift. Usually, the shift is due to quantum confinement effects (Niu et al., 2004a). For NWs with a diameter of 15 nm, a shift of the TO peak by 13 cm^{-1} was reported. Shifts of 10 cm^{-1} for 20 nm diameter (Niu et al., 2004a) and 9 cm^{-1} for NWs of 10–100 nm diameter (Pan et al., 2005) have also been reported. However, there are some reports indicating that even for NWs with diameters ranging from 3 to 43 nm, this shift was not observed (Zhang et al., 1998). It was observed that the SiNW peak's being down shifted in frequency, in comparison to the peak of c-Si, has a broadened line width with the line shape of asymmetric (Li et al., 1999). The asymmetric broadening in Raman spectra in comparison with bulk silicon is often due to the small diameters of the NWs and the presence of defects within the SiNWs' (Zhang et al., 1998) variations in the diameter of the NWs. Scheel and coworkers (2006) have shown that the material around the NWs affect the Raman spectra of NWs, due to their low thermal conductivity. They found that the exhibited phonon frequency in spectra is related to the thermal conductivity of gases such as He and air surrounding the NWs (Scheel et al., 2006).

11.4.2.2 Optical Properties

Optical methods provide a sensitive and simple experiment for characterization of the electronic structures of NWs, which is related to the fact that they need a minimal sample preparation (the contacts not being required), and the optical measurements are sensitive to quantum effects. Nevertheless, the explanation of these measurements is not always straightforward. Various optical techniques have indicated that the properties of NWs are different from those of their bulk counterparts. An optical graph of 1D materials, such as carbon nanotubes, often indicates an intense specification at particular energies near uniqueness in the common density of states formed with strong quantum confinement conditions. Usually, the light wavelength used to probe the sample is smaller than the length of the wire, but larger than the diameter of the wire. Hence, the probe light used in an optical measurement cannot be focused particularly onto the wire. This means that in an optical measurement, the wire and its substrate are simultaneously probed. However, in transmission and reflection characterizations, even a non-absorbing substrate can modify the measured spectra of NWs. The bandgap or energy gap is an important property of SiNWs, which is dependent on the NW's diameter. Ma and coworkers (2003) have reported that the bandgap augments with the decrease in diameter from 1.1 eV for 7 nm to 3.5 eV for 1.3 nm. Their results are in line with the theoretical calculations made by other groups (Ma et al., 2003). The energy gap was measured using a scanning tunneling spectroscopy (STS) system. The increase in bandgap indicates quantum confinement effect in the SiNWs at low diameters (Ma et al., 2003). In another report, the energy-gap magnitude of the SiNWs approximates that of bulk silicon for larger diameter NWs (Scheel et al., 2005). This means that the energy gap of NWs with diameter approaching 7–10 nm is approaching energy gaps of bulk silicon. Scheel et al. (2005) have also demonstrated that the value of the energy gap is due to quantum confinement and depends not only on the wire diameter but also on the direction of the wire axis.

11.4.2.3 Electrical Properties

Some researchers have measured the electrical properties of SiNWs. To connect the NWs to an external circuit, we often carry out nanolithography and e-beam lithography techniques since it is difficult to establish contact between individual NWs. Electrical measurements of single SiNWs have been carried out, in which the SiNWs were found to be slightly of p-type (Chung et al., 2000). The ohmic contact on the NWs is important as it connects the NWs with devices or an external circuit. Mohney et al. (2004) have measured the contact resistance for p-type NWs of 78 and 104 nm diameters. They found the contact resistance to be about 5×10^{-4} Ω cm^2 for Ti/Au contacts to p-type SiNWs. The measurements were carried out using a four-point technique (Mohney et al., 2005). Cui and Liber (2001) have fabricated devices based on SiNWs using intercrossed p-doped and n-doped SiNWs, in which the cross showed good current reflection. High density SiNWs grown on ITO substrates via the VLS and VS mechanisms have also been examined and doped longitudinally (Goncher et al., 2006). The measured characteristics of the NWs in bulk showed a standard diode characteristic. The synthesized NWs via the VS mechanism were found to have a more pronounced diode characteristic than those grown via the VLS mechanism with Au as a catalyst (Goncher et al., 2006). As reported in the literature, the properties of SiNWs are strongly dependent on the size of the NW. The bandgap of SiNWs for NWs with small diameters changes greatly from that of bulk silicon. The properties of larger diameter NWs tend to approximate those of bulk silicon.

11.5 GOLD CATALYST

Among the metal catalysts, Au has been selected as a catalyst material to grow SiNWs. It is the most widely used, the best, and the simplest catalyst for SiNW growth. Why is gold a favorable catalyst? To answer this basic question, it is necessary to look at the Au–Si system to get informative pieces of information. Using the Au–Si system, one can notice general standards being utilized in other catalyst materials. There are many favorable reasons and advantages in using Au as a catalyst material. Availability is the first reason. Gold is one of the standard metals used in solid state research and electrical contacts. There are many semiconductor research laboratories with evaporation systems equipped with Au. A thin layer of Au can be deposited onto a typical sample substrate with this system. To a thin evaporated layer, one can also use Au colloid nanoparticles with diameters ranging from 2 to 250 nm. Further, Au has a high chemical stability. In particular, considering the point that Au does not get oxidized in air to make an in situ deposition unnecessary, it is an important advantage for the pregrowth sample preparation. Furthermore, the technical requirements of the growth system are reduced due to the high chemical stability of Au, especially with regard to the maximum tolerable oxygen background pressure. Besides, regarding a work safety point of view, we can mention Au being a nontoxic metal as one of its advantages. However, the main reason why Au is used to grow SiNW lies in the binary phase diagram of Au–Si system shown in Section 11.7.

11.6 SUBSTRATES USED FOR THE GROWTH OF SiNWs

SiNWs can be synthesized on a variety of substrates. Therefore, the substrate used to grow SiNWs is often a matter of interest as it can affect the quality of the SiNWs. Selective patterning of the substrate being developed by using various masks or lithographic techniques can also control the growth location of the NWs.

11.6.1 SUBSTRATES

Sometimes, NWs simply fabricated in bulk are placed on the walls of a tube furnace (Shi et al., 2005); however, most of them are synthesized onto c-Si substrates with various orientations. As discussed earlier, usually the VLS mechanism is one of the growth mechanisms used with good catalysts to grow NWs. In the VLS mechanism, a metal catalyst is located on the surface silicon and forms a eutectic alloy. SiNWs are synthesized by injecting a vapor phase of silicon into the system. In the VLS mechanism and other growth mechanisms, the silicon substrate can also be considered as a source of silicon to grow SiNWs. Wu and his coworkers have shown that the appropriate c-Si substrates affect the direction of growth of the NWs in relation to the substrate, using an epitaxial growth mechanism (Wu et al., 2004). Usually, there is a thin native oxide layer on the silicon substrates. Removing this native layer from the substrate is advantageous in the epitaxial growth mechanism. However, this is not always the best approach for the growth of SiNWs since the oxide layer is advantageous for a number of growth techniques. In some growth techniques, the oxygen from the substrate plays a role in the seeding and nucleation of the NWs. The growth mechanisms are the stress-driven growth and the oxide-assisted growth mechanisms of SiNWs. In both these growth mechanisms, the oxide layer is used to synthesize the SiNWs (Prokes and Arnold, 2005; Pan et al., 2005). The HF solution is used to remove the native oxide layer before the deposition of the gold colloid or a gold layer as it is critical for epitaxial growth. Formation of SiO_2 on the gold-deposited Si (111) substrates can be disadvantageous to the epitaxial growth of SiNWs (Jagannathan et al., 2006). Jagannathan et al. (2006) found that removing the silicon oxide layer from the silicon substrate through HF etching, depositing the gold deposition (3 nm), and immediately growing the SiNWs remarkably improved both the growth density and the epitaxy of the SiNWs. Other substrates such as SOI (Gotza et al., 1998), highly oriented pyrolytic graphite (Marsen and Sattler, 1999), and ITO on glass have also been used for the growth of NWs.

11.6.2 PATTERNED GROWTH

SiNWs have been grown in a selected region of a substrate using numerous patterning techniques. A variety of lithographic techniques can be used to aid position control of SiNWs. These include photolithography and electron beam lithography. Both pattern the gold film used for the catalyzed growth of SiNWs by the VLS mechanism (Wu et al., 2002). Regular arrays of vertically aligned NWs have been grown using photographic techniques (Ishida et al., 2003). A SiO_2 mask was created and patterned on a Si (111) substrate. The resist itself was patterned by electron beam lithography and gold was deposited onto the SiO_2 window mask. The gold on the resist could then be removed leaving a regular array of gold dots (Ishida et al., 2003). This leads to growth of single SiNWs in a regular array that can be used for field emission devices. NWs produced by VLS are known to grow where the gold is deposited. By exploiting this, NWs can be grown in a patterned region of substrate, by placing gold nanoparticles on the substrate surface or by e-beam deposition of gold (Salhi et al., 2006). A simple technique for patterning SiNWs into large regular arrays of NW clusters is by using a shadow mask during the deposition of the gold film onto the substrate (Wu et al., 2002; Hofmann et al., 2003). A TEM grid, or fine metal mesh, can be used to produce several micron-sized squares in a regular array.

11.7 VLS GROWTH MECHANISM

The VLS mechanism reflects the pathway of silicon, which coming from the vapor phase passes through a liquid droplet, and finally ends up as a solid. There are also other growth mechanisms proposed and named similar to the VLS mechanism. The key in the growth of wire, the so-called vapor–solid–solid (VSS) mechanism, comes into play when the wire is catalyzed by the growth of catalyst particles in place of solid catalyst liquid droplets. The catalyst material and temperature are very important in sustaining the wire growth via the VLS or VSS mechanism. The VLS mechanism was originally suggested by Wagner and Ellis (1964) to produce micrometer-sized whiskers, and was later reexamined by Lieber's group (Gudekssen and Lieber, 2000; Lauhon et al., 2004) to generate NWs from a rich variety of inorganic materials. There are several reports on 1D nanostructure fabrication from vapor phase precursors without a metal catalyst or obvious VLS evidence, in which the Au clusters are oxidation resistant and have unusual catalytic properties. The basic principle underlying the VLS process is the precipitation of one material from a supersaturated liquid alloy. The more detailed process in the VLS growth is schematically illustrated in Figure 11.6 for Au-catalyzed SiNWs. Four stages can be

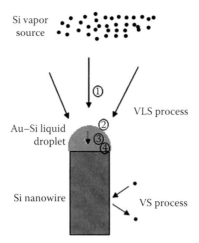

FIGURE 11.6   The VLS process for Au catalyzed Si.

distinguished in the steady-state growth (Cleland and Roukes, 2002): (1) diffusion of species from the vapor source to the vapor/Au–Si droplet interface; (2) diffusion and/or surface reaction on the vapor/droplet interface; (3) diffusion in the liquid droplet; (4) precipitation of Si at the droplet/NW interface.

   Direct deposition of Si from vapor on the side surfaces of NWs takes place in parallel with the VLS growth. The accommodation coefficient of a solid surface, however, is of several orders of magnitude lower than that of a liquid surface (Weaver et al., 1990) resulting in a negligible vapor–solid (VS) deposition rate. It is the large accommodation coefficient of the liquid alloy that ensures the unidirectional growth of NWs and makes the metal catalyst meaningful in most cases. The metal could also reduce the activation energy for precursor decomposition in some systems, such as Au-catalyzed decomposition of SiH_4. The growth process begins when the catalyst becomes supersaturated with reactant, and terminates when the vapor phase precursors pass out of the reaction zone. The resulting wire morphology depends on experimental parameters such as temperature, pressure, flow rate, and the nature of the metal of the catalyst. As a rule, the size limit of wires is controlled by both thermodynamics and growth kinetics (Gudekssen and Lieber, 2000). The liquid droplet and the solid NW together form a symmetrical linearized system. If the size of the catalyst nanoparticles is known, according to equilibrium thermodynamics, the following formula derived by Lieber's group (Gudekssen and Lieber, 2000; Lauhon et al., 2004) can be used to calculate the minimum radius r_{min} of a liquid metal cluster:

$$r_{min} = \frac{2\sigma_{LV}V_L}{RT\ln\sigma} \qquad (11.1)$$

where
   σ_{LV} is the liquid–vapor surface free energy
   V_L is the molar volume

R is the Avogadro gas constant ($R = 8.32441$ J K⁻¹ mol⁻¹)

T is the temperature

σ is the vapor phase supersaturation

Wastwater contributed another formula based on the Gibbs–Thomson equation:

$$r_{min} = \frac{2\sigma_{LV}V_L}{\Delta\mu_{wire} - \Delta\mu_{bulk}} \tag{11.2}$$

where $\Delta\mu_{wire} - \Delta\mu_{bulk}$ is the effective chemical potential change in wire with respect to bulk. The Gibbs–Thomson effect relates surface curvature to vapor pressure and chemical potential. Small liquid droplets exhibit a higher effective vapor pressure, and concentration gradients direct small precipitates to dissolve and larger ones to grow (Westwater et al., 1997). Therefore, the NW diameter is limited by the size of the metallic seed particles and the Gibbs–Thomson equation (Westwater et al., 1997; Gudekssen and Lieber, 2000). Au and Si form a simple eutectic system. Figure 11.7 displays a binary phase diagram of Si–Au. Theoretically, VLS growth occurs only at temperatures above the eutectic point. In many cases, the catalytic nature is still obscure. Au is a good choice for growing many compound NWs because many semiconductors alloy with Au. At a temperature lower than the melting temperature of the components, a liquid alloy, implied as a eutectic alloy, forms. The word "eutectic" is usually used to explain an alloy with the lowest possible melting point. For example, pure Au and pure Si are solid up to 1064°C and 1410°C, respectively, while the Au–Si alloy with 18.6 at.% Si and 81.4 at.% Au melts at $T = 363°C$.

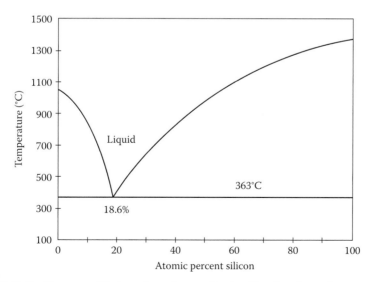

FIGURE 11.7 Binary equilibrium phase diagram for Au–Si and a schematic representation of the VLS growth of single crystal NWs.

11.8   GROWTH TECHNIQUE

11.8.1   CVD

CVD is one of the popular methods used to synthesize certain nanostructures such as NWs. CVD includes all the deposition techniques using chemical reactions in a gas phase. The energy required for the chemical reaction is usually supplied by keeping the substrate at raised temperatures. In the CVD method, a vaporizable gaseous precursor, such as SiH_4 or $SiCl_4$, is used as the silicon source. It is conveyed onto the deposition surface as the precursor reacts, and is cracked into its selectors. Originally, the CVD method was used for high-purity film deposition. Epitaxial growth of SiNWs can occur with the various growth rates by the CVD method depending on the type of Si precursor, pressure, and growth temperature. Moreover, CVD offers, by a controlled method, a broad range of silicon wires with improved properties. Since silicon is easily oxidized when exposed to oxygen at high temperatures, reduction of the oxygen pressure is crucial for the epitaxial growth of uniform SiNWs. When one uses an oxygen-sensitive catalyst, it is useful to merge the deposition of catalyst and the NW growth in one system without breaking the vacuum in between (Wang et al., 2006). Anyway, in the CVD reactor, it is beneficial to use lower base pressure for high or even ultrahigh vacuum since it reduces unwanted impurity, and raises growth at low temperatures (Akhtar et al., 2008a). The pressures in CVD during growth of SiNWs are mainly dependent on the gaseous silicon precursor and its cracking probability at the catalyst surface. For instance, the growth with Si_2H_6 can be carried out at low partial pressures around 10^{-6} mbar. These low pressures enable to combine the CVD with TEM, allowing the *in situ* observation of the NW growth (Hofmann et al., 2008). In contrast to Si_2H_6 gas, the partial pressures required for wire growth with SiH_4 gas are about five orders of magnitude higher. One of the problems of SiNWs grown using the CVD method is that especially for a diameter of about 50 nm or smaller, they exhibit a certain shift of the growth direction (Schmidt et al., 2005). However, when the NWs are grown in a template, such as anodic aluminum oxide (AAO), this impassability can be dominated (Zhang et al., 2001; Lew et al., 2002; Shimizu et al., 2007). In this case, the catalyst is involved in the pores of an AAO membrane, so that the wire growth is limited to the pores of AAO. Therewith, the wire is confined to grow along the pore direction. In this way, epitaxial growth of NWs can be achieved in the <100> orientation. After the growth of NWs, we can remove the template with phosphoric acid as only the upright NWs stand on the substrate. An absence of molecular-level control on bonding of covalent within the films is another disadvantage of the CVD method. The next disadvantage of the CVD method is related to the fact that it is very difficult to integrate the functional organic molecules in the thin film deposition because evaporation at high temperature causes the degradation of the organic species.

11.8.2   PECVD

PECVD is another common method to produce nanostructures. The PECVD method was introduced in the early 1970s by which Soviet researchers were able to increase the concentration of atomic hydrogen by using an electrical discharge. This technology was then developed to be used in the semiconductor IC industry. Today, the PECVD technique is the most commonly used technique for the deposition of dielectric films and other materials like metals and semiconductors (Steiner, 2004). Considering the fact that the deposition temperatures needed to deposit PECVD oxide films, which is referred to as using plasma to dissociate the gaseous precursors, is lower compared to that of low pressure chemical vapor deposition (LPCVD) films, PECVD oxides are commonly used as protective layers for masking and passivation, specifically on devices that have been deposited with metals. In the PECVD process, glow-discharge plasmas are sustained within chambers where simultaneous vapor-phase chemical reactions and deposition occur. Just as the demands of semiconductor technology provided the driving force for advances in sputtering and plasma etching, PECVD processing arose from similar imperatives. A major early commercial application of PECVD was the low-temperature deposition of silicon nitride films for the passivation and encapsulation of completely fabricated microelectronic devices. In the majority of the PECVD processing activities, glow-discharge plasmas are excited by an RF field. It is related to the fact that most of the depositions done by this method are dielectrics, in which DC discharges are not feasible. Generally, the employed RF frequencies range from about 40 kHz to 100 MHz. A conventional PECVD operates at a typical plasma frequency of 13.56 MHz. The highest frequency regime of the PECVD that has been reported is 144 and 150 MHz (Oda and Nishiguchi, 2001). In the reduced gas pressure environment, typically sustained between 50 mtorr and 5 torr, electron and positive ion densities rate range between 109 and 10 n cm^{-3}, and average electron energies range from 1 to 10 eV. This energetic discharge is sufficient to decompose gas molecules into a variety of component species, that is, ions, atoms, and molecules in ground and excited states, molecular fragments, free radicals, etc. The PECVD process can occur in plasma systems. Using the RF energy to create species with high reactivity in the plasma, the process can operate at lower temperatures at the substrates (150–350°C).

11.8.3 Very High Frequency Plasma Enhanced Chemical Vapor Deposition

For the cost-effective mass production of micro/nanostructures, it is needed to provide large width and high rate deposition technologies using the PECVD methods. For μc-Si, especially, the coefficient of absorption for infrared (IR) and red is very low such that layer thicknesses of more than 1.5 μm are needed for sufficient absorption. Therefore, researchers focused on the very high frequency plasma enhanced chemical vapor deposition (VHF-PECVD) method. This method has advantages due to high plasma density and less ion bombardment. For high quality and high rate deposition using the VHF-PECVD method, some researchers tried in the late 1980s (Curtins et al., 1987; Oda and Yasukawa, 1991; Heintze et al., 1993; Kuske et al., 1995). The deposition rates of only a few angstroms per second at a plasma excitation frequency of 13.56 MHz to 2–3 nm s^{-1} at VHF were reported (Schwarzenbach et al., 1996). However, a group from Neuchatel University in Switzerland in the mid-1990s reported a study on a-Si/μc-Si tandem solar cells using VHF plasma enhanced in the 70–100 MHz range, after which the VHF-PECVD began to receive widespread attention (Meier et al., 1996; Kroll et al., 1997). The VHF-PECVD is more proper

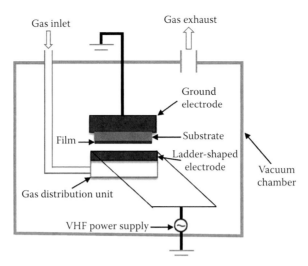

FIGURE 11.8 A diagram of the VHF-PECVD using the ladder-shaped electrode.

for high quality and high rate deposition than the conventional RF plasma. However, for large area deposition, the shorter VHF wavelength nullifies the drawback of interelectrode voltage inhomogeneities. This problem was overcome by developing an appropriate ladder-shaped electrode (Figure 11.8) with a VHF phase-controlled power supply (Murata et al., 1996; Takeuchi et al., 2001).

This was achieved in the subsequent development of a VHF-PECVD instrument and deposition technologies for a-Si films with a deposition rate as high as 1 nm s^{-1} onto a glass substrate as large as 1.1 × 1.4 m^2. This is the largest substrate used in the world for solar cell deposition. Lower plasma potentiality and higher electron density, higher film qualities, resulting in higher deposition rates, are the advantages of the VHF-PECVD. Positively, VHF power is correlated with plasma potentiality and electron density. Therefore, when VHF power increases, electron density augments and the plasma potential increases as well. The ion damage and sequential film quality degradation can be due to the increase of plasma potentiality. Affirmatively, plasma excitation frequency is correlated with electron density and negatively correlated with plasma potentiality. This means that higher plasma excitation frequencies are desirable for higher film qualities and higher deposition rates. The film uniformity of the large area deposition and electrical qualifications such as VHF feeding determine the upper limit of the VHF frequency. Here, it is notable that the gas pressure has similar effects with respect to the frequency. Since the VHF plasma at higher pressure should allow for higher film qualities and higher rates, the VHF has the advantages of higher qualities and higher rates in PECVD, which was shown by theoretical and experimental consideration. However, since the reactor dimensions become comparable to the wavelength associated with the excitation frequency, achieving a large area homogeneous deposition is difficult using VHF plasma. Hence, the substantiation of VHF-PECVD requires the improvement of an innovative technology for large area homogeneous deposition using VHF plasma.

11.9 FABRICATION OF ULTRA-SHARP TIP POINTED Si NANONEEDLE

1D nanostructures have attracted extensive interest due to their potential uses in nanoelectronic and optoelectronic devices and being of fundamental importance for the study of size-dependent chemical and physical phenomena (Morales and Lieber, 1997; Cao et al., 2010). Among these 1D nanomaterials, SiNWs have attracted much attention due to their suitability for use with conventional silicon-based IC technology. SiNWs are particularly favorable to be used as building blocks to fabricate nanoscale devices such as nanotransistors (Shan and Fonash, 2008), metal–oxide–semiconductor field effect transistors (MOSFETs) (Najmzadeh et al., 2010), sensors (Wanekaya et al., 2006), logic device (Li et al., 2007), and solar cells (Kelzenberg et al., 2008). Small diameter SiNWs are fronted as attractive nanomaterials due to their quantum size effect for future devices. Theoretical calculations indicate that essential quantum size effects will happen for a SiNW with a diameter less than 10 nm (Olinga et al., 2000). In the past few years, several researchers have reported the synthesis of SiNWs using various methods, including laser ablation (Zhang et al., 1998); supercritical fluid solution phase (Holmes et al., 2000), thermal vapor deposition (Zhang et al., 2000), rapid chemical etching technique (Peng et al., 2005), CVD (Wu et al., 2004), and PECVD (Hofmann et al., 2003). Among the chemical techniques, PECVD is one of the most advanced techniques. The PECVD technique is a directional deposition method. In the PECVD process, a gas such as silane and the subsequent metal catalyst moderates the growth of SiNWs. This involves a mechanism that is known as the VLS, and was first proposed by Wagner and Ellis (1964). The VLS fabrication mechanism is the most successful synthesis method for fabricating the single crystalline SiNWs with high production yields (Paulo et al., 2007). In the VLS mechanism, a metal catalyst, such as Au, is deposited onto a substrate, such as silicon. This catalyst can be of thin layers or nanoparticles. Usually, Au nanoparticle catalysts have a remarkable capacity to control SiNW sizes. In the present research, the VHF-PECVD method was used for the growth of Au-catalyzed SiNWs on Si substrate. The theoretical and experimental consideration indicates that the VHF-PECVD has advantages such as high deposition speed, good quality, higher electron density, lower plasma potentiality, and photovoltaic performance due to reduced ion bombardment compared to the conventional RF (13.56 MHz) PECVD (Takatsuka et al., 2004; Shah et al., 2005). In this technique, NWs can be grown with good crystallinity and high growth rate at lower substrate temperature compared to other methods. The morphology of these Si nanowires is different from the Si nanowires grown previously by the conventional VLS methods (Gentile et al., 2008; Niu and Wang, 2008). Our Si nanowires are needle-like with a unique sharpness below 3 nm. These nanoneedles are sharper than the nanoneedles fabricated in the last report (Červenka et al., 2010). The SiNWs were synthesized with an Au catalyst on a Si (111) wafer by the VHF-PECVD method in vacuum. In this study, the Si wafer was coated with a 30 nm sized gold colloid particle solution as a catalyst. The size of the Au catalyst is important for the diameter of the SiNWs in the VLS mechanism because the Au catalyst takes place at the droplet interface (Cui et al., 2001). To obtain small diameter catalysts, the Au-coated substrate was heated at a temperature of 500°C for 5 min. For the synthesis of the SiNWs, pure SiH_4 (99.9995%) gas as the Si source with 10 sccm was introduced to the VHF-PECVD vacuum reactor at the pressure of 77 mtorr for 10 min. The power of RF plasma and frequency were 15 W and 150 MHz, respectively. During the deposition, the substrate temperature was maintained at 370°C. The as-grown SiNWs were then analyzed using a field emission scanning electron microscope (FESEM, JEOL, JSM-6701F) and high-resolution transmission electron microscopy (HRTEM, JEOL, JEM-2100). The elemental composition of the ultrasharped SiNWs and their crystal phase were analyzed through an x-ray diffraction technique (XRD), with Cu Kα radiation and an energy dispersive x-ray spectrometer (EDX). Raman spectroscopy was performed using a spectrum GX (NIR, FT-Raman) system with an Nd crystal laser source and 1 μm spot size.

11.10   RESULT AND DISCUSSION ON PROPERTIES OF ULTRA-SHARP TIP POINTED Si NANONEEDLES

Figure 11.9 shows the FESEM image of SiNWs synthesized on Si (111) substrates using a gold-catalyzed VHF-PECVD method under the experimental conditions described in the figure caption. According to Figure 11.9, a population of needle-like ultrasharp pointed NWs with lengths ranging from 4 to 5 μm were grown on the Si wafer. Adjusting the RF power and the temperature led to the controlled growth of ultrasharp pointed SiNWs, as seen in Figure 11.9.

Figure 11.10 shows the EDX spectra of the nanowires. EDX shows that these nanowires are composed mainly of Si and Au (93.37 and 6.62 at.%, respectively).

Figure 11.11 displays low-magnification TEM and HRTEM images of the uppermost part of the generated SiNWs. The low-magnified TEM (Figure 11.11a) provides further information regarding the structure of grown SiNWs. From this image, it can be seen that the wire has a needle-like structure with an ultrasharp pointed tip. This result corresponds to the results obtained with the FESSEM image. In Figure 11.11b, a very small Au catalyst particle is also observed at the tips of ultrasharp SiNWs.

FIGURE 11.9   FESEM image of ultrasharp pointed SiNWs synthesized with gold colloid as catalyst on Si (111)-oriented using VHF-PECVD method. The nanowires were grown for 10 min at a temperature of 370°C and RF power of 15 W.

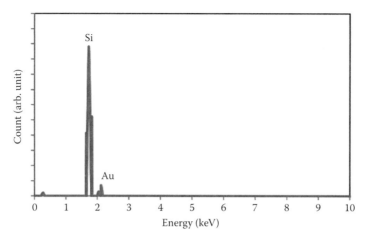

FIGURE 11.10   The EDX spectra of the ultrasharp pointed SiNWs.

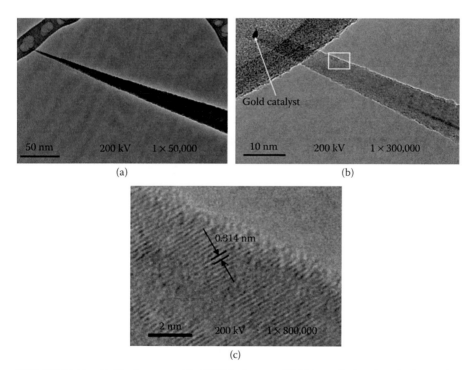

FIGURE 11.11   (a) The low-magnified TEM of grown SiNW reveals that the wire is needle-like structure with ultrasharp pointed tip. (b) A very small Au catalyst particle (black spot) is placed at the tip of the ultrasharp SiNW. The size of Au particle and tip of wire are 1.85 nm and 2.408 nm, respectively. (c) The HRTEM image of a section of SiNW shown in white box in image (b) reveals that each nanowire consists of crystalline structure with interplanar spacing of crystals about 0.314 nm, matching well with the {111} plane of silicon.

The morphology confirms that the SiNWs were obtained by the Au-catalyzed VLS growth mechanism. Before the growth of nanowires, the size of the gold colloid solution deposited on the silicon substrate was 30 nm. However, the Au catalyst size on the tip of the nanowires at the end of synthesis decreased 10 to 15 times of the nominal size at the beginning. Figure 11.11c shows the HRTEM image of SiNW grown in this experiment. The structure is further confirmed by HRTEM characterization of Si nanowire. The HRTEM image reveals that each nanowire consists of a single crystalline structure. According to the HRTEM measurement and the following calculation by the software of the HRTEM digital micrograph, the interplanar spacing of crystals is about 0.314 nm, matching well with the {111} plane of silicon and the growth of the nanowire along the <111> direction.

Figure 11.12 shows the XRD spectrum of the ultrasharp pointed SiNWs grown on the Si substrate, which displays a high-intensity peak of the Si (111) indicating that the SiNWs were the well single crystalline structure. The spectrum also exhibited the growth direction of the SiNWs in the <111> orientation, which is consistent with the results of the HRTEM (Figure 11.11). Calculated from the interplanar spacing of the most intense (111) peak ($d = 0.314$ nm), the lattice parameter of the SiNWs was obtained as $a = 0.5452$ nm, which is larger than the standard value $a = 0.5430$ nm for bulk silicon. This reveals that there is a slight lattice expansion and distortion in the SiNWs structure. There is no trace of oxygen peaks in the XRD spectrum. Au peak was also detected because of the gold nanoparticles serving as a catalyst being located on the top of the tapered SiNWs as shown in the HRTEM images.

Most of the Si nanowires produced by the Au catalyst have had uniform diameters of higher than 10 nm (Salhi et al., 2006; Gentile et al., 2008). The tip radius of nanocones produced by other researchers via the VLS mechanism were also in the range of 10–200 nm (Jeon and Kamisako, 2009; Woo et al., 2009; Červenka et al., 2010). Thus, our ultrasharp pointed Si nanowires with an unprecedented sharpness of below 3 nm are unique in the category of Au-catalyzed grown Si nanowires via

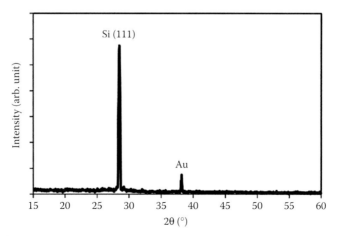

FIGURE 11.12   The XRD spectrum of the ultrasharp pointed SiNWs grown on the Si substrate.

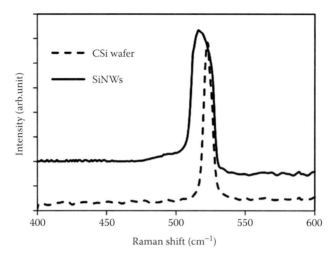

FIGURE 11.13 Raman spectrum of a single crystalline of ultrasharp pointed SiNW (solid line) and silicon wafer (dashed line) at room temperature. The ultrasharp pointed SiNW has an asymmetry peak centered at 516 cm^{-1}. The sharp peak at 522 cm^{-1} corresponds to crystalline Si substrate.

the VLS mechanism using the PECVD method. Figure 11.13 shows the Raman spectra of the as-grown SiNWs (solid line) and a single crystalline Si wafer (dashed line). Raman spectroscopy is very sensitive to the lattice structure and the crystal symmetry. The peak energy, the peak width, and the symmetry in Raman spectra change with reduction of the Si nanowires' size (Adu et al., 2006). As seen in Figure 11.13, the as-grown SiNWs have an asymmetry Raman peak centered at 516 cm^{-1} where this peak is remarked to be the first-order TO photon mode. The high symmetric peak centered at 522 cm^{-1} is related to bulk c-Si. The important downshift of the asymmetrical TO peak shown from 522 cm^{-1} of bulk silicon to 516 cm^{-1} of SiNWs is due to the diameter decrease of the SiNWs (Yu et al., 1998b). Here, size confinement or the nanoscale size of wires is the reason for the asymmetry and broadened width of peak in the spectrum. This is a sign of phonon confinement (Adu et al., 2005). Raman spectroscopy taken from SiNWs reveals that their structure is crystalline. This is consistent with the HRTEM analysis in Figure 11.11c.

A comparison between the VHF-PECVD method and those of some other researchers for the growth of Si nanoneedles and nanowires is shown in Table 11.3. According to the table, the tip diameter and the growth temperature of nanowires obtained by the VHF-PECVD method are remarkably lower than those of the nanoneedles and nanowires grown using other systems. Furthermore, grown nanoneedles using the VHF-PECVD method in this research have only single crystalline structure, while other results show nanoneedles and nanowires having crystalline core and amorphous shell structures.

It should be noted that in the present work, the growth rate of Si nanowires, around 500 nm min^{-1}, was approached during 10 min deposition. This value is significantly higher than the usual growth rates (12–50 nm min^{-1}) of Si nanowires grown by classical CVD, low pressure CVD (LPCVD), and conventional medium

TABLE 11.3
Characteristic of Nanowires and Nanoneedles Grown Using Various Methods

Method	Tip Diameter (nm)	Structure	Growth Rate (nm min⁻¹)	Growth Temperature (°C)
VHF-PECVD (present research)	<3	Crystalline	500	370
PECVD (Červenka et al., 2010)	<10	Crystalline/amorphous	1000	400
LPCVD (Salhi et al., 2006)	40	Crystalline/amorphous	50	500
PECVD (Woo et al., 2009)	<30	Crystalline/amorphous	12	380
CVD (Adu et al., 2006)	50–60	–	50	400–500

frequency of 13.56 MHz PECVD methods (Hofmann et al., 2003; Salhi et al., 2006; Becker et al., 2008). This fast growth rate is probably due to plasma power, which causes high decomposition of the SiH_4 gas. The existence of Au peak in the spectra confirmed that the Au catalyst-assisted VLS mechanism occurred during the growth process. The VLS mechanism correctly describes the gold-catalyzed CVD growth of NWs. In this mechanism, the silane (SiH_4) molecules at first decompose onto the surface of gold catalyst nanoparticles in the form of silicon. Then, the silicon atoms dissolve into the liquid gold particles resulting in the formation of Au–Si alloy. The continuing adsorption of the Si in Au–Si liquid droplets leads to the alloy being supersaturated, which brings about the growth of a solid silicon core in the bottom of the droplet. The Au–Si droplet remains at the top of the solid silicon, and can accept more silicon atoms from the source. As a result, the solid silicon grows in the liquid–solid interface leading to the growth of the NW. Thus, the VLS reaction includes three distinct stages: (a) diffusion of silicon atoms from the vapor source to the vapor/Au–Si droplet interface; (b) diffusion of silicon atoms into the liquid droplet; and (c) precipitation of Si atoms at the droplet/NW interface. The decomposition of SiH_4 gas is a thermally activated process in which the activation energy of the CVD-grown SiNWs with an Au-catalyzed type was found in compliance with the activation energy of SiH_4 decomposition (Kikkawa et al., 2005). With utilizing the plasma activation to CVD system for growth of SiNWs, the VLS mechanism is no longer valid since the plasma deposition is a non-equilibrium process thermodynamically. Moreover, VHF-PECVD causes SiH_4 to predissociate to an ionized gas. Accordingly, the silicon atoms have a greater diffusion and adsorption potential in VHF-PECVD than in CVD. As the SiH_4 molecules are dissociated in the plasma, the growth process is not limited by the SiH_4 decomposition rate as in CVD, and other factors start playing a role. The other main effect of the VHF-PECVD system is the non-catalyzed Si sidewall deposition. This type of deposition causes tapering of Si nanowires in VHF-PECVD, where it allows production of Si nanoneedles with ultrasharp tips. As shown in the HRTEM image (Figure 11.11c), the Si nanoneedle consists of a crystalline Si structure. The VHF-PECVD growth conditions promoted the growth of crystalline Si layers on the surface of the nanoneedle sidewall without gold catalytic nanoparticles during the growth of the NWs.

Synthesizing needle-like SiNWs below 3 nm in tip will open up the door to fabricate worthwhile devices where quantum effects play a significant role. These ultrasharp nanoneedles can be used as tips of special microscopes, for instance atomic force microscopy (AFM), due to their low apex angle (<5°). These Si nanowires would also be beneficial for field emission and solar cell applications as good light absorbers (Chuen et al., 2005; Najmzadeh et al., 2010).

11.11 SYNTHESIS OF Si NANONEEDLES VIA VSS MECHANISM

Over the last few years, self-assembled semiconductor nanowires have been considered as one of the promising candidates for fabricating nanoelectronics, optoelectronics, and nanosensor devices (Samuelson, 2003; Yu et al., 2010). This is due to the difference between their electronic, optical, chemical, and mechanical properties with respect to their bulk matches. SiNWs have attracted extensive research among semiconductor nanowires (Akhtar et al., 2008b; Swain et al., 2010; Wolfsteller et al., 2010; Shiu et al., 2011). Different methods have been employed to synthesize SiNWs. However, in most of these methods, the growth temperature was very high. Among these methods, PECVD with various frequencies has some advantages such as excellent quality, lower growth temperatures, and higher length of NWs compared to other methods (Shiu et al., 2011). In the PECVD process, the precursor gas dissociates by the plasma into highly active radicals (Yu et al., 2008), which can be moderated for SiNW growth. SiNW growth by means of PECVD using a metal catalyst via the so-called VLS mechanism was extensively investigated by Wagner and Ellis (1964). However, temperatures above the eutectic point were usually required for the growth of SiNWs. For instance, temperatures over 363°C, which corresponds to a eutectic transformation of the Au–Si alloy, are needed when Au is used as the catalyst. Silicon atoms resulting from the decomposition of vapor-phase source materials such as silane are absorbed by the Au nanoparticles. The absorbed Si atoms and the Au nanoparticles form an alloy (Au–Si) in the form of liquid droplet at the growth temperature. Si precipitates onto the substrate when the liquid alloy reaches supersaturation, to form SiNWs. Meanwhile, some researchers (Persson et al., 2004; Wang et al., 2006; Kodambaka et al., 2007) identified a very interesting growth mechanism, that is, the VSS mechanism, to grow NWs below the eutectic temperature using metal nanoparticles as catalysts, that is solid-phase, during NW growth. In this research, synthesis of SiNWs at temperatures lower than the Au–Si eutectic temperature and grown by the VHF-PECVD (150 MHz) method was applied. As a result, the VSS process using pure silane precursor gas and Au colloid nanoparticles as the catalyst was investigated. SiNWs were grown on one-sided polished n-type (111) silicon wafers with a resistivity ranging from 0.001 to 0.002 Ω cm by using the VHF-PECVD (150 MHz) system via the VSS mechanism. First, the substrate surface was cleaned by immersing it in a mixture of HCl, deionized (DI) water, and H_2O_2 solution. At the next stage, the cleaned substrate was dipped into 0.1% w/v poly-L-lysine, then rinsed with DI water, and dried by nitrogen. After being dried, the substrate was deposited with a colloidal solution of 30 nm sized gold nanoparticles, and finally rinsed with DI water and dried by nitrogen. The Si substrate with the gold nanoparticles colloid catalyst on top was then placed in a VHF-PECVD chamber to carry out the SiNW growth. The chamber was degassed under vacuum and was purged using Ar gas. Finally, the substrate was heated to growth temperature ranging from 363°C to 230°C. A pure SiH$_4$ (99.9995%) gas with 10 sccm total flow rate was then introduced into the chamber to begin the SiNW growth. The RF power and the pressure were kept at 15 W and 78 mtorr, respectively, during the SiNW growth. After growth completion, the samples were cooled down slowly to room temperature in the vacuum chamber. To investigate the synthesis of SiNWs at temperature lower than the Au–Si eutectic temperature, regardless of different growth temperatures starting from 363°C to 230°C, the same above-mentioned experimental procedure was repeated. The grown SiNWs were then analyzed using a FESEM (FESEM, JEOL, JSM-6701F) and HRTEM (HRTEM, JEOL, JEM-2100) with TEM operation voltage at 200 kV. The elemental composition of the SiNWs and their crystal phase were analyzed through an XRD, with Cu Kα radiation and an EDX.

11.12   STRUCTURAL AND PHOTOLUMINESCENCE PROPERTIES OF Si NANONEEDLES GROWN VIA VSS MECHANISM

SiNWs were synthesized at below Au–Si eutectic temperature by the VHF-PECVD via the VLS mechanism. Figure 11.14 shows the FESEM images of SiNW morphology obtained at different growth temperatures. As can be seen, most of the wires possess long and tapered morphology with a smaller diameter at their tip compared to the bottom. Moreover, it was observed that by decreasing the growth temperature, heterogeneous distribution of SiNWs was achieved, which is clearly shown in Figure 11.14. Furthermore, decreasing the temperature to 230°C prevented SiNW growth (Figure 11.14f) indicating the fact that 250°C was the lowest temperature for the growth of SiNWs by the VHF-PECVD method with other experimental conditions in this research. The growth prevention can be attributed to the lack of SiH$_4$ dissociation during the process at temperatures lower than 250°C. This significantly reveals that temperature is an important parameter to be considered for SiNW growth.

The EDX analysis of Au-catalyzed SiNWs grown by the VSS mechanism at 320°C is shown in Figure 11.15. The EDX spectrum indicates that the nanoneedles are composed of Si, Au, and O. Figure 11.15a displays the EDX measurement on the tip of the SiNW where Au is located. This spectrum and the relative table display Au as the dominant element on the tip, while Si dedicates only a little fraction of 4.62 wt.%. Such a small fraction is a result of low solubility of Si in solid Au. Figure 11.15b shows the EDX spectra of a SiNW stem. It indicates that no trace of Au is found, confirming that Au was not diffused along the SiNW surface, while a high percentage of Si, 80.8%, is found in the stem. It reveals that in the growth of Au-catalyzed SiNWs by the VSS mechanism, deposition of silicon on the sidewall of the NW took place. Besides, the detected O in the spectrum was mostly due to the reaction of Si with the oxygen in the air. Cu and C were also detected in the EDX analysis, which were not considered as nanowire. The detected Cu peak in the spectra is the effect of using

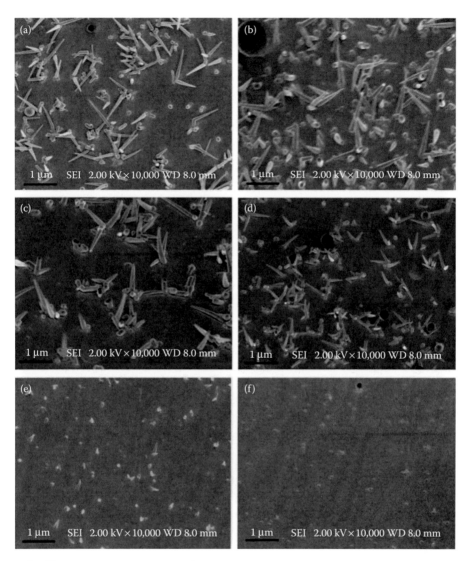

FIGURE 11.14 FESEM images of SiNWs at various growth temperatures: (a) 363°C, (b) 320°C, (c) 290°C, (d) 270°C, (e) 250°C, and (f) 230°C.

a copper grid in the EDX sample preparation, and the detected C is because of the carbon tape used to reduce the charging effect on the EDX system.

For further investigation in the relationship between the morphology and microstructure, the TEM and HRTEM were employed on individual SiNWs. Figure 11.16a shows a TEM image of an interesting NW with bulk quantity of Au catalyst particles at the tip of the NW. It clearly can be seen that the SiNW consists of straight, tapered, and cone-like structures. From this image, it can be observed that the tip of the wire is composed of an Au–Si droplet with a smooth and curved surface. The tapering

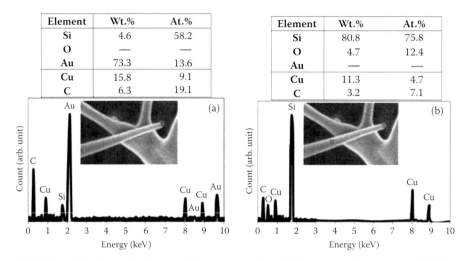

Element	Wt.%	At.%
Si	4.6	58.2
O	—	—
Au	73.3	13.6
Cu	15.8	9.1
C	6.3	19.1

Element	Wt.%	At.%
Si	80.8	75.8
O	4.7	12.4
Au	—	—
Cu	11.3	4.7
C	3.2	7.1

FIGURE 11.15 **(See color insert.)** The corresponding EDX spectra of Au-catalyzed SiNWs grown by VSS mechanism at 320°C (a) tip and (b) stem of the nanowire.

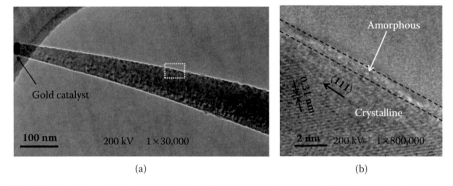

FIGURE 11.16 (a) The low-magnified TEM image of a grown SiNW reveals that a gold nanoparticle terminated the wire tip and (b) HRTEM image of a middle part of SiNW stem (white square box in image (a)) reveals that NW consists of a crystalline Si core and an amorphous sheath with growth direction of <111>.

shape of the NWs is probably due to the uncatalyzed deposition of Si on the sidewall of the NWs (vapor–solid mechanism) (Wagner and Ellis, 1965; Givargizov, 1975; Hannon et al., 2006). To investigate the details of the crystal structure, the corresponding HRTEM image, shown in Figure 11.16b, was taken from the middle part of the SiNW stem located at about 400 nm from the tip.

According to this image, the SiNW consists of a c-Si core surrounded by a thin a-Si oxide layer. The NW surface is oxidized because the sample has been transferred through air. The silicon oxide thicknesses are about 1–2 nm. NW growth directions are generally determined using the HRTEM that shows lattice fringe spacing. The spacing of the lattice plane is about 0.314 nm, which matched well with the (111) plane of the cubic diamond silicon structure. From Figure 11.16b, it

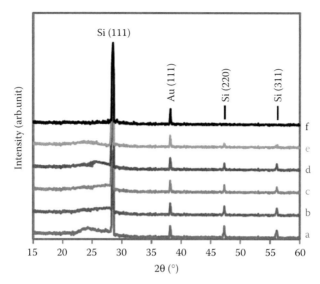

FIGURE 11.17 XRD patterns of the as-synthesized SiNWs at various temperatures. (a) 363°C, (b) 320°C, (c) 290°C, (d) 270°C, (e) 250°C, and (f) 230°C.

can also be found that the growth direction of this NW is along the <111> direction. Figure 11.17 shows the XRD patterns of the SiNWs synthesized at different temperatures. The three main peaks of Si located at 2θ of 28.501°, 47.405°, and 56.247° corresponding to (111), (220), and (311) crystallographic planes, respectively, confirm that the synthesized NWs are crystallized SiNWs. However, the pattern corresponding to the sample treated at 230°C does not show the peaks at 2θ of 47.405° and 56.247°, indicating that no wire growth has taken place.

It should be noted that the intensity of the Si (111) peak mainly belonging to the Si substrate is obviously higher than the other peaks in all spectra. This discloses that the peak originating from the Si substrate is in the (111) direction. Besides the mentioned diffraction peaks, a weak peak in the form of a hump located at ~25° indicates that amorphous SiO_2 contributes to the XRD patterns. The XRD pattern and HRTEM images confirm that SiNWs consist of a c-Si core with a thin a-Si oxide shell. Raman spectroscopy was applied to investigate more closely the crystalline structure of SiNWs. Figure 11.18 shows the Raman spectrum of the SiNWs growth at different temperatures. It consists of six peaks representing six different growth temperatures. The two peaks at 519 cm⁻¹ correspond to the SiNWs shown in Figure 11.14a and b, whereas the rest at 516, 517, 518, and 522 cm⁻¹ refer to those displayed in Figure 11.14c–f, respectively. These peaks are regarded to be the first-order TO phonon mode. Owing to the fact that the size of the grown SiNWs at different temperatures is invariant, there are downshifts for the peaks. The different growth temperatures can be distinguished through the variation in intensity and symmetricity of the peaks.

The peak *f* is approximately symmetric due to the nonexistence of SiNW growth at this temperature, but the rest are asymmetric. Usually, two factors contribute to asymmetric Raman peaks: nanoscale size and defects (Nolsson and Nelin, 1972;

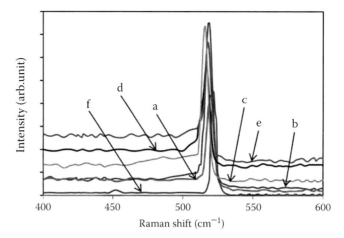

FIGURE 11.18 Raman spectrum of the SiNWs fabricated by PECVD at (a) 363°C, (b) 320°C, (c) 290°C, (d) 270°C, (e) 250°C, and (f) 230°C.

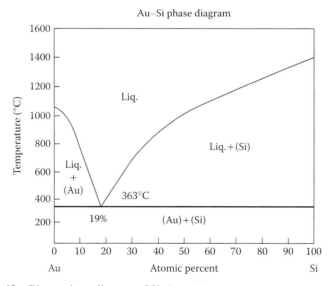

FIGURE 11.19 Binary phase diagram of Si–Au system.

Zhang et al., 1999). Some researchers have previously reported Raman spectroscopy measurements of SiNWs (Niu et al., 2003a, 2004b; Marczak et al., 2009) and it is similar with these results. It should be noted that both the VLS and VSS mechanisms are likely to happen in the growth of SiNWs based on the applied temperature. Theoretically, VLS growth occurs at the temperatures above the eutectic point when an alloy droplet is formed (Figure 11.19). In the VLS mechanism, the growth temperature is higher than the Au–Si eutectic temperature (363°C) in which the Au catalyst particle is a liquid.

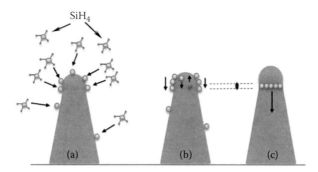

FIGURE 11.20 A schematic picture of the VSS mechanism for SiNW growth. (a) decomposition and transfer of SiH_4 species to the Si substrate, (b) delivery of the Si atoms to the Au catalyst/SiNW interface via surface or bulk diffusion reactions, and (c) incorporation of adatoms into the Au catalyst/SiNW interface leading to VSS growth. (Adapted from Kovalev, D., Heckler, H., Polisski, G., and Koch, F., *Physica Status Solidi B*, 215, 1999.)

On the other hand, the VSS (Falk et al., 2007; Li et al., 2009; Zhang et al., 2009) mechanism is more likely to take place at the lower temperatures (<363°C). Figure 11.20 shows a schematic picture of the VSS mechanism for the growth of tapered NWs. First, a proper source material is delivered to the Si substrate for VSS growth to occur. During the process, the vapor phase is deposited on the surface of the catalyst or NW. SiH_4 molecules in the VHF-PECVD system are decomposed by RF power at the growth temperature to provide adatoms such as Si to the surfaces of the catalyst and NW. Growth of Si nanowires below the nominal decomposition temperatures is possible because solid and liquid metal particles accelerate decomposition of SiH_4. At the next step (Figure 11.20b), Si adatoms are delivered to the growth interface via bulk or surface diffusion. If the metal or semiconductor atoms are more moveable, bulk or surface diffusion must be fast enough to support the growth of NWs. Usually, surface diffusion is much faster than bulk diffusion. Finally, as shown in Figure 11.20c, the inclusion of adatoms at the interface is preferred in VSS growth (Wacaser et al., 2009). In the VSS mechanism, during the growth of NWs, the catalyst remains solid at the top of the wires as a spherical cap, and increases the adsorption and decomposition of the precursor gas. In summary, Au-catalyzed growth of SiNWs can be divided into three steps: (1) decomposition and transfer of SiH_4 species to the Si substrate, (2) diffusion of Si atoms into the Au nanoparticle and/or precipitation of the atoms on the surface reaction, and (3) incorporation of adatoms into the growing SiNWs. All the foregoing steps should take place in order to make the VSS mechanism dominant.

In contrast to the VLS growth mechanism, the growth through the VSS mechanism consumes much more energy due to the diffusion of Si atoms into the NW through a solid catalyst. Furthermore, the inadvertent incorporation of impurities, particularly originated from the catalyst, is reduced in the wires grown by the VSS mechanism. This is related to lower solid solubility and atom diffusivity associated with the lower temperatures in the VSS mechanism. It seems that precipitation of silicon in the solid Au is much slower than in the liquid. It was found that Au–Si

NWs can be synthesized at lower temperatures than that of Au–Si eutectic if appropriate pressure and RF power are chosen. Higher plasma excitation frequencies are capable of producing atomic hydrogen with high density and high hydrogen coverage, thus providing more sources of favorable gas phase reactive species: SiH_1, SiH_2, and SiH_3 radicals. This promoted growth surface to increase surface diffusion of precursors. As a result, higher deposition rate and high qualities of crystallized SiNWs were achieved. Generally, the NW growth rate by the VSS process is much slower because the rate of precipitation of the supersaturated silicon in the solid Au catalyst is much lower than the rate in the liquid one. Furthermore, VHF-PECVD causes SiH_4 to predissociate to an ionized gas. Accordingly, the silicon atoms have a greater diffusion and adsorption potential in the VHF-PECVD than in other methods such as the CVD. The higher diffusion and adsorption increase the growth rate of SiNWs unlike the VSS mechanism, as it is found, that is, ~67 nm min^{-1} at 320°C; however, the rate of 250°C achieved in this research is considerably much lower. In comparison with the previous works, it is clear that the operating temperatures of the present investigation for the synthesis of SiNWs using a Au catalyst was much lower than the temperature (340°C) used by other researchers (Aella et al., 2007; Qi et al., 2008) with a similar method. However, their applied temperature is lower than the Au–Si eutectic temperature. Since the growth temperatures in this research were less than the Au–Si eutectic temperature, NWs were expected to grow via the VSS mechanism.

REFERENCES

Adu, K. W., Gutierrez, H. R., and Eklund, P. C. 2006. Raman-active phonon line profiles in semiconducting nanowires. *Vibrational Spectroscopy.* 42(1), 165–175.

Adu, K. W., Gutierrez, H. R., Kim, U. J., Sumanasekera, G. U., and Eklund, P. C. 2005. Confined phonons in Si nanowires. *Nano Letters.* 5(3), 409–414.

Aella, P., Ingole, S., Petuskey, W. T., and Picraux, S. T. 2007. Influence of plasma stimulation on Si nanowire nucleation and orientation dependence. *Advanced Materials.* 19, 2603–2607.

Akhtar, S., Tanaka, A., Usami, K., Tsuchiyaa, Y., and Oda, S. 2008a. Influence of the crystal orientation of substrate on low temperature synthesis of silicon nanowires from Si_2H_6. *Thin Solid Films.* 517(1), 317–319.

Akhtar, S., Usami, K., Tsuchiya, Y., Mizuta, H., and Oda, S. 2008b. Vapor–liquid–solid growth of small- and uniform-diameter silicon nanowires at low temperature from Si_2H_6. *Applied Physics Express.* 1, 014003.

Albuschies, J., Baus, M., Winkler, O., Hadam, B., Spangenberg, B., and Kurz, H. 2006. High density silicon nanowire growth from self-assembled Au nanoparticles. *Microelectronic Engineering.* 83, 1530–1533.

Becker, M., Sivakov, V., Gösele, U., Stelzner, T., Andrä, G., Reich, H. J., Hoffmann, S., Michler, J., and Christiansen, S. H. 2008. Nanowires enabling signal-enhanced nanoscale Raman spectroscopy. *Small.* 4, 398–404.

Cabarrocas, P. R. I. 2004. New approaches for the production of nano-, micro-, and polycrystalline silicon thin films. *Physica Status Solid C.* 1(5), 1115–1130.

Cao, L., Fan, P., Barnard, E. S., Brown, A. M., and Brongersma, M. L. 2010. Tuning the color of silicon nanostructures. *Nano Letters.* 10(7), 2649–2654.

Carius, R., Merdzhanova, T., Fingers, F., Klein, S., and Vetterl, O. 2003. A comparison of microcrystalline silicon prepared by plasma-enhanced chemical vapor deposition and hot-wire chemical vapor deposition: Electronic and device properties. *Journal of Material Science: Materials in Electronics*. 14, 625–628.

Červenka, J., Ledinský, M., Stuchlík, J., Stuchlíková, H., Bakardjieva, S., Hrůska, K., Fejfar, A., and Kŏcka, J. 2010. The structure and growth mechanism of Si nanoneedles prepared by plasma-enhanced chemical vapor deposition. *Nanotechnology*. 21, 415604.

Chuen, Y. L., Chou, L. J., Hsu, C. M., and Kung, S. C. 2005. Synthesis and characterization of taper- and rodlike Si nanowires on $Si_{(x)}Ge_{(1-x)}$ substrate. *Journal of Physical Chemistry B*. 109(46), 21831–21835.

Chung, S. W., Yu, J.-Y., and Heath, J. R. 2000. Silicon nanowire devices. *Applied Physics Letters*. 76, 2068–2070.

Cleland, A. N. and Roukes, M. L. 2002. Noise processes in nanomechanical resonators. *Journal of Applied Physics*. 92, 2758–2769.

Cui, Y. and Lieber, C. M. 2001. Functional nanoscale electronic devices assembled using silicon nanowire building blocks. *Science*. 291, 851.

Cui, Y., Lauhon, L. J., Gudiksen, M. S., Wang, J., and Lieber, C. M. 2001. Diameter-controlled synthesis of single crystal silicon nanowires. *Applied Physics Letters*. 78, 2214–2216.

Curtins, H., Wyrsch, N., Favre, M., and Shah, A. V. 1987. Influence of plasma excitation frequency for a-Si:H thin film deposition. *Plasma Chemistry and Plasma Process*. 7, 267–273.

Dailey, E., Madras, P., and Drucker, J. 2010. Au on vapor–liquid–solid grown Si nanowires: Spreading of liquid AuSi from the catalytic seed. *Journal of Applied Physics*. 108, 064320.

Das, D. and Jana, M. 2004. Hydrogen plasma induced microcrystallization in layer-by-layer growth scheme. *Solar Energy Materials and Solar Cells*. 81, 169–181.

Droz, C., Vallat-Sauvain, E., Bailat, J., Feitknecht, L., Meier, J., Niquille, X., and Shah, A. 2003. *In Procceedings of World Conference on Photovoltaic Energy Conversion*, vol. 2. Arisumi Printing Inc., Osaka, Japan. 12–16 May, pp. 1544–1547.

Elechiguerra, J. L., Manriquez, J. A., and Yacaman, M. J. 2004. Growth of amorphous SiO_2 nanowires on Si using a Pd/Au thin film as a catalyst. *Applied Physics A*. 79, 461–467.

Falk, J. L. L., Hemesath, E. R., Lopez, F. J., and Lauhon, L. J. 2007. Vapor–solid–solid synthesis of Ge nanowires from vapor-phase-deposited manganese germanide seeds. *Journal of the American Chemical Society*. 129, 10670–10671.

Fan, X. H., Xu, L., Li, C. P., Zheng, Y. F., Lee, C. S., and Lee, S. T. 2001. Effects of ambient pressure on silicon nanowire growth. *Chemical Physics Letters*. 334, 229–232.

Feng, S. Q., Yu, D. P., Zhang, H. Z., Bai, Z. G., and Ding, Y. 2000. The growth mechanism of silicon nanowires and their quantum confinement effect. *Journal of Crystal Growth*. 209, 513–517.

Fontcuberta i Morral, A., and Roca i Cabarrocas, P. 2001. Study on fabrication and characterization of thin film GaAs solar cells with mirror reflector on thermal dissipation substrate. *Thin Solid Films Third Symposium on Thin Film for Large Aria Electronics*, vol. 383, 30 May–2 June 2000, pp. 161–164.

Fukata, N., Oshima, T., Murakami, K., Kizuka, T., Tsurni, T., and Ito, S. 2005a. Phonon confinement effect of silicon nanowires synthesized by laser ablation. *Applied Physics Letters*. 86, 213112-1.

Gentile, P., David, T., Dhalluin, F., Buttard, D., Pauc, N., Hertog, M. D., Ferret, P., and Baron, T. 2008. The excitation of one-dimensional plasmons in Si and Au–Si complex atom wires. *Nanotechnology*. 19(35), 1–5.

Givargizov, E. I. 1975. Fundamental aspects of VLS growth. *Journal of Crystal Growth*. 31, 20–30.

Goncher, G., Solanki, R., Carruthers, J. R., Conley, J., Jr., and Ono, Y. 2006. p–n junctions in silicon nanowires. *Journal of Electronic Materials.* 35(7), 1509–1512.

Gotza, M., dutoit, M., and Ilegems, M. 1998. Fabrication and photoluminescence investigation of silicon nanowires on silicon-on-insulator material. *Journal of Vacuum Science & Technology B (Microelectronics and Nanometer Structures).* 16, 582–588.

Gourbilleau, F., Achiq, A., Voivenel, P., and Rizk, R. 1999. Thin solid films. *E—Materials Research Society Spring Conference, Symposium E: Thin Film Materials for Large Area Electronics,* vol. 337. Strasbourg, France, 16–19 June 1998, pp. 74–77.

Griffiths, H., Xu, C., Barrass, T., Cooke, M., Iacopi, F., Vereecken, P., and Esconjauregui, S. 2007. Plasma assisted growth of nanotubes and nanowires. *Surface and Coatings Technology.* 201, 9215–9220.

Gudekssen M. S. and Lieber C. M. 2000. Diameter-selective synthesis of semiconductor nanowires. *Journal of the American Chemical Society.* 122, 8801–8802.

Hannon, J. B., Kodambaka, S., Ross, F. M., and Tromp, R. M. 2006. The influence of the surface migration of gold on the growth of silicon nanowires. *Nature.* 440, 69–71.

Heintze, M., Zedlitz, R., and Bauer, G. H. 1993. Analysis of high rate a-Si:H deposition in a VHF plasma. *Journal of Physics D: Applied Physics.* 26, 1781–1786.

Hofmann, S., Ducati, C., Neill, R. J., Piscanec, S., Ferrari, A. C., Geng, J., Dunin-Borkowski, R. E., and Robertson, J. 2003. Gold catalyzed growth of silicon nanowires by plasma enhanced chemical vapor deposition. *Journal of Applied Physics.* 94(9), 6005.

Hofmann, S., Sharma, R., Wirth, C. T., Cervantes-Sodi, F., Ducati, C., Kasama, T., Dunin-Borkowski, R. E., Drucker, J., Bennett, P., and Robertson, J. 2008. Ledge-flow-controlled catalyst interface dynamics during Si nanowire growth. *Nature Materials.* 7, 372–375.

Holmes, J. D., Johnston, K. P., Doty, R. C., and Korgel, B. A. 2000. Control of thickness and orientation of solution-grown silicon nanowires. *Science.* 287, 1471–1473.

Hu, J. and Lieber, C. M. 1999. Nanowires for integrated multicolor nanophotonics. *Accounts of Chemical Research.* 32, 435.

Hu, J., Wang, T., and Lieber, Ch. 1999. Chemistry and physics in 1D: Synthesis and properties of nanowires and nanotubes. *Accounts of Chemical Research.* 32, 435–445.

Ishida, M., Kawano, T., Futagawa, M., Aria, Y., Takao, H., and Sawada, K. 2003. A Si nano-micro-wire array on a Si(111) substrate and field emission device applications. *Superlatices and Microstructures.* 34, 567–575.

Jagannathan, H., Nishi, Y., Reuter, M., Copel, M., Tutuc, E., Guha, S., and Pezzi, R. P. 2006. Effect of oxide overlayer formation on the growth of gold catalyzed epitaxial silicon nanowires. *Applied Physics Letters.* 88, 103113–103115.

Jeon, M. and Kamisako, K. 2009. Synthesis of Au-catalyzed silicon nanowires by hydrogen radical-assisted deposition method. *Metals and Materials International.* 15(1), 83–87.

Kayali, S., Ponchak, G., and Shaw, R. (Eds). 1996. *GaAs MMIC Reliability Assurance Guideline for Space Applications.* Pasadena, CA: Jet Propulsion Laboratory California Institute of Technology.

Kelzenberg, M. D., Turner-Evans, D. B., Kayes, B. M., Filler, M. A., Putnam, M. C., Lewis, N. S., and Atwater, H. A. 2008. Photovoltaic measurements in single-nanowire silicon solar cells. *Nano Letters.* 8(2), 710–714.

Kikkawa, J., Ohno, Y., and Takeda, S. 2005. Growth rate of silicon nanowires. *Applied Physics Letters.* 86, 123109.

Kodambaka, S., Tersoff, J., Reuter, M. C., and Ross, F. M. 2007. Germanium nanowire growth below the eutectic temperature. *Science.* 316, 729–732.

Kovalev, D., Heckler, H., Polisski, G., and Koch, F. 1999. Optical properties of Si nanocrystals. *Physica Status Solidi B.* 215, 871–932.

Kroll, U., Shah, A., Keppner, H., Meier, J., Torres, P., and Fischer, D. 1997. Potential of VHF-plasmas for low-cost production of a-Si:H solar cells. *Solar Energy Materials Solar Cell.* 48, 343–350.

Kulkarni, N. N., Bae, J., Shih, C.-K., Stanley, S. K., Coffee, S. S., and Ekerdt, J. G. 2005. Low-threshold field emission from cesiated silicon nanowires. *Applied Physics Letters.* 87, 213115–213117.

Kuske, J., Stephan, U., Steinke, O., and Rohlecke, S. 1995. Power feeding in large area PECVD of amorphous silicon. *Materials Research Society Symposium Proceedings.* 27, 377.

Lauhon, L. J., Gudiksen, M. S., and Lieber, C. M. 2004. Semiconductor nanowire heterostructures: Organizing atoms: manipulation of matter on the sub-10 nm scale. *Philosophical Transactions of the Royal Society A: Mathematical, Physical and Engineering Sciences.* 362(1819), 1247–1260.

Lerose, D., Bechelany, M., Philippe, L., Michler, J., and Christiansen, S. 2010. Ordered arrays of epitaxial silicon nanowires produced by nanosphere lithography and chemical vapor deposition. *Journal of Crystal Growth.* 312, 2887–2891.

Lew, K.-K., Reuther, C., Carim, A. H., Redwing, J. M., and Martin, B. R. 2002. Template directed vapor–liquid–solid growth of silicon nanowires. *Journal of Vacuum Science & Technology B.* 20, 389–392.

Li, B., Yu, D., and Zhang, S.-L. 1999. Raman spectral study of silicon nanowires. *Physical Review B (Condensed Matter).* 59, 1645–1648.

Li, C. B., Usami, K., Mizuta, H., and Oda, S. 2009. Vapor–solid–solid radial growth of Ge nanowires. *Journal of Applied Physics.* 106, 046102.

Liu, W. L., Hsieh, S. H., Chen, C. H., and Chen, W. J. 2009. Effect of Fe metal on the growth of silicon oxide nanowires. *International Journal of Minerals, Metallurgy and Materials.* 16(3), 317.

Liu, Z. Q., Pan, Z. W., Sun, L. F., Tang, D. S., Zhou, W. Y., Wang, G., Qian, L. X., and Xie, S. S. 2000. Synthesis of silicon nanowires using AuPd nanoparticles catalyst on silicon substrate. *Journal of Physics and Chemistry of Solids.* 61, 1171–1174.

Liu, Z. Q., Zhou, W. Y., Sun, L. F., Tang, D. S., Zou, X. P., Li, Y. B., Wang, C. Y., Wang, G., and Xie, S. S. 2001. Growth of amorphous silicon nanowires. *Chemical Physics Letters.* 341, 523–528.

Lu, X. and Korgel, B. A. 2003. Growth of single crystal nanowires. *Nanoletters.* 3, 93.

Ma, D. D. D., Lee, C. S., Au, F. C. K., Tong, S. Y., and Lee, S. T. 2003. Small-diameter silicon nanowire surfaces. *Science.* 299, 1874–1877.

Marczak, M., Judek, J., Kozak, A., Gebicki, W., Jastrzebski1, C., Adamowicz, L., Luxembourg, D., Hourlier, D., and Mélin, T. 2009. The individual core/shell silicon nanowire structure probed by Raman spectroscopy. *Physica Status Solidi C.* 6(9), 2053–2055.

Marsen, B. and Sattler, K. 1999. Fullerene structured nanowires of silicon. *Physics Review B (Condensed Matter).* 60, 11593–11600.

McIlroy, D. N., Alkhateeb, A., Zhang, D., Aston, D. E., Marcy, A. C., and Norton, M. G. 2004. Nanospring formation unexpected catalyst mediated growth. *Journal of Physics. Condensed Matter.* 16, 415–440.

Meier, J., Kroll, U., Vallat-Sauvain, E., Spitznagel, J., Graf, U., and Shah, A. 2004. Amorphous solar cells, the micromorph concept and the role of VHF-GD deposition technique. *Solar Energy.* 77, 983–993.

Meier, J., Torres, P., Platz, R., Dubail, S., Kroll, U., Anna Selvan, J. A., Pellaton-Va ucher, N., Hof, C., Fischer, D., Keppner, H., Shah, A., Ufert, K.-D., Giannoulès, P., and Köhler, J. 1996. On the way towards high-efficiency thin film silicon solar cells by the micromorph concept. *In: Proceedings of Materials Research Society Symposium. Spring Meeting,* vol. 420. San Francisco, pp. 3–14.

Mohney, S. E., Wang, Y., Cabassi, M. A., Lew, K. K., Dey, S., Redwing, J. M., and Mayer, T. S. 2005. Measuring the specific contact resistance of contacts to semiconductor nanowires. *Solid State Electronics.* 49(2), 227–232.

Morales, A. M. and Lieber, C. M. 1997. SiO-enhanced synthesis of Si nanowires by laser ablation. *ACS Meeting,* vol. 213. San Francisco, CA, p. 651.

Morales, A. M. and Lieber, C. M. 1998. A laser ablation method for the synthesis of crystalline. *Science*. 279, 208–211.

Murata, M., Takeuchi, Y., Sasagawa, E., and Hamamoto, K. 1996. Inductively coupled radio frequency plasma chemical vapor deposition using a ladder-shaped antenna. *Review of Scientific Instruments*. 67, 1542–1545.

Najmzadeh, M., Michielis, L. D., Bouvet, D., Dobrosz, P., Olsen, S., and Ionescu, A. M. 2010. Silicon nanowires with lateral uniaxial tensile stress profiles for high electron mobility gate-all-around MOSFETs. *Microelectronic Engineering*. 87(5–8), 1561–1565.

Niu, J. J. and Wang, J. N. 2008. A study in the growth mechanism of silicon nanowires with or without metal catalyst. *Materials Letters*. 62, 767–771.

Niu, J., Sha, J., Ma, X., Xu, J., and Yang, D. 2003a. Array-orderly single crystalline silicon nano-wires. *Chemical Physics Letters*. 367, 528–532.

Niu, J., Sha, J., Wang, Y., Ma, X., and Yang, D. 2003b. Crystallization and disappearance of defects of the annealed silicon nanowires. *Microelectronic Engineering*. 66, 65–69.

Niu, J., Sha, J., and Yang, D. 2004a. Silicon nanowires fabricated by thermal evaporation of silicon monoxide. *Physica E*. 23, 131–134.

Niu, J., Sha, J., Yang, Q., and Yang, D. 2004b. Crystallization and Raman shift of array-orderly silicon nanowires after annealing at high temperature. *Japanese Journal of Applied Physics A*. 43(7), 4460–4461.

Nolsson, G. and Nelin, G. 1972. Study of the homology between silicon and germanium by thermal-neutron spectrometry. *Physical Review B*. 6(10), 3777–3786.

Oda, S. and Nishiguchi, K. 2001. Nanocrystalline silicon quantum dots prepared by VHF plasma enhanced chemical vapor deposition. *Thirteenth European Conference on Chemical Vapor Deposition. Journal de PhysiqueIV France*. 11. 1065–1071.

Oda, S. and Yasukawa, M. 1991. High quality a-Si:H films and interfaces prepared by VHF plasma CVD. *Journal of Non-Crystalline Solids*. 137–138, 677–680.

Olinga, T. E., Frayasse, J., Travers, J. P., Dufresne, A., and Pron, A. 2000. Highly conducting and solution-processable polyaniline obtained via protonation with a new sulfonic acid containing plasticizing functional groups. *Macromolecules*. 33(6), 2107–2113.

Pan, H., Lim, S., Poh, C., Sun, H., Wu, X., Feng, Y., and Lin, J. 2005. Growth of Si nanowires by thermal evaporation. *Nanotechnology*. 16, 417–421.

Parlevliet, D. and Cornish, J. 2006. Pulsed PECVD for the growth of silicon nanowires. *International Conference on Nanoscience and Nanotechnology. ICONN*. Brisbane, Qld. Art. no. 4143323, pp. 35–38.

Paulo, A. S., Arellano, N., He, R., Carraro, C., Maboudian, R., Howe, R., Bokor, J., and Yang, P. 2007. Suspended mechanical structures based on elastic silicon nanowire arrays. *Nano Letters*. 7, 1100–1104.

Pei, L. Z., Tang, Y. H., Chen, Y. W., Guo, C., Zhang, W., and Zhang, Y. 2006. Silicon nanowires grown from silicon monoxide under hydrothermal conditions. *Journal of Crystal Growth*. 289, 423–427.

Peng, K., Xu, Y., Wu, Y., Yan, Y., Lee, S. T., and Zhu, J. 2005. Aligned single-crystalline Si nanowire arrays for photovoltaic applications. *Small*. 1(11), 1062.

Persson, A. I., Larsson, M. W., Stenstrom, S., Ohlsson, B. J., Samuelson, L., and Wallenberg, L. R. 2004. Solid-phase diffusion mechanism for GaAs nanowire growth. *Nature Materials*. 3, 677–681.

Prokes, S. M. and Arnold, S. 2005. Stress-driven formation of Si nanowires. *Applied Physics Letters*. 86, 193105–193107.

Qi, P., Wong, W. S., Zhao, H., and Wang, D. 2008. Low-temperature synthesis of Si nanowires using multizone chemical vapor deposition methods. *Applied Physics Letters*. 93, 163101.

Ross, F. M., Tersoff, J., and Reuter, M. C. 2005. Saw tooth faceting in silicon nanowires. *Physics Review Letters*. 95, 146104–146107.

Salhi, B., Gelloz, B., Koshida, N., Patriarche, G., and Boukherrou, R. 2007. Synthesis and photoluminescence properties of silicon nanowires treated by high-pressure water vapor annealing. *Physica Status Solidi A.* 204(5), 1302–1306.

Salhi, B., Grandidier, B., and Boukherroub, R. 2006. Controlled growth of silicon nanowires on (111) silicon surfaces. *Journal of Electroceramics.* 16, 15–21.

Samuelson, L. 2003. Self-forming nanoscale devices. *Materials Today.* 6(10), 22.

Srivastava, S. K., Singh, P. K., Singh, V. N., Sood, K. N., Haranath, D., and Kumar, V. 2009. Large-scale synthesis, characterization and photoluminescence properties of amorphous silica nanowires by thermal evaporation of silicon monoxide. *Physica E.* 41, 1545–1549.

Sato, K., Izumi, T., Iwase, M., Show, Y., Morisaki, H., Yaguchi, T., and Kamino, T. 2003. Nucleation and growth of nanocrystalline silicon studied by TEM, XPS and ESR. *Applied Surface Science Fourth International Symposium on the Control of Semiconductor Interface,* vol 216. Karuizawa, Japan, 21–25 October, pp. 376–381.

Scheel, H., Reich, S., Ferrari, A. C., Cantoro, M., Colli, A., and Thomsen, C. 2006. Raman scattering on silicon nanowires: The thermal conductivity of the environment determines the optical phonon frequency. *Applied Physics Letters.* 88, 233114-1–233114-4.

Schmidt, V., Senz, S., and Gösele, U. 2005. Diameter-dependent growth direction of epitaxial silicon nanowires. *Nano Letters.* 5, 931–935.

Schwarzenbach, W., Howling, A. A., Fivaz, M., Brunner, S., and Hollenstein, Ch. 1996. Sheath impedance effects in very high frequency plasma experiments. *Journal of Vacuum Science & Technology A.* 14, 132–138.

Shah, A., Meier, J., Vallat-Sauvain, E., Wyrsch, N., Kroll, U., Droz, S., and Graf, U. 2005. Material and solar cell research in microcrystalline silicon. *Solar Energy Materials and Solar Cells.* 78, 469.

Shan, Y. and Fonash, S. J. 2008. Self-assembling silicon nanowires for device applications using the nanochannel-guided "Grow-in-Place" approach. *ACS Nano.* 2(3), 429–434.

Shan, Y., Kalkan, A. K., and Fonash, S. J. 2004. From Si source gas directly to positioned, electrically contacted Si nanowires: The self-assembling "grow-in-place" approach. *Nano Letters.* 4, 2085.

Sharma, S. and Sunkara, M. K. 2004. Direct synthesis of single–crystalline silicon nanowires using molten gallium and silane plasma. *Nanotechnology.* 15, 130–134.

Sharma, S., Kamins, T. I., and Stanley Williams, R. 2004. Diameter control of Ti-catalyzed silicon nanowires. *Journal of Crystal Growth.* 267: 613–618.

Shi, Y., Hu, Q., Araki, H., Suzuki, H., Gao, H., Yang, W., and Nada, T. 2005. Long Si nanowires with millimeter-scale length by modified thermal evaporation from Si powder. *Applied Physics A.* 80(8), 1733–1736.

Shimizu, T., Xie, T., Nishikawa, J., Shingubara, S., Senz, S., and Gösele, U. 2007. *Advanced Materials.* 19, 917.

Shiu, S. C., Lin, S. B., Hung, S. C., and Lin, C. F. 2011. Influence of pre-surface treatment on the morphology of silicon nanowires fabricated by metal-assisted etching. *Applied Surface Science.* 257(6), 1829–1834.

Steiner, T. 2004. *Semiconductor Nanostructures for Optoelectronic Applications.* London: Artech House.

Swain, B. S., Swain, B. P., and Hwang, N. 2010. Chemical surface passivation of silicon nanowires grown by APCVD. *Current Applied Physics.* 10, S439–S442.

Takatsuka, H., Noda, M., Yonekura, Y., Takeuchi, Y., and Yamauchi, Y. 2004. Development of high efficiency large area silicon thin film modules using VHF-PECVD. *Solar Energy.* 77(6), 951–960.

Takeuchi, Y., Nawata, Y., Ogawa, K., Serizaw A., Yamauchi, M., and Murata, M. 2001. Preparation of large uniform amorphous silicon films by VHF-PECVD using a ladder shaped antenna. *Thin Solid Films.* 386, 133–136.

Tan, T. Y., Lee, S. T., and Gösele, U. 2002. A model for growth directional features in silicon nanowires. *Applied Physics A: Materials Science & Processing.* 74, 423–432.

Wacaser, B. A., Dick, K. A., Johansson, J., Borgström, M. T., Deppert, K., and Samuelson, L. 2009. Preferential interface nucleation: An expansion of the VLS growth mechanism for nanowires. *Advanced Materials.* 21(2), 153–165.

Wagner, R. S. and Ellis, W. C. 1964. Vapor-liquid-solid mechanism of single crystal growth. *Applied Physics Letters.* 4, 89–90.

Wagner, R. S. and Ellis, W. C. 1965. The vapor–liquid–solid mechanism of crystal growth and its application to silicon. *Transactions of the Metallurgical Society of AIME.* 233, 1053.

Wanekaya, A. K., Chen, W., Myung, N. V., and Mulchandani, A. 2006. Nanowire-based electrochemical biosensors. *Electroanalysis.* 18(6), 533–550.

Wang, C., Jiang, Y., Li, G., and Zhang, Z. 2008. A Wurtz-like reaction to silicon nanowires. *Materials Letters.* 62, 2497–2499.

Wang, L. C., Feng, D., Li, Q., He, Y. L., and Chu, Y. M. 1992. Microstructures and characteristics of nano-size crystalline silicon films. *Journal of Physics: Condensed Matter.* 4, 509–512.

Wang, N., Tang, Y. H., and Zhang, Y. F. 1998. A simple route to annihilate defects in silicon nanowires. *Chemical Physics Letters.* 283, 368.

Wang, R. P., Zhou, G. W., Pan, Y. L. S. H., Zhang, H. Z., Yu, D. P., and Zhang, Z. 2000. Raman spectral study of silicon nanowires: High-order scattering and phonon confinement effects. *Physical Review B.* 61(24), 16827–16832.

Wang, Y., Schmidt, V., Senz, S., and Gösele, U. 2006. Epitaxial growth of silicon nanowires using an aluminium catalyst. *Nature Nanotechnol.* 1(3), 186–189.

Weaver, J. W., Timoshenko, S. P., and Young, D. H. 1990. *Vibration Problems in Engineering.* New York: Wiley, pp. 417–432.

Westwater, J., Gosain, D. P., Tomiya, S., Usui, S., and Ruda, H. 1997. Growth of silicon nanowires via gold/silane vapor–liquid–solid reaction. *Journal of Vacuum Science & Technology B.* 15, 554–557.

Westwater, J., Gosain, D. P., and Usui, S. 1998. Si nanowires grown via the vapour–liquid–solid reaction. *Physics Status Solidi A.* 165(1), 37–42.

Wilson, J. I. B. 1980. Amorphous silicon. *Sunworld*, 4, 14–15.

Wolfsteller, A., Geyer, N., Duca, T. K. N., Kanungo, P. D., Zakharov, N. D., Reiche, M., Erfurth, W., Blumtritt, H., Kalem, S., Werner, P., and Gösele, U. 2010. Comparison of the top-down and bottom-up approach to fabricate nanowire-based silicon/germanium heterostructures. *Thin Solid Films.* 518(9), 2555–2561.

Woo, R. L., Gao, L., Goel, N., Hudait, M. K., Wang, K. L., Kodambaka, S., and Hicks, R. F. 2009. Kinetic control of self-catalyzed indium phosphide nanowires, nanocones, and nanopillars. *Nano Letters.* 9(6), 2207–2211.

Wu, Y., Chi, Y., Huynh, L., Barrelet, C. J., Bell, D. C., and Lieber, C. M. 2004. Controlled growth and structures of molecular-scale silicon nanowires. *Nano Letters* 4(3), 433–436.

Wu, Y., Fan, R., and Yang, P. 2002. Block-by-block growth of single-crystalline Si/SiGe superlattice nanowires. *Nano Letters.* 2(2), 83–86.

Xing, Y. J., Yu, D. P., Xi, Z. H., and Xue, Z. Q. 2003. Silicon nanowires grown from Au-coated Si substrate. *Applied Physics A: Materials Science & Processing.* 76, 551–553.

Yan, H. F., Xing, Y. J., Hang, Q. L., Yu, D. P., Wang, Y. P., Xu, J. Xi, Z. H., and Feng, S. Q. 2000. Growth of amorphous silicon nanowires via a solid–liquid–solid mechanism. *Chemical Physics Letters.* 323, 224–228.

Yan, X. Q., Zhou, W. Y., Sun, L. F., Gao, Y., Liu, D. F., Wang, J. X., Zhou, Z. P., Yuan, H. J., Song, L., Liu, L. F., Wang, G., and Xie, S. S. 2005. The influence of hydrogen on the growth of gallium catalyzed silicon oxide nanowires. *Journal of Physics and Chemistry of Solids.* 66, 701–705.

Yao, Y., Li, F., and Lee, S.-T. 2005. Oriented silicon nanowires on silicon substrates from oxide-assisted growth and gold catalysts. *Chemical Physics Letters.* 406, 381–385.

Yu, D. P., Bai, Z. G., Ding, Y., Hang, Q. L., Zhang, H. Z., Wang, J. J., Zou, Y. H., Qian, W., Xiong, G. C., Zhou, H. T., and Feng, S. Q. 1998a. Nanoscale silicon wires synthesized using simple physical evaporation. *Applied Physics Letters.* 72(26), 3458–3460.

Yu, D. P., Hang, Q. L., Ding, Y., Zhang, H. Z., Bai, Z. G., Ding, Wang, J. J., Zou, Y. H., Qian, W., Xiong, G. C., and Feng, S. Q. 1998b. Amorphous silica nanowires: Intensive blue light emitters. *Applied Physics Letters.* 73(21), 3076–3078.

Yu, L., Alet, P. J., Picardi, G., Maurin, I., and Cabarrocas, P. R. I. 2008. Synthesis, morphology and compositional evolution of silicon nanowires directly grown on SnO_2 substrates. *Nanotechnology.* 19(48), 485605.

Yu, L., Donnell, B. O., Alet, P. J., and Cabarrocas, P. R. 2010. All-in-situ fabrication and characterization of silicon nanowires on TCO/glass substrates for photovoltaic application. *Solar Energy Materials & Solar Cells.* 94(11), 1855–1859.

Yue, G., Yan, B., Ganguly, G., Yang, J., Guha, S., and Teplin, C. W. 2006. Material structure and metastability of hydrogenated nanocrystalline silicon solar cells. *Applied Physics Letters.* 88, 263507-3.

Zardo, I., Yu, L., Conesa-B., Estradé, S., Alet, P. J., Rössler, J., Frimmer, M., Roca, C. P., Peiró, F., Arbiol, J., Morante, J. R., and Fontcuberta, M. A. 2009. Gallium assisted plasma enhanced chemical vapor deposition of silicon nanowires. *Nanotechnology.* 20(15), 155602.

Zhang, B. Z., Shimizu, T., Chen, L., Senz, S., and Gösele, U. 2009. Bottom-imprint method for VSS growth of epitaxial silicon nanowire arrays with an aluminium catalyst. *Advanced Materials.* 21(46), 4701–4705.

Zhang, S. L., Zhu, B. F., Huang, F. M., Yan, Y., Shang, E. Y., Fan, S. S., and Han, W. G. 1999. Effect of defects on optical phonon Raman spectra in SiC nanorods. *Solid State Communications.* 111(11), 647–651.

Zhang, X.-W. and Han, G.-R. 2002. Microstructure and conductivity of large area nanocrystalline silicon films grown by specially designed thermal-assisted chemical vapor deposition. *Thin Solid Films.* 415, 5–9.

Zhang, X.-Y., Zhang, L.-D., Meng, G.-W., Li, G.-H., Jin-P., N.-Y., and Phillipp, F. 2001. Synthesis of ordered single crystal silicon nanowire arrays. *Advanced Materials.* 13(16), 1238–1241.

Zhang, Y. F., Tang, Y. H., Lam, C., Wang, N., Lee, C. S., Bello, I., and Lee, S. T. 2000. Bulk-quantity Si nanowires synthesized by SiO sublimation. *Journal of Crystal Growth.* 212, 115–118.

Zhang, Y. F., Tang, Y. H., Wang, N., Lee, C. S., Bello, I., and Lee, S. T. 1998. One-dimensional growth mechanism of crystalline silicon nanowires. *Journal of Crystal Growth.* 197, 136–140.

Zhang, Y. F., Tang, Y. H., Wang, N., Yu, D. P., Lee, C. S., Bello, I., and Lee, S. T. 1998. Silicon nanowires prepared by laser ablation at high temperature. *Applied Physics Letters.* 72(15), 1835.

12 Top-Down Fabrication of ZnO NWFETs

Suhana Mohamed Sultan, Peter Ashburn, and Harold M. H. Chong

CONTENTS

12.1 INTRODUCTION

Currently, ZnO nanowires are gaining much attention for their applications in thin film transistors (TFTs), display electronics, biosensors, and optoelectronics, which is due to these direct wide bandgap materials being highly transparent in the visible light with a high surface conductivity. ZnO nanowires are generally fabricated using bottom-up technology and require the use of tedious pick and place method and electron beam lithography, which limit their uses in large area and low-cost applications. This chapter provides a new perspective of the top-down fabrication technique to produce highly oriented and reproducible nanowires with different channel lengths in defined locations on a large-scale processing mode. All the fabrication and characterizations were conducted at the Southampton Nanofabrication Center, University of Southampton.

12.1.1 ZnO Material Properties

ZnO is one of the semiconductor materials that is known for its versatile applications. It is a wide and direct bandgap semiconductor material (E_g = 3.4 eV at 300 K) that has garnered the focus of intense materials research and device development among the established II–VI wide bandgap compounds such as ZnS, ZnSe, ZnTe, CdS, and CdSe. Distinguished advantages of ZnO include the following: (a) the availability of large area native substrates [1], (b) high transparency in the visible region [2], (c) an exciton binding energy of 60 meV ensuring excitonic emission at temperatures above room temperature [3], (d) its resistance to high energy radiation damage enabling it a potential semiconductor to be applied in hostile environments [4], and (e) its susceptibility to wet chemical processing [5]. Table 12.1 gives a comparison of ZnO material properties with other wide bandgap compounds.

ZnO crystallizes in the wurtzite structure with a hexagonal lattice and has an energy bandgap similar to GaN. Furthermore, with a lattice mismatch of less than 1.8%, it is suitable as an alternative substrate for growth of high quality GaN epilayers. ZnO is intrinsically an n-type semiconductor due to vacancies and interstitials. P-type doping has still been a challenge due to the instability of device performance.

ZnO has high transparency from the ultraviolet (UV) region to the near-infrared (NIR) region, which enables transparent electronics applications such as in flat panel displays and solar cells. Additional advantages of ZnO and other wide bandgap semiconductors are its thermal stability, which facilitates high-temperature operation and the high breakdown voltage making it ideal for high-power switching devices [6]. However, so far there are few experimental evidences on the high breakdown voltage on ZnO-based devices.

The research on ZnO started in the 1930s and peaked in the 1980s but faded away because it was not possible then to dope ZnO of both the n- and p-types, which is crucial for applications in electronics and optoelectronics. The area of interest at that time was on bulk samples covering topics like growth, doping, band structure, and excitons [7].

The present interest on ZnO material is because of its application in high mobility electronic devices for flat panel display and flexible electronic applications, light-emitting and laser diodes that can replace GaN-based structures, and also a transparent

TABLE 12.1
Comparison of ZnO Properties with Other Wide Bandgap Semiconductors

Material	Crystal Structure	Bandgap (eV)	Lattice a (Å)	Constant c (Å)	Exciton Binding Energy (meV)
ZnO	Wurtzite	3.37	3.25	5.21	60
ZnS	Wurtzite	3.68	3.82	6.26	39
ZnSe	Zinc blende	2.7	5.67	–	20
GaN	Wurtzite	3.4	3.19	5.19	21
6H-SiC	Wurtzite	2.9	3.08	15.1	–

Source: Chen, Y., Bagnall, D.M., Koh, H.-J., Park, K.-T., Hiraga, K., and Zhu, Z. *J. Appl. Phys.* 84, 3912, 1998.

conducting oxide, which is a cheaper alternative to indium tin oxide (ITO). Due to its high surface conductivity, it is also well researched in sensor applications [8].

12.1.2 ZnO: From Thin Film to Nanowires

There has been a growing interest in using ZnO as a new material for TFT applications over the past decade due to its versatile film properties. Since the first demonstration of transparent ZnO TFTs in 2003 [9–12], there have been considerable advances in the scientific understanding of the properties of ZnO thin films. ZnO TFTs are already seen as potential alternative to the currently polysilicon-based TFT as a select transistor for active matrix liquid crystal display (AMLCD) technology. This is due to their high transparency in the visible region and high channel mobility compared to amorphous and polysilicon material, which only achieved a mobility of 1–10 cm^2/Vs [7].

Figure 12.1 shows the picture image of ZnO-based AMLCD, which was demonstrated by Furuta et al. [13]. This shows the viability of ZnO material to be implemented in LCD applications. Another special characteristic of ZnO thin film is that it can be grown at/near room temperature, which opens a new avenue in flexible electronic applications as shown in Figure 12.1 [14].

From thin film development, there is an increase in attention toward ZnO nanowires. ZnO nanostructure is a widely researched area compared to other materials. Figure 12.2 shows some of the ZnO nanostructures demonstrated for the past years [15]. Due to an increase of surface to volume ratio and high surface conductance, it is an ideal material for field-effect transistor (FET), a device which can be functionalized as TFTs for high resolution display applications, gas sensors, and UV photodetectors [16].

It has also been demonstrated that ZnO nanowires assembled in logic gate application such as NOT, OR, AND, and NOR gates [17]. With a combination of low growth temperature along with high reported mobility of 4000 cm^2/Vs [1], ZnO nanowires have high potential in TFT electronics applications.

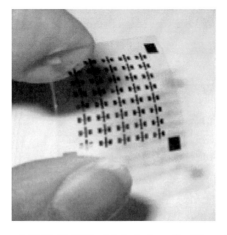

FIGURE 12.1 Picture of 40 ZnO TFTs fabricated on flexible polyethylene naphthalate (PEN) polyester substrate. (From Carcia, P.F., et al., *J. Appl. Phys.*, 102, 074512, 2007.)

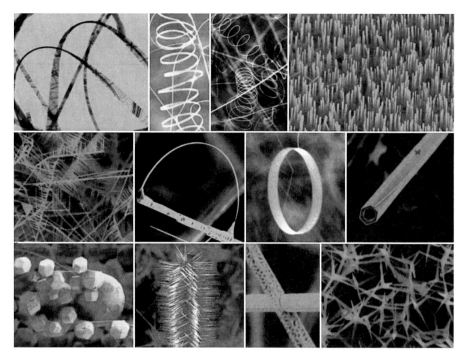

FIGURE 12.2 A collection of nanostructures of ZnO synthesized by thermal evaporation of solid powders. (From Wang, Z.L., *Mater. Today*, 7, 26, 2004.)

12.1.3 CONCEPT OF NANOWIRE TFTs

As potential application of nanowires in electronics is tremendous, it has been a very popular research area since Wagner and Ellis (1964) published the novel method way back in 1964 of growing silicon whiskers by using gold catalysts via chemical vapor deposition (CVD) method [18]. A nanowire is a structure that has large aspect ratio or length-to-width ratio. It is considered as a one-dimensional nanostructure, and the fabrication process has been carried out by a host of bottom-up and top-down techniques. Bottom-up method of fabrication such as chemical synthesis, that is, vapor–liquid–solid (VLS) or vapor–solid (VS), proves to be a more flexible approach to produce high-quality nanowires. In particular, carrier mobility values of 300 cm^2/Vs have been demonstrated for p-type Si nanowires [19], 730 cm^2/Vs for Si/Ge nanowire heterostructures [20], 2000–4000 cm^2/Vs for n-type indium phosphide (InP) nanowires (NWs) [21], and up to 20,000 cm^2/Vs for single walled carbon nanotubes (CNTs) [22]. These nanostructures can often be processed in solution and assembled onto a wide range of substrates. The main drawback of this method is related to the point that after the growth, nanowires will be scattered on the substrate making it necessary to employ a good pick-and-place method to place the grown nanowire on the device substrate. Although the bottom-up approach is most commonly used in the research area to construct a nanowire field-effect transistor (NWFET), too many variables such as difference in sizes of the grown nanowires and issues on the alignment prove that realization of an NWFET structure is a difficult task.

Meanwhile, top-down techniques such as nanolithography (electron beams and ion beams), scanning probe lithography (SPL), and extreme UV or x-ray lithography have been widely used to study the basic properties of nanowire devices. Although these techniques have made great contributions to nanowire technology, their use on a large scale remains a significant challenge due to their high cost and low throughput. In 2002, Choi et al. had demonstrated sub-7 nm structures with conventional lithography and dry etching [23] called spacer patterning technique. This technique exploits a combination of lithography, anisotropic etchings, and excellent homogeneity and reproducibility of conformal growth processes. This method has been used to pattern silicon fins for double-gate metal–oxide–semiconductor field-effect transistors (MOSFETs), FinFETs [23], crossbar array of memory [16], and biosensors [24].

In short, there is need for highly oriented nanowire TFT fabrication techniques, which can be applied to any materials that could form TFTs with performance comparable to or better than polycrystalline, and which can be processed at low temperatures over a large area of substrate and also enable new applications in flexible and wearable electronics. Therefore, a broad range of materials including group III–V and II–VI compound semiconductors can be exploited as the TFT channel materials, opening new avenues to further improve device performance or create new device functions. One of the new semiconductor materials that have been extensively researched in recent years for various applications is ZnO.

12.2   THIN FILM GROWTH AND PROPERTIES

There have been many reports on ZnO thin film growth and the relationship between the crystallography and electrical properties. The results are tabulated in Table 12.2. Generally, ZnO thin film can be grown by various methods on different substrates. However, the room temperature mobilities are lower (typically below 100 cm^2/Vs) compared to bulk ZnO, which is about 200 cm^2/Vs [5]. This is because of the grain boundaries that exist in the thin film.

From Table 12.2, it is observed that pulsed laser deposition (PLD), metal–organic chemical vapor deposition (MOCVD), and metal oxide vapor phase epitaxy (MOVPE) deposition normally requires high temperature deposition. There has been research on room temperature sputtering but there is lack of control of impurities in this technique and therefore requires post-deposition annealing at high temperature. Atomic layer deposition (ALD) has proven to deposit ZnO material at a low temperature (<150°C) with reasonable carrier concentration and mobility [25]. This enables depositions on flexible substrates such as polyethylene terephthalate (PET) [26], TEFLON (a brand name made of polytetrafluoroethylene), and polyphthlate carbonate (PPC). The use of these polymer substrates enables applications such as flexible displays, sensor arrays, as well as other large area electronics [19] and are receiving much attention recently mainly due to their attractive properties such as lightweight and high resistance to impact damage.

The above-mentioned growth techniques produce ZnO film that is highly n-typed. The unintentional n-type conductivity has conventionally been attributed to point defects like zinc interstitials (Zn_i) and oxygen vacancies (V_o). However, there is still

TABLE 12.2
ZnO Thin Film Deposition Techniques

Sample	Growth Temperature (°C)	Carrier Concentration (cm^{-3})	Mobility (cm^2/Vs)	References
ZnO thin film on c-plane sapphire substrates grown by PLD	750	2×10^{16}	155	[27]
ZnO thin film grown on MgZnO-buffered ScAlMgO$_4$ substrates by PLD	350	1×10^{16}	440	[28]
ZnO thin film grown by PLD on glass	450	3×10^{16}	0.7	[10]
ZnO thin film on MgO buffered c-plane sapphire substrates grown by MBE	700	1.2×10^{17}	130	[29]
ZnO thin film grown on sapphire by MOCVD	600	4×10^{17}	24	[30]
H-doped and N-doped ZnO thin film by radio frequency (RF) magnetron sputter (H = 6.5%) (N = 60%) on ZnO, GaN, and sapphire	300–700	H: 2.1×10^{20}	27.8	[31]
ZnO thin film by RF magnetron sputtering on glass	Room temperature	3×10^{16}	2.0	[32]
Ga-doped ZnO thin film by RF sputter on SiO$_2$/p-Si	Room temperature	6.9×10^{21}	0.62	[33]
ZnO thin film grown by ALD on c-plane sapphire substrate	>300 <300	10^{18}–10^{19} 1×10^{16}	N/A	[34]
ZnO thin film grown on ZnO buffer layer Si (111) substrate by ALD	Buffer layer = 180 Main = 270	N/A	N/A	[35]
ZnO thin film grown on silicon by ALD	155–220	1×10^{20}	160	[36]
Characteristics of ZnO TFT by ALD at various temperatures	70–130	10^{14}–10^{19}	8.82×10^{-3}– 6.11×10^{-3}	[37]

controversy over which of these defects is the dominant cause of the n-type characteristics. Zn_i occurs as a shallow donor (~25–30 meV) while V_o is a relatively deep level donor [38]. However, the formation energy of Zn_i has been reported to be high; therefore, it is not expected to be the dominant cause of background concentrations (10^{15}–10^{17} cm^{-3}) of as-grown, undoped ZnO. Hydrogen also is usually considered the main cause of n-type doping [39]. References [38–41] discuss theoretically in detail about the possible causes of the n-type characteristics of ZnO.

However, achieving p-type doping has proven to be a very difficult task. One reason is that ZnO is intrinsically n-typed, in which there is not much knowledge to understand the conductivity. Another reason is that the defects such as oxygen vacancies and zinc interstitials do play a role as compensating centers in p-type doping [42]. A third reason is that there are a few candidate shallow acceptors in ZnO. Elements like Li, Na, K, Cu, Ag, and Au are proven to be deep acceptors and do not contribute to p-type conductivity [43]. There have also been reports on p-type doping in nitrogen-doped ZnO, but these have not been followed up by reports of reproducible p–n junctions and stability of the p-type doping.

Meanwhile, recently, A. Janotti et al. have proposed fluorine impurities as an alternative to achieve a p-type doping [44]. Despite this, there is no experimental evidence to prove this p-type conductivity by F-doping.

12.3 ZnO NANOWIRE FABRICATION PROCESS

Many techniques have been used to fabricate nanowires. These can be categorized into two major methods. The first method uses a bottom-up process, which is widely used, and the second uses a top-down process.

12.3.1 BOTTOM-UP PROCESS

ZnO has been fabricated for various applications using mostly catalyst-assisted VLS growth and non-catalyst VS method. For ZnO nanowires grown via VLS process, the commonly used catalyst is Au. The size of the catalyst defines the diameter of the nanowires [45,46]. In the VS process, no catalyst is required resulting in higher purity of the ZnO, because there will be no impurities. There have been reports demonstrating the growth of ZnO nanowires without catalysts using MOCVD, MOVPE, pulsed laser ablation, and thermal evaporation [47].

However, the bottom-up process requires expensive and the low throughput e-beam lithography process to define the source and drain area. It is suitable only to study the performance of individual nanowires. Although the bottom-up process can produce highly crystalline nanowires, which have been proven by x-ray diffraction and/or high-resolution transmission electron microscopy (HRTEM) [48], the reproducibility and stability of these nanowire devices are still questionable (see Figures 12.3 and 12.4).

12.3.2 TOP-DOWN PROCESS

The drawbacks of the bottom-up process are related to the tedious "pick and place" method with its time-consuming processing steps. It is an impractical method to fabricate planar nanowire devices in large scale. Therefore, researchers are looking for a complementary process to fabricate nanowire using a top-down process, which involves deposition, patterning, and etching processes.

There are not many top-down processes to produce ZnO nanowires. Ra et al. have used the nanoscale spacer lithography (NSL) method, which has been implemented in Si nanowires previously [49]. In this method, ZnO thin film is deposited through ALD and followed by dry etching using inductively coupled plasma, (ICP). The advantages of this top-down technique are the compatibility with the Si process and the precise control of electrical contacts. The resolution of this NSL technique is determined by the ZnO film thickness deposition process.

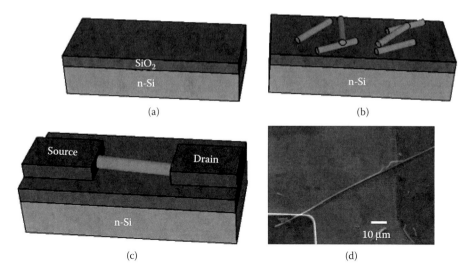

FIGURE 12.3 ZnO nanowire fabrication from bottom-up process.

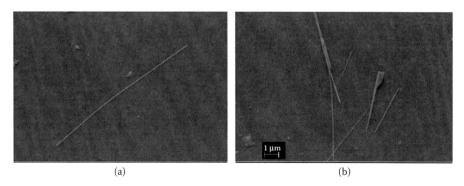

FIGURE 12.4 Scanning electron microscopy (SEM) images of (a) single ZnO nanowire and (b) multiple ZnO nanowires on Si/n-Si substrate. (From P. Singjai, Chiang Mai University.)

12.3.3 BENCHMARKING ZnO NWFETs

In this section, several published performances of ZnO NWFETs will be analyzed and compared. Table 12.3 shows the characteristics of some of the published NWFET work, and Figure 12.5 plots some of the results based on mobility and the ZnO nanowire dimensions.

From Figure 12.5, the mobility of ZnO NWFET increases by more than two orders after passivation by either polymer as demonstrated by Park et al. [31] *or* by dielectrics such as SiO_2 and Si_3N_4 as shown by Chang et al. [33]. Song et al. had also demonstrated that passivated ZnO NWFET exhibits stable electrical characteristics under different environments [34]. Apart from passivation, some have used different dielectric materials as the gate insulator as shown by Ju et al. and hence the high mobility achieved in their work [35]. All works presented here are based on

TABLE 12.3

Characteristics of Some of the Published NWFET Work

References	Method of Growth	μ_{FE} (cm²/Vs)	L_g (μm)	d_{ZnO} (nm)	Insulator	t_{ins} (nm)	V_t (V)	V_{ds} (V)	S (V/dec)	I_{on}/I_{off}
[1]	Vapor trap (untreated)	30	4	80	SiO_2	500	−18	1	3	–
	Passivate with SiO_2/Si_3N_4 coat	3118	4	90	SiO_2	500	−18	0.005	0.15	1×10^4
[49]	Top-down ZnO by ALD	80	5	70	SiO_2	400	−7.5	1	0.7	1×10^5
[50]	Catalyst-free MOVPE without passivation (100 nm)	75	5	100	SiO_2	250	−21	1	3.37	–
	With polyimide passivation	1000	5	100	SiO_2	250	−0.33	1	0.20	1×10^5
[51]	Use organic dielectric, d, of ZnO = 120 nm	196	2.1	120	Organic	16	−0.4	0.1	0.4	1×10^3
	After ozone treatment		2	120	Organic	16	0.2	0.1	0.15	–
[52]	Ga-doped VLS technique Carbon used to reduce Zn vapor	42	2	100	SiO_2	300	−12	0.5		Not shown
[53]	Vapor trap (untreated)	17.2	7	120	SiO_2	500	−8.37	0.1	5.56	Not shown
[54]	CVD grown untreated		4	85	SiO_2	100	−1.73 (air) 0.05 (dry O_2) −0.1 (vacuum)	0.1		1×10^5
[55]	Treated with PMMA	60	4	85	SiO_2	100	6	0.1	0.38	1×10^5
	Top gate: ZnO suspended in air. VLS Au coated Si substrate (Nb as S/D)	928	0.968	60	Air	26	0.4	0.8	0.129	1×10^6
[56]	All devices passivated with PMMA Grown on Au coated sapphire substrate	20	4	220	SiO_2	100	−4.14	0.1	0.4	1×10^6

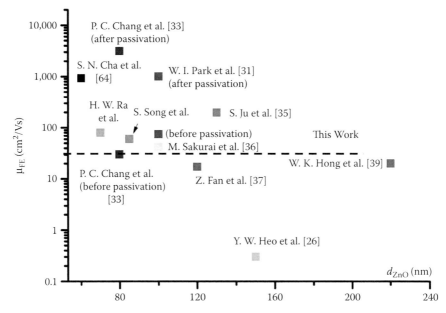

FIGURE 12.5 Field effect mobilities, μ_{FE}, versus ZnO diameters for all published ZnO NWFET characteristics.

bottom-up grown ZnO nanowires except for the work of Ra et al. in which they have achieved a mobility of 80 cm²/Vs even for top-down fabrication [32]. This work is inspired to achieve a mobility of at least 100 cm²/Vs with the top-down method as well as well-behaved transistor characteristics suitable for electronics applications.

12.4 TOP-DOWN FABRICATION OF ZNO NWFET

In this section, we discuss the fabrication of ZnO NWFETs using the top-down method. To begin with, we start with the top-down fabrication techniques using conventional optical lithography process. The ZnO film is deposited by remote plasma atomic layer deposition (RP-ALD).

12.4.1 Fabrication Process

Figure 12.6 shows fabrication process steps. The process starts with a p-type Si wafer. The wafer is cleaned in fumic nitric acid for 15 min and dipped in hydro-fluoric acid before a layer of SiO_2 thermally grown by wet oxidation. The SiO_2 is then etched anisotropically by photolithography pattern transfer and reactive ion etching (RIE). The etched depth was 100 nm. The SiO_2 etch is executed in CHF_3 and Ar mixture at a pressure of 30 mTorr. Figure 12.6b shows the SiO_2 trench formed after the RIE etch. A layer of ZnO film is deposited on top of the trenches. The ZnO film is deposited using the RP-ALD, which will not be discussed in this chapter, at deposition temperature of 150°C, and oxygen plasma time of 5 s, plasma pressure of 57.5 mTorr, and plasma power of 100 W.

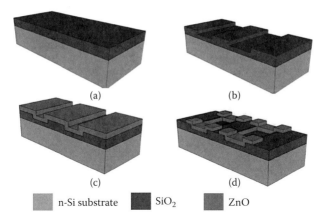

n-Si substrate   SiO$_2$ ZnO

FIGURE 12.6 Top-down fabrication process of ZnO NWFET. (a) SiO$_2$ thermally grown through wet oxidation ~200 nm, (b) SiO$_2$ dry etched in the RIE to form 100 nm trench, (c) ZnO thin film deposited in the ALD, and (d) anisotropic RIE etch to obtain nanowires at the side of the oxide trenches. (From Sultan, S.M., Sun, K., Partridge, J., Allen, M., Ashburn, P., and Chong, H.M.H., *12th International Conference on ULIS 2011 Ultimate Integration on Silicon*, Cork, Ireland, March 14–16, 2011.)

After deposition, the films were etched in RIE anisotropically. The gas used was CHF$_3$ at a pressure of 20 mTorr. Aluminum source and drain contacts were evaporated and lifted off in 1-methyl-2-pyrrolidinone (NMP) solution. The contacts were annealed in rapid thermal annealer (RTA) for 2 min at 350°C.

The nanowire device structure with Al pad is shown in the optical microscopy image in Figure 12.7a. The nanowires fabricated have length ranging from 2 to 20 µm. The width and height of the ZnO nanowires are 40 and 80 nm, respectively. These dimensions can be controlled by adjusting the thickness of the ZnO layer and the height of the SiO$_2$ trench. Figure 12.7b shows 16 identical wires across the two Al electrodes demonstrating that ZnO nanowire arrays were successfully formed at the side of the oxide spacers. Figure 12.7c shows the cross section of the ZnO film deposited in ALD before the RIE etch. The film deposited is uniform and almost conformal over the SiO$_2$ trench. The ZnO layer was slightly thicker (about 20%) on the pillar sidewall than on the planar surfaces. Figure 12.7d shows a single nanowire using field-emission SEM (FESEM) after the RIE etch.

12.4.2 ETCHING

Apart from deposition techniques, etching of these ZnO films play an important role in producing a high conductivity ZnO NWFET. Since these nanowires are not covered by the photoresist, they are prone to be bombarded by plasma ions and radicals from the etching gas. Therefore, it is important to monitor the various etching methods on a ZnO film. Table 12.4 shows a review of how ZnO can be etched using various dry etch equipments. All results shown are based on room temperature etching. Since nanowires are formed from anisotropic etching, only dry etch techniques have been used.

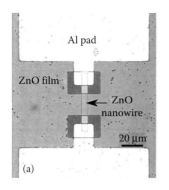
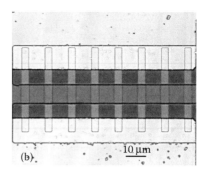

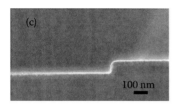
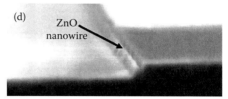

FIGURE 12.7 (a) Optical microscope view of a pair of ZnO nanowire in contact with Al pad, (b) optical image of 16 nanowire array contacted with Al pad, (c) cross section of as-deposited ZnO film before etch, and (d) cross section of a single ZnO nanowire after RIE etch. (From Sultan, S.M., Sun, K., Clark, O.D., Masaud, T.B., Fang, Q., Gunn, R., Partridge, J., Allen, M.W., Ashburn, P., and Chong, H.M.H., *Electron Device Letters*, 33, 201, 2012.)

TABLE 12.4
ZnO Etch Overview from Previous Work

Machine	Power (W)	Gas	Pressure (mTorr)	Etch Rate (nm/min)	Mask	References
RIE	200	CHF_3	20	5	Cr	[58]
ECR plasma	80	$CF_4 + CH_4$	0.075	55	Resist	[59]
ICP	RF = 150	Ar/Cl_2	5	15	SiNx	[60]
	ICP = 300	Ar/CH_4				
ICP	RF = 400	CF_4/Ar	15	45.2	–	[61]
	ICP = 800					

Basically, the mechanism of all these techniques is the same, which is utilizing the plasma to etch the material. RIE etch is typically performed at 10–100 mTorr pressure in an asymmetric parallel plate reactor and is highly anisotropic. The ion density is in the range of 10^{10}–10^{11} cm^{-3} [57]. In comparison, electron cyclotron resonance (ECR) etching also produces high anisotropic etch with lower gas pressures of 0.2–10 mTorr and the etch rate is faster than in the RIE. This is due to higher ion density (up to 10^{12} cm^{-3}). However, ECR technology is expensive; etch uniformity is not good due to the presence of a magnetic field. An inductive coupled plasma (ICP) tool provides the

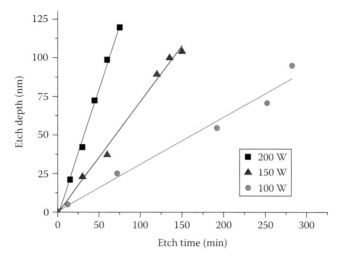

FIGURE 12.8 The etch rate of ZnO film with pressure = 20 mTorr and CHF_3 = 25 sccm. (From Sultan, S.M., Sun, K., Partridge, J., Allen, M., Ashburn, P., and Chong, H.M.H., *12th International Conference on ULIS 2011 Ultimate Integration on Silicon*, Cork, Ireland, March 14–16, 2011.)

same advantage as ECR, that is, low pressure (1 mTorr), and higher etch rate because of high ion density. In ICP, the ion flux is controlled by the ICP power and the gas pressure while the ion energy is controlled by the RF power of the substrate holder.

Based on this review, the initial experiment of this work was executed in RIE in CHF_3 gas flow with a pressure of 20 mTorr following Schuler et al. [58]. The mask used was a positive resist. Note that the mask is used for the thin film area and not for the nanowire. Figure 12.8 shows the etch rate of ZnO film with increasing RF power. For this fabrication, highest power is used due to higher etch rate as well as higher selectivity toward the underlying SiO_2 film. However, the selectivity toward the photoresist mask was not good (<1). The etch rate results demonstrate highest etch rate of 1.64 nm/min for RF power of 200 W, followed by 0.72 and 0.3 nm/min for RF power of 150 and 100 W, respectively.

However, ICP etching of ZnO thin film shows a much higher etch rate and higher selectivity toward the underneath SiO_2 and over the photoresist mask. Figure 12.9 shows the etch rate of ZnO in ICP in CHF_3/Ar gas flow and selectivity over the photoresist mask as CHF_3 gas increased from 0% to 100%. In this experiment, the total gas flow rates, RF power, and ICP power were fixed at 25 sccm, 300 W, and 1000 W, respectively with a pressure of 10 mTorr.

The etch rates and the selectivity were measured using a surface profiler and ellipsometer. The etch rate of ZnO at 100% Ar was 19.51 nm/min. With increasing the CHF_3 content further, the etch rate of ZnO increased and maintained at about 50 nm/min. The selectivity of ZnO to the photoresist shows an increase from 0.05 at 100% Ar to 0.353 at 40% of CHF_3 and reduced to 0.26 at 100% of CHF_3.

At 100% Ar, the etching is considered as pure physically and is also known as Ar sputtering. It lacks selectivity among different materials because the ion energy required to eject the surface atoms is very high compared to the surface

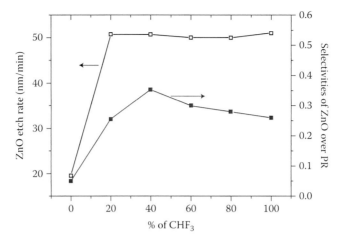

FIGURE 12.9 ICP etch rate and selectivities of ZnO film as a function of CHF_3/Ar ratios.

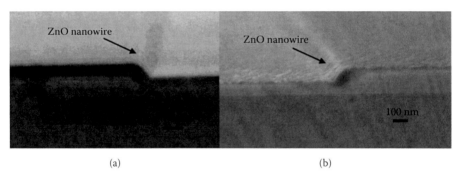

FIGURE 12.10 Raman image mapping at E_2 phonon mode after ICP etch. (a) Incomplete etching, (b) complete etching, and (c) SEM image of a complete etch of ZnO nanowire.

bond energies. Thus, the etch rate is low. At 100% CHF_3, it is considered as purely chemical, and etching in this manner needs to be carefully monitored as it could be etching isotropically.

Figure 12.10 shows the SEM image of the ZnO nanowire after the ICP etch. Figure 12.10a shows the ZnO nanowire formed after 1 min ICP etch with only CHF_3, while Figure 12.10b shows the ZnO nanowire forming at the SiO_2 pillar with 20% of CHF_3 gas and 80% of argon gas flow. Etching in pure CHF_3 produces a smoother surface compared with Ar-assisted etch as shown.

12.4.3 RAMAN MAPPING

Figure 12.11 shows the results from Raman mapping of the ZnO nanowires at the side of the oxide trenches after RIE etch. Figure 12.11a shows a bright region, indicating the presence of ZnO, and this is verified by the Raman spectrum shown in Figure 12.11d. The laser source used for the Raman spectroscopy map was 532 nm.

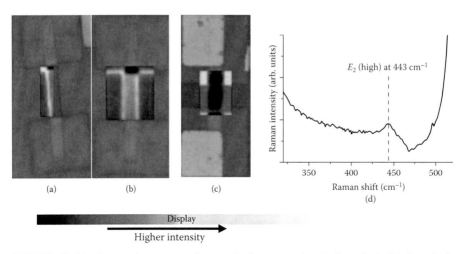

FIGURE 12.11 Raman image mapping at E_2 phonon mode. (a) Sample 1, (b) Sample 2, (c) Sample 3, and (d) Raman spectra obtained from one of the mapped point along ZnO nanowire device. (From Sultan, S.M., Sun, K., Partridge, J., Allen, M., Ashburn, P., and Chong, H.M.H., *12th International Conference on ULIS 2011 Ultimate Integration on Silicon*, Cork, Ireland, March 14–16, 2011.)

The peaks at 437 cm^{-1} should be assigned to the vibration modes of E_2. The E_2 mode was related to band characteristics of the wurtzite phase, and it can be shifted due to residual stress in the nanowire. As shown, the E_2 (high) peak is at 443 cm^{-1} which indicates that the nanowire is under compressive stress. This can be due to the prolonged RIE etch, Al metallization process or contact annealing, which exposed the nanowires to different ion bombardments and high temperatures, respectively. Note that the E_2 (high) peak is broader compared to the bottom-up ZnO nanowire Raman results (not shown). Basically, the full width at half maximum (FWHM) of the E_2 (high) peak is narrower for bottom-up nanowire, which indicates that the crystal quality is better compared to top-down fabricated nanowires.

Figure 12.11a of Sample 1 and Figure 12.11b of Sample 2 show high intensity (yellow) at the side of the oxide trench, which indicates the existence of ZnO nanowires. Meanwhile, Figure 12.11c of Sample 3 does not show peaks at E_2 mode near the oxide trench, which suggests that the ZnO film has been overetched from prolonged RIE etch.

12.4.4 Contact Annealing

The electrical characterization is performed using a semiconductor parameter analyzer HP4155. Figure 12.12 shows the *I–V* characteristics of the ZnO nanowire device measured before and after contact annealing in RTA at 350°C. The current–voltage characteristics were measured at room temperature. The nanowires measured were 10 µm long. The voltage is biased from −1 to 1 V. Before annealing, the ZnO nanowire device shows asymmetrical behavior with a maximum current of 1×10^{-8} A at 1 V. This result indicates the existence of surface barriers between the ZnO nanowire device and the Al metal electrode. The black curve shows the same nanowire

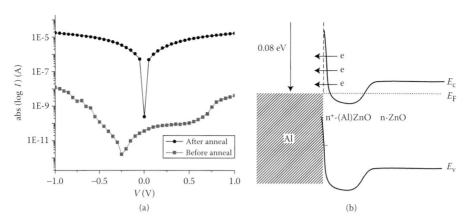

FIGURE 12.12 (a) The *I–V* characteristics of Al-ZnO nanowire before and after contact anneal [57] and (b) schematic band diagram of Al-based ohmic contact with ZnO (Reproduced from Liao, Z., Liu, K., Zhang, J., Xu, J., and Yu, D., *Phys. Lett. A*, 367, 207, 2007.)

device after contact annealing with RTA in vacuum. The curve shows a distinctive characteristic of ohmic behavior from the symmetric *I–V* result. Contact annealing also improves the current of the ZnO nanowire device by threefold in magnitude with maximum current achieved at 1 V is 2×10^{-5} A. This result using an Al electrode is comparable to results obtained using Ti/Au metal [63]. This is due to the dissociation of the ZnO at the surface region and strong reaction between Al and O in the ZnO layer with barrier heights as shown in Figure 12.12b. As a result, it causes the accumulation of oxygen vacancies near the ZnO surface hence lowering the n-type barrier height.

12.4.5 FIELD-EFFECT CHARACTERISTICS

Ensuring ohmic contact between ZnO and the source and drain electrodes is vital for the electrons to flow without any contact barriers. From Figure 12.12, it is observed that ZnO and Al can form an excellent ohmic contact after the RTA anneal. Transistor characteristics were measured using the measurement setup shown in Figure 12.13. Three probes were used with two probes connected to the source and drain terminal via Al electrodes. The third probe was connected to the substrate by a connection through the chuck of the probe station. Source electrode was kept at zero potential. The voltage at the Si substrate, which serves as bottom gate, was varied from 0 to 20 V at a constant bias of $V_D = 1$ V for linear region and $V_D = 10$ V for saturation region.

The nanowire length in these measurements was 8.6 µm. Figure 12.13 shows the I_D–V_G plot of the NWTFT device in log scale at $V_D = 0.5$ V after RIE etch. From this plot, as V_G increases positively, I_D increases, indicating that these NWFETs are typical n-type semiconductors. The threshold voltages are measured by extrapolating the linear part of the I_D–V_G plot at the peak of the transconductance plot and it is found to be 4.8 V. Since the threshold voltage is greater than 0, this device operates in an enhancement mode.

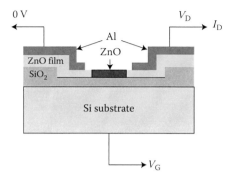

FIGURE 12.13 Schematic of probe connections for electrical measurements at room temperature.

The calculated capacitance value is 3.8×10^{-16} F, and the radius of the nanowire is taken as 20 nm. These dimensions are based on SEM images taken after the RIE etch process. The field-effective channel mobility is calculated from Equation 12.1 and it is found to be 0.5 cm²/Vs. From this plot, the subthreshold swing (SS) = $(dV_G)/[d(\log I_D)]$ is determined to be 0.69 V/dec. Subthreshold slope measurements were taken at the steepest linear line of the I_D–V_G plot.

$$\mu FE = \frac{g_m L}{Z C_{ins} V_D} \tag{12.1}$$

where
g_m is the transconductance from the I_D–V_G plot
L is the nanowire length
Z is the nanowire width
C_{ins} is the capacitance per unit area

The capacitance is given in Equation 12.2 for cylinders using an infinite plate model [50]

$$C = \frac{2\pi\varepsilon_0\varepsilon_{Si}L}{\ln\dfrac{2t}{r}} \tag{12.2}$$

where
t is the SiO₂ thickness
r is the nanowire radius
ε is the dielectric constant of the SiO₂

The I_{ON}/I_{OFF} ratio is measured from the lowest positive measured current to the highest measured current at a drain voltage of 0.5 V. This is to give a rough estimate of the values since some of the I_D data points are missing in the negative gate voltage region due to leakage in the gate oxide. The leakage can be attributed to the overetch of the SiO₂ layer during the ZnO RIE etch.

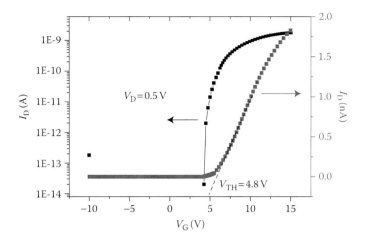

FIGURE 12.14   Transfer characteristics of ZnO NWTFT with $L = 8.6$ μm.

Figure 12.14 shows the output characteristics of the ZnO NWTFT. Well-behaved output characteristics are obtained with clearly defined linear and saturation regions. The output characteristics obtained with our top-down process are also significantly better behaved than those reported for ZnO nanowire transistors fabricated by bottom-up self-assembly. These devices tend to show either a lack of saturation or very poor saturation [64,50]. From this plot, the carrier concentration is estimated to be in the order of 10^{17} cm^{-3}.

From these results, the top-down fabricated NWTFT exhibits well-behaved transistor characteristics. The field-effect mobility value is consistent with the range of values from 0.13 to 17 cm^2/Vs reported for thin film ZnO transistors fabricated using ALD [37,65,11]. There is considerable scope to improve the mobility value of these nanowire TFTs by optimizing the ZnO ALD process, passivation techniques, and using different metals such as Ti/Au or Ti/Al as source and drain electrodes.

Another interesting property of these NWTFTs is the high breakdown voltage exhibited in the output characteristics, which has been reported to be the first demonstration of high voltage operation [66]. A breakdown voltage of 75 V is obtained for $V_G = 0$ V and $I_D = 5$ nA for this ZnO nanowire transistors as shown in Figure 12.15. A typical breakdown field for SiO$_2$ is 600 V/μm. Therefore, the breakdown voltage of a 100 nm SiO$_2$ layer is 60 V. This value suggests that the breakdown voltage obtained at 75 V is due to SiO$_2$ and that the ZnO nanowires have the potentiality to withstand even higher voltages than 75 V. The transistor operation up to a voltage of 75 V shown in Figure 12.15 is significantly better than the 3 V operation reported by Park et al. [50] and the 1.5 V operation reported by Cha et al. [55] for ZnO nanowire transistors fabricated by bottom-up self-assembly. This high operating voltage in small dimension nanowire devices indicates the suitability of this device for applications such as high resolution flat panel displays [67,68], which require a supply voltage of 15–20 V.

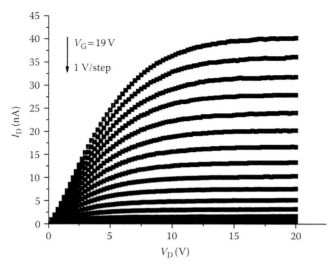

FIGURE 12.15 The output characteristics of ZnO NWTFT with $L = 8.6$ μm after aniso-tropic RIE etch. (From Sultan, S.M., Sun, K., Clark, O.D., Masaud, T.B., Fang, Q., Gunn, R., Partridge, J., Allen, M.W., Ashburn, P., and Chong, H.M.H., *Electron Device Lett.*, 33, 203, 2012.)

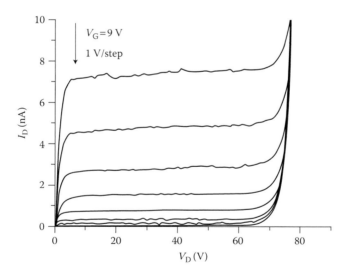

FIGURE 12.16 Breakdown voltage measurement for ZnO NWFET, $L = 8.6$ μm. (From Sultan, S.M., Sun, K., Clark, O.D., Masaud, T.B., Fang, Q., Gunn, R., Partridge, J., Allen, M.W., Ashburn, P., and Chong, H.M.H., *Electron Device Lett.*, 33, 203, 2012.)

12.4.6 SCALABILITY AND REPRODUCIBILITY OF TOP-DOWN FABRICATION

To test the scalability of our top-down ZnO nanowire process, Figure 12.16 shows the results for NWFETs with channel lengths of 1.3, 8.6, and 18.6 μm processed at plasma time of 5 s. Well-behaved output characteristics are obtained at all channel lengths. The poor saturation seen in Figure 12.16a for the 1.3 μm transistor suggests the presence of short-channel effects. Similar behavior has been observed in short-channel ZnO TFTs [69]. At $V_G = 10$ V and $V_D = 5$ V, drain currents of 110 and 8.8 nA are obtained for $L = 1.3$ and 18.6 μm, respectively. This factor of 12.4 differences in drain current is consistent with the expected value of 14.3 obtained from the $1/L$ scaling behavior of long-channel FETs [70]. Breakdown voltages of 79 and 78 V were measured for channel lengths of 1.3 and 18.6 μm, respectively.

To test the reproducibility of the fabrication process, Figure 12.17 shows the transfer characteristic of a device with 16 parallel nanowires. This device has a drain current of 35.7 nA at $V_D = 5$ V and $V_G = 10$ V, which compares with a drain current of 4.3 nA for the transistor with two parallel nanowires. The 8.3 difference in current demonstrates that the drain current scales in a reproducible way. Figure 12.17 also shows subthreshold characteristic for NWFETs with different channel lengths at $V_D = 5$ V. Subthreshold slopes of 1.02, 0.69, and 0.9 V/dec are obtained for transistors with channel lengths of 1.3, 8.6, and 18.6 μm, respectively. The poor value of subthreshold slope for the 1.3 μm device again can be explained by short-channel effects due to poor control of the channel by the gate as a result of the thick gate

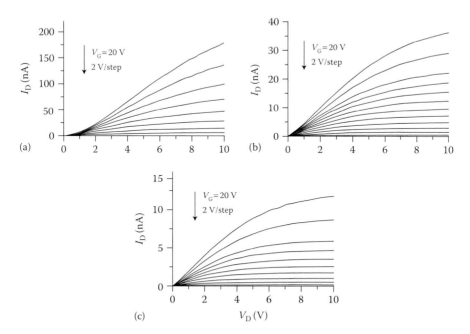

FIGURE 12.17 Output characteristics of ZnO NWFET comprising two parallel nanowires with channel lengths L of (a) 1.3, (b) 8.6, and (c) 18.6 μm. (From Sultan, S.M., Sun, K., Clark, O.D., Masaud, T.B., Fang, Q., Gunn, R., Partridge, J., Allen, M.W., Ashburn, P., and Chong, H.M.H., *Electron Device Lett.*, 33, 203, 2012.)

oxide. The 8.6 μm transistor shows a respectable I_{ON}/I_{OFF} ratio of 2×10^6, measured between $V_G = 20$ and 3 V.

12.4.7 COMPARISON WITH THEORY

Based on Equation 12.3 of a typical MOSFET device operating in a saturation region, measured I_D values were compared with calculated values to obtain a better understanding of the NWFET operation. From the plots in Figure 12.18, it is shown for channel length of 8.6 and 18. 6 μm, the transistor current matches very well with the calculated values. This shows the reliability of this device operation without the influence of other issues such as series resistances, short channel effects, etc.

$$I_D = \frac{Z}{2L} \mu C_{ins} \left(V_G - V_{TH} \right)^2 \tag{12.3}$$

However, for $L = 1.3$ μm, the measured current is higher than the theoretical prediction. As can be observed in Figure 12.18, the subthreshold slope is poor for this channel length due to thick gate oxide. Due to poor control of the gate on the channel, the drain current increases rapidly and thus the discrepancy with the calculated values (see Figure 12.19).

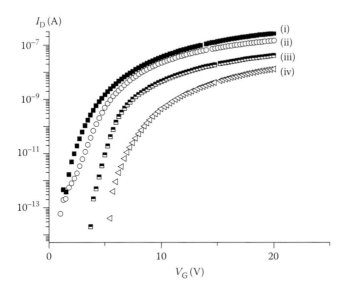

FIGURE 12.18 Transfer characteristics for devices with different channel lengths and different numbers of parallel nanowires: (i) 16 NWFETs with $L = 8.6$ μm, (ii) 2 NWFETs with $L = 1.3$ μm, (iii) 2 NWFETs with $L = 8.6$ μm, and (iv) 2 NWFETs with $L = 18.6$ μm. (From Sultan, S.M., Sun, K., Clark, O.D., Masaud, T.B., Fang, Q., Gunn, R., Partridge, J., Allen, M.W., Ashburn, P., and Chong, H.M.H., *Electron Device Lett.*, 33, 203, 2012.)

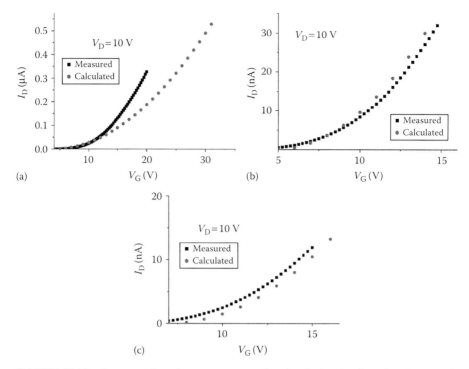

FIGURE 12.19 I_D comparison between measured and calculated values for (a) $L = 1.3$, (b) 8.6, and (c) 18.6 μm.

12.4.8 CONCLUSIONS

In conclusion, a top-down fabrication of ZnO NWFETs was successfully demonstrated using remote plasma ALD and anisotropic RIE. Etching makes a significant contribution in controlling the surface roughness of the nanowires. Despite the preliminary work suggesting well-behaved transistor characteristics using RIE etch, it suffers from low mobility in the range of 0.5 cm²/Vs. However, with this top-down technique, nanowire transistors with different channel lengths could be fabricated in defined locations. A breakdown voltage of ≥75 V has been achieved at all channel lengths, which is ample for applications such as active matrix display electronics. It is proved that the device is scalable and reproducible where the drain current scales as $1/L$, as expected for long-channel FETs. The current conduction also adheres well with typical MOSFET theory for $L = 8.6$ μm.

ACKNOWLEDGMENTS

S.M. Sultan would like to thank the Faculty of Electrical Engineering, Universiti Teknologi Malaysia, Johor Bahru and the Ministry of Higher Education of Malaysia for supporting her doctoral studies.

REFERENCES

1. P.-C. Chang, Z. Fan, C.-J. Chien, D. Stichtenoth, C. Ronning, and J. G. Lu. High-performance ZnO nanowire field effect transistors. *Applied Physics Letters*. 89(3), 133113, 2006.

2. W.-H. Luo, T.-K. Tsai, J.-C. Yang, W.-M. Hsieh, C.-H. Hsu, and J.-S. Fang. Enhancement in conductivity and transmittance of zinc oxide prepared by chemical bath deposition. *Journal of Electronic Materials*. 38(11), 2264–2269, 2009.

3. D. M. Bagnall et al. Optically pumped lasing of ZnO at room temperature. *Applied Physics Letters*. 70(17), 2230, 1997.

4. F. Tuomisto, K. Saarinen, D. Look, and G. Farlow. Introduction and recovery of point defects in electron-irradiated ZnO. *Physical Review B*. 72(8), 1–11, 2005.

5. D. C. Look. Recent advances in ZnO materials and devices. *Materials Science and Engineering B*. 80(1–3), 383–387, 2001.

6. J. D. Albrecht, P. P. Ruden, S. Limpijumnong, W. R. L. Lambrecht, and K. F. Brennan. High field electron transport properties of bulk ZnO. *Journal of Applied Physics*. 86(12), 6864, 1999.

7. K. Ellmer, A. Klein, and B. Rech. *Transparent Conductive Zinc Oxide—Basics and Applications in Thin Film Solar Cells*. Springer, Berlin, Germany. 2008, p. 203.

8. P.-C. Chang and J. G. Lu. ZnO nanowire field-effect transistors. *IEEE Transactions on Electron Devices*. 55(11), 2977–2987, 2008.

9. Y. Chen, D. M. Bagnall, H.-J. Koh, K.-T. Park, K. Hiraga, and Z. Zhu. Plasma-assisted molecular beam epitaxy of ZnO on c-plane sapphire: Growth and characterization. *Journal of Applied Physics*. 84(7), 3912–3918, 1998.

10. K. Nomura, H. Ohta, K. Ueda, T. Kamiya, M. Hirano, and H. Hosono. Thin-film transistor fabricated in single-crystalline transparent oxide semiconductor. *Science*. 300(5623), 1269–1272, 2003.

11. S. Masuda, K. Kitamura, Y. Okumura, S. Miyatake, H. Tabata, and T. Kawai. Transparent thin film transistors using ZnO as an active channel layer and their electrical properties. *Journal of Applied Physics*. 93(3), 1624, 2003.

12. R. L. Hoffman, B. J. Norris, and J. F. Wager. ZnO-based transparent thin-film transistors. *Applied Physics Letters*. 82(5), 733, 2003.

13. P. F. Carcia, R. S. McLean, M. H. Reilly, and G. Nunes. Transparent ZnO thin-film transistor fabricated by RF magnetron sputtering. *Applied Physics Letters*. 82(7), 1117, 2003.

14. P. F. Carcia et al. A comparison of zinc oxide thin-film transistors on silicon oxide and silicon nitride gate dielectrics. *Journal of Applied Physics*. 102(7), 074512, 2007.

15. Z. L. Wang. Nanostructures of zinc oxide. *Materials Today*. 7(6), 26–33, 2004.

16. G. F. Cerofolini et al. A hybrid approach to nanoelectronics. *Nanotechnology*. 16(8), 1040–1047, 2005.

17. W. I. Park, J. S. Kim, G. C. Yi, and H. J. Lee. ZnO nanorod logic circuits. *Advanced Materials*. 17(11), 1393–1397, 2005.

18. R. S. Wagner and W. C. Ellis. Vapor-liquid-solid mechanism of single crystal growth. *Applied Physics Letters*. 4(5), 89–90, 1964.

19. X. Duan. Nanowire thin-film transistors: A new avenue to high-performance macro-electronics. *IEEE Transactions on Electron Devices*. 55(11), 3056–3062, 2008.

20. J. Xiang, W. Lu, Y. Hu, Y. Wu, H. Yan, and C. M. Lieber. Ge/Si nanowire hetero-structures as high-performance field-effect transistors. *Nature*. 441(7092), 489–493, 2006.

21. X. Duan, Y. Huang, Y. Cui, J. Wang, and C. M. Lieber. Indium phosphide nanowires as building blocks for nanoscale electronic and optoelectronic devices. *Nature*. 409(6816), 66–69, 2001.

22. S. Rosenblatt, Y. Yaish, J. Park, J. Gore, V. Sazonova, and P. L. McEuen. High performance electrolyte gated carbon nanotube transistors. *Nano Letters*. 2(8), 869–872, 2002.

23. Y.-K. Choi, T.-J. King, and C. Hu. A spacer patterning technology for nanoscale CMOS. *IEEE Transactions on Electron Devices*. 49(3), 436–441, 2002.

24. W. Lu and C. M. Lieber. Nanoelectronics from the bottom up. *Nature Materials*. 6(11), 841–850, 2007.

25. N. Huby, S. Ferrari, E. Guziewicz, M. Godlewski, and V. Osinniy. Electrical behavior of zinc oxide layers grown by low temperature atomic layer deposition. *Applied Physics Letters*. 92(2), 023502, 2008.

26. X. G. Zheng et al. Photoconductive ultraviolet detectors based on ZnO films. *Applied Surface Science*. 253(4), 2264–2267, 2006.

27. E. M. Kaidashev et al. High electron mobility of epitaxial ZnO thin films on c-plane sapphire grown by multistep pulsed-laser deposition. *Applied Physics Letters*. 82(22), 3901, 2003.

28. A. Ohtomo and A. Tsukazaki. Pulsed laser deposition of thin films and superlattices based on ZnO. *Semiconductor Science and Technology*. 20(4), S1–S12, 2005.

29. H. Kato, M. Sano, K. Miyamoto, and T. Yao. Effect of O/Zn flux ratio on crystalline quality of ZnO films grown by plasma-assisted molecular beam epitaxy. *Japanese Journal of Applied Physics*. 42 (Part 1, No. 4B), 2241–2244, 2003.

30. X. Wang et al. Structural and optical properties of ZnO film by plasma-assisted MOCVD. *Optical and Quantum Electronics*. 34(9), 883–891, 2002.

31. S. Eisermann et al. Hydrogen and nitrogen incorporation in ZnO thin films grown by radio-frequency (RF) sputtering. *Thin Solid Films*. 518(4), 1099–1102, 2009.

32. E. M. C. Fortunato et al. Wide-bandgap high-mobility ZnO thin-film transistors produced at room temperature. *Applied Physics Letters*. 85(13), 2541, 2004.

33. H. Jeon et al. Characteristics of gallium-doped zinc oxide thin-film transistors fabricated at room temperature using radio frequency magnetron sputtering method. *Japanese Journal of Applied Physics*. 47(1), 87–90, 2008.

34. A. Wojcik, M. Godlewski, E. Guziewicz, R. Minikayev, and W. Paszkowicz. Controlling of preferential growth mode of ZnO thin films grown by atomic layer deposition. *Journal of Crystal Growth*. 310(2), 284–289, 2008.

35. S. Lee, Y. Im, S. Kim, and Y. Hahn. Structural and optical properties of high quality ZnO films on Si grown by atomic layer deposition at low temperatures. *Superlattices and Microstructures*. 39(1–4), 24–32, 2006.

36. S.-Y. Pung, K.-L. Choy, X. Hou, and C. Shan. Preferential growth of ZnO thin films by the atomic layer deposition technique. *Nanotechnology*. 19(43), 435609, 2008.

37. S. Kwon et al. Characteristics of the ZnO thin film transistor by atomic layer deposition at various temperatures. *Semiconductor Science and Technology*. 24(3), 035015, 2009.

38. C. Van de Walle. Defect analysis and engineering in ZnO. *Physica B: Condensed Matter*. 308–310(1–2), 899–903, 2001.

39. A. Janotti and C. Vandewalle. New insights into the role of native point defects in ZnO. *Journal of Crystal Growth*. 287(1), 58–65, 2006.

40. A. Janotti and C. G. Van de Walle. Native point defects in ZnO. *Physical Review B*. 76(16), 1–22, 2007.

41. D. Look. Electrical and optical properties of defects and impurities in ZnO. *Physica B: Condensed Matter*. 340–342, 32–38, 2003.

42. U. Özgür et al. A comprehensive review of ZnO materials and devices. *Journal of Applied Physics*. 98(4), 041301, 2005.

43. C. Park, S. Zhang, and S. Wei. Origin of p-type doping difficulty in ZnO: The impurity perspective. *Physical Review B*. 66(7), 1–3, 2002.

44. A. Janotti, E. Snow, and C. G. Van de Walle. A pathway to p-type wide-band-gap semi-conductors. *Applied Physics Letters*. 95(17), 172109, 2009.
45. P.-C. Chang, C.-J. Chien, D. Stichtenoth, C. Ronning, and J. G. Lu. Finite size effect in ZnO nanowires. *Applied Physics Letters*. 90(11), 113101, 2007.
46. Y. W. Heo et al. Depletion-mode ZnO nanowire field-effect transistor. *Applied Physics Letters*. 85(12), 2274, 2004.
47. M. H. Huang et al. Room-temperature ultraviolet nanowire nanolasers. *Science*. 292(5523), 1897–1899, 2001.
48. D. S. Kim, J. P. Richters, R. Scholz, T. Voss, and M. Zacharias. Modulation of carrier density in ZnO nanowires without impurity doping. *Applied Physics Letters*. 96(12), 79–81, 2010.
49. H.-W. Ra, K.-S. Choi, J.-H. Kim, Y.-B. Hahn, and Y.-H. Im. Fabrication of ZnO nanowires using nanoscale spacer lithography for gas sensors. *Small Weinheim an der Bergstrasse Germany*. 4(8), 1105–1109, 2008.
50. W. I. Park, J. S. Kim, G.-C. Yi, M. H. Bae, and H. J. Lee. Fabrication and electrical characteristics of high-performance ZnO nanorod field-effect transistors. *Applied Physics Letters*. 85(21), 5052, 2004.
51. S. Ju, K. Lee, D. B. Janes, M.-H. Yoon, A. Facchetti, and T. J. Marks. Low operating voltage single ZnO nanowire field-effect transistors enabled by self-assembled organic gate nanodielectrics. *Nano Letters*. 5(11), 2281–2286, 2005.
52. M. Sakurai, Y. G. Wang, T. Uemura, and M. Aono. Electrical properties of individual ZnO nanowires. *Nanotechnology*. 20(15), 155203, 2009.
53. Z. Fan, D. Wang, P.-C. Chang, W.-Y. Tseng, and J. G. Lu. ZnO nanowire field-effect transistor and oxygen sensing property. *Applied Physics Letters*. 85(24), 5923, 2004.
54. S. Song, W.-K. Hong, S.-S. Kwon, and T. Lee. Passivation effects on ZnO nanowire field effect transistors under oxygen, ambient, and vacuum environments. *Applied Physics Letters*. 92(26), 263109, 2008.
55. S. N. Cha et al. High performance ZnO nanowire field effect transistor using self-aligned nanogap gate electrodes. *Applied Physics Letters*. 89(26), 263102, 2006.
56. W.-K. Hong et al. Realization of highly reproducible ZnO nanowire field effect transistors with n-channel depletion and enhancement modes. *Applied Physics Letters*. 90(24), 243103, 2007.
57. S. M. Sultan, K. Sun, J. Partridge, M. Allen, P. Ashburn, and H. M. H. Chong. Fabrication of ZnO nanowire device using top-down approach. *12th International Conference on ULIS 2011 Ultimate Integration on Silicon*. Cork, Ireland. pp. 1–3, 14–16 March 2011.
58. S. J. Pearton and D. P. Norton. Dry etching of electronic oxides, polymers, and semiconductors. *Plasma Processes and Polymers*. 2(1), 16–37, 2005.
59. L. P. Schuler, M. M. Alkaisi, P. Miller, R. J. Reeves, and A. Markwitz. Comparison of DC and RF sputtered zinc oxide films with post-annealing and dry etching and effect on crystal composition. *Japanese Journal of Applied Physics*. 44(10), 7555–7560, 2005.
60. K. Ogata et al. Electron-cyclotron-resonance plasma etching of the ZnO layers grown by molecular-beam epitaxy. *Growth Lakeland*. 4, 531–533, 2004.
61. W. Lim et al. Dry etching of bulk single-crystal ZnO in CH4/H2-based plasma chemistries. *Applied Surface Science*. 253(2), 889–894, 2006.
62. J. Woo, G. Kim, J. Kim, and C. Kim. Etching characteristic of ZnO thin films in an inductively coupled plasma. *Surface and Coatings Technology*. 202(22–23), 5705–5708, 2008.
63. Z. Liao, K. Liu, J. Zhang, J. Xu, and D. Yu. Effect of surface states on electron transport in individual ZnO nanowires. *Physics Letters A*. 367(3), 207–210, 2007.

64. S. N. Cha et al. High performance ZnO nanowire field effect transistor. *Proceedings of 35th European Solid-State Device Research Conference 2005 ESSDERC 2005.* 1, 3–6, 2005.

65. P. F. Carcia, R. S. McLean, and M. H. Reilly. High-performance ZnO thin-film transistors on gate dielectrics grown by atomic layer deposition. *Applied Physics Letters.* 88(12), 123509, 2006.

66. S. M. Sultan, K. Sun, O. D. Clark, T. B. Masaud, Q. Fang, R. Gunn, J. Partridge, M. W. Allen, P. Ashburn, and H. M. H. Chong. Electrical characteristics of top-down ZnO nanowire transistors using remote plasma ALD. *Electron Device Letters.* 33, 203–205, 2012.

67. S.-H. K. Park et al. Transparent and photo-stable ZnO thin-film transistors to drive an active matrix organic-light-emitting-diode display panel. *Advanced Materials.* 21(6), 678–682, 2009.

68. T. Kamiya, K. Nomura, and H. Hosono. Present status of amorphous In–Ga–Zn–O thin-film transistors. *Science and Technology of Advanced Materials.* 11(4), 044305, 2010.

69. H.-H. Hsieh and C.-C. Wu. Scaling behavior of ZnO transparent thin-film transistors. *Applied Physics Letters.* 89(4), 041109, 2006.

70. S. M. Sze and K. K. Ng. *Physics of Semiconductor Devices,* vol. 6, no. 1. Wiley-Interscience, New York. 2007, p. 815.

13 Quantum Mechanical Effects in Nanometer Scale Strained Si/Si$_{1-x}$Ge$_x$ MOSFETs

Kang Eng Siew and Razali Ismail

CONTENTS

13.1 INTRODUCTION

Si-based complementary metal–oxide–semiconductor (CMOS) devices have experienced aggressive scaling over the past three decades to enhance their performance. However, continuous shrinking of the device dimensions into the nano regime induces gate leakage and suppresses short channel effects (SCEs). As devices continue to be scaled down, the tunneling through the thin oxide limits the MOS performance. The current leaks across the gate without any applied voltage. Consequently, alternative technologies, namely new materials and new device architectures, are introduced to further boost the device performance.

One of the most popular new material technologies is strained technology. Strained technology changes the device materials rather than device geometry. When a crystal layer is pseudomorphically epitaxially grown on top of another layer, a strain layer is induced owing to the 4.2% lattice mismatch (Rim et al., 2003). The strain induced is applicable to alter the band structure at the channel, namely bandgap, electron affinity, and heterointerface offsets. With the alteration of the band structure in the channel layer due to the strain, the effective mass is lowered while suppressing the intervalley scattering. The major interest in this process is attributed to the enhanced carrier mobility, drive current, high field velocity, and carrier velocity overshoot (Hoyt et al., 2002) without scaling down the device dimensions.

However, in today's technology, where the channel length has been reduced to 22 nm, the metal–oxide–semiconductor field effect transistors (MOSFETs) have now entered a regime where the quantum mechanical effects (QMEs) have become very dominant. The classical expression is no longer adequate to model an accurate and precise device characteristic. QMEs describe many phenomena where classical mechanics drastically fails to explain with great accuracy and precision. Thus, in this chapter, besides discussing the incorporation of strain into Si MOS devices, the effects of QMEs on the strained Si threshold voltage will also be considered.

13.2   PHYSICAL PROPERTIES OF BULK Si AND Ge

Silicon (Si) and germanium (Ge) are both group IV elements in the periodic table, and both elements have the diamond crystal structure. Some of the characteristics of bulk Si and Ge are listed in Table 13.1.

Experiments performed by Braunstein et al. (1958) show the values for the minimum indirect energy gap (E_g) for unstrained $Si_{1-x}Ge_x$. It is observed that indirect E_g

---

TABLE 13.1
Properties of Si and Ge

Properties	Silicon (Si)	Germanium (Ge)
Crystal structure	Diamond	Diamond
Lattice constant (Å)	5.43	5.66
Atoms (cm^3)	5.0×10^{22}	4.42×10^{22}
Atomic weight	28.09	72.60
Bandgap (eV)	1.12	0.67
Dielectric constant	11.9	16.2
Electron affinity	4.05	4
Effective electron mass, me/mo	m (longitudinal) = 0.98	m (longitudinal) = 1.64
	m (transverse) = 0.19	m (transverse) = 0.08
Effective hole mass, mh/mo	m (light hole) = 0.16	m (light hole) = 0.04
	m (heavy hole) = 0.49	m (heavy hole) = 0.28

is divided into two regions, following a quadratic expression where x represents the germanium mole fraction:

$$E_g(x) = (1.12 - 0.43x + 0.206x^2) \text{ eV} \quad \text{for } 0 < x < 0.85 \tag{13.1}$$

$$E_g(x) = (2.010 - 1.27x) \text{ eV} \quad \text{for } 0.85 < x < 1 \tag{13.2}$$

The lattice constant for unstrained $Si_{1-x}Ge_x$ is determined by the following quadratic expression, where x represents the germanium mole fraction:

$$a_0(x) = (0.002733x^2 + 0.01992x + 0.5431) \text{ nm} \tag{13.3}$$

13.3 PHYSICS OF STRAINED MOSFETs

Strained Si MOSFET is a device architecture that takes advantage of strain-induced enhancement of carrier transport in Si. When a thin layer of Si is pseuodomorphically grown on a thick relaxed $Si_{1-x}Ge_x$, the Si layer experiences a strain effect due to the mismatch of the lattice constant. The strain modifies the band structure, provides lower effective masses, and suppresses the interval-ley scattering, consequently resulting in the enhancement of carrier mobility and drive current.

There are two possible types of strain in a SiGe/Si heterostructure, including tensile strain and compressive strain. When a thin layer of Si is pseudomorphically grown on top of a thick relaxed $Si_{1-x}Ge_x$ layer, the silicon atoms are stretching far apart from each other to get aligned with the $Si_{1-x}Ge_x$ atoms. Thus, the Si layer is under a biaxial tensile strain. Alternately, growing a thin layer of $Si_{1-x}Ge_x$ on top of Si layer causes the larger constant lattice of $Si_{1-x}Ge_x$ layer to be under compressive strain. This is attributed to the 4.2% of lattice mismatch between Si and Ge as shown in Table 13.1. Both the tensile and compressive strain layers are illustrated in Figure 13.1. The tensile and compressive strains significantly alter the alignment of conduction and valence bands.

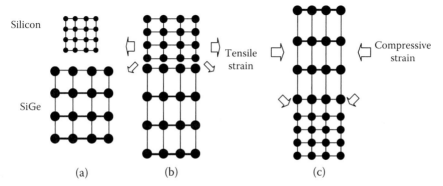

FIGURE 13.1 Schematic illustration of (a) silicon and SiGe atom, (b) tensile strained Si, and (c) compressive strained $Si_{1-x}Ge_x$.

13.3.1 BAND ALIGNMENT IN COMPRESSIVE STRAINED $Si_{1-x}Ge_x$ ON RELAXED SI

For unstrained $Si_{1-x}Ge_x$, the valence band is maximum composed of three bands located at $k = \Gamma$ (Pollak and Cardona, 1968), namely heavy holed (HH), lightly holed (LH), and the split off (SO) bands as shown in the qualitative energy k-diagram as in Figure 13.2a. As can be observed, both the HH and LH bands are at the same energy level, $k = 0$, while the SO band is located at a slightly lower energy. As discussed earlier, the interatomic distance in this layer is larger than that of pure Si. Thus, when a thin layer of $Si_{1-x}Ge_x$ is pseudomorphically grown on top of relaxed Si, the $Si_{1-x}Ge_x$ layer is under the compressive strain to match with the Si lattice constant. Once the strain is induced in the $Si_{1-x}Ge_x$ layer, the compressive strain splits the HH and LH degeneracy at the Γ point. The SO band is also lifted toward lower energy as illustrated in Figure 13.2b. The effective masses subsequently become highly anisotropic with the strain (Lime et al., 2006). Holes are preferred to stay at the HH band as it is at a higher energy level compared to the other two bands. With lighter effective masses, the holes, mobility is definitely enhanced. The mobility enhancement is also due to the suppression of the intervalley scattering.

At room temperature, all the six degeneracy valleys in the conduction band are located at the equivalent energy level. Two of the six valleys are in the plane of Si and the other four valleys are out of plane direction. However, the compressive strain in $Si_{1-x}Ge_x$ splits the sixfold degenerate conduction band into two diverse energy levels: Δ_2 at the higher energy and Δ_4 toward lower energy (Lang et al., 1985). The effect of the compressive strain on the conduction band is illustrated in Figure 13.3.

There are two types of band alignment in the lattice, namely type I and type II band alignments. When strained $Si_{1-x}Ge_x$ is grown on top of relaxed Si, it is

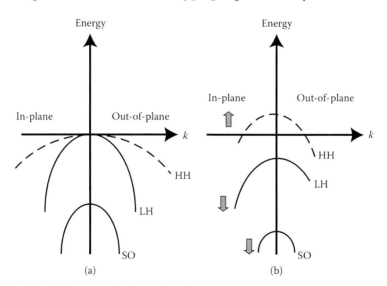

FIGURE 13.2 Schematic of valence band energy versus k: (a) conventional bulk Si and (b) compressive strained $Si_{1-x}Ge_x$ on relaxed Si.

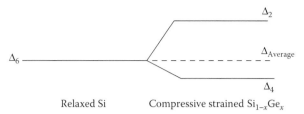

FIGURE 13.3 Schematic of conduction band splitting due to the compressively strained $Si_{1-x}Ge_x$. Sixfold degeneracy conduction band splits into Δ_2 at the higher energy and Δ_4 toward lower energy.

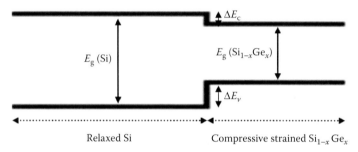

FIGURE 13.4 The effect of strain on the energy band diagram: Type I band alignment for compressive strained $Si_{1-x}Ge_x$.

called type I band alignment. The band alignment for $Si_{1-x}Ge_x$ grown on relaxed Si is shown in Figure 13.4. From the figure, it is observed that the majority of the band alteration occurs in the valence band and negligible conduction band alteration (Chaudhry et al., 2010; People and Bean, 1986), making it ideal for hole confinement. Thus, the valence band discontinuity of the type I band alignment has been fully carried out in the fabrication of p-channel SiGe MOSFETs. Another possible heterostructure that experiences type I band alignment is dual channel strained heterostructure, where strained $Si_{1-y}Ge_y$ is grown on relaxed $Si_{1-x}Ge_x$ with $y > x$.

13.3.2 BAND ALIGNMENT IN TENSILE STRAINED SI ON RELAXED $SI_{1-x}GE_x$

Tensile strain can be induced when a thin layer of Si is grown on a relaxed $Si_{1-x}Ge_x$. The effect of the compressive strain on both the valence and conduction bands was discussed in Section 13.3.1. Similarly, the tensile strain in the Si layer splits the degeneracy of HH and LH bands. The SO band is also lifted to the lower energy. However, for tensile strain, the LH band is lifted higher than the HH bands, a phenomenon different from that of compressive strain. The effect of the tensile strain on the valence band is illustrated in Figure 13.5b. The curvature of the band determines the hole effective mass (Hensel and Feher, 1963). The splitting of the

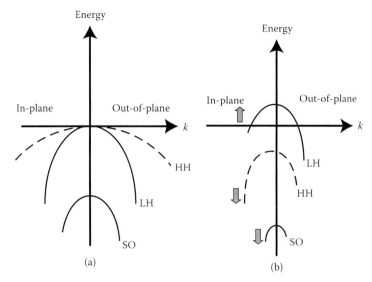

FIGURE 13.5 Schematic of valence band energy versus k: (a) conventional bulk Si and (b) tensile strained Si.

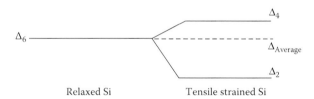

FIGURE 13.6 Schematic of conduction band splitting due to the tensile strained Si. Sixfold degeneracy conduction band splits into Δ_4 at the higher energy and Δ_2 toward lower energy.

valence band reduces in-plane effective masses. As a result, the holes are preferable to stay at the LH band compared to the other two bands.

When tensile strain is induced in the Si layer, the strain splits the sixfold degeneracy in the conduction band into two energy levels. The conduction band shifts the Δ_2 degeneracy toward the lower energy, and Δ_4 degeneracy is shifted to higher energy as can be seen from Figure 13.6. The electrons prefer to occupy the lower Δ_2 degeneracy, which is attributed to lower electron effective masses.

Figure 13.7 shows the band alignment for tensile strained Si on relaxed $Si_{1-x}Ge_x$, which is a type II band alignment. For type II band alignment, major alterations occur at both the conduction band and the valence band, making it ideal for both the electron and hole confinement. The continuities of the type II band alignment for conduction band and valence band have been fully implemented in both n-channel and p-channel strained MOSFETs.

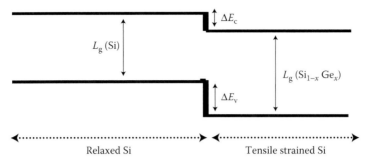

FIGURE 13.7 The effect of strain on the energy band diagram: Type II band alignment for tensile strained Si.

13.4 HOLE/ELECTRON MOBILITY

The main contribution of strain technology is the increase of carrier mobility without aggressively reducing the gate length. Works have been carried out by several researchers to quantify the mobility enhancement in strained silicon over bulk Si, in terms of both the n-channel and p-channel MOSFETs.

13.4.1 HOLE MOBILITY IN COMPRESSIVE STRAINED $Si_{1-x}Ge_x$

The improvement of hole mobility in compressive strained $Si_{1-x}Ge_x$ was fully extended to the fabrication of p-channel MOSFETs. Several researchers reported an improvement in hole mobility, at least a factor of four for a modulation doped Si/SiGe heterostructure compared to the conventional p-channel MOSFETs (Ismail et al., 1993). The increase of hole mobility in compressive strained $Si_{1-x}Ge_x$ is gaining most interest of the researchers too. Experimental results from Shima (2003) demonstrated an increase of 25% of the hole mobility in sub-100 nm compressive strained $Si_{1-x}Ge_x$, while for long channel devices, the increase of hole mobility can reach up to three times compared to the bulk Si MOSFETs (Kaya et al., 2000). The large hole mobility enhancement in compressive strained $Si_{1-x}Ge_x$ is considered to be resulting from the large hole population at the splitting of the valence band degeneracy and reduced effective masses in strained $Si_{1-x}Ge_x$. The hole mobility enhancement is also considered to be due to various scattering processes, especially the alloy scattering. Equation 13.4 shows the relationship among mobility, carrier effective mass, electronic charge, and scattering time:

$$\mu = \frac{q\tau}{m^*} \tag{13.4}$$

13.4.2 ELECTRON MOBILITY IN TENSILE STRAINED SI

As discussed earlier, the tensile strained Si experiences a splitting of the sixfold degeneracy into two set of energy ladders: Δ_2 degeneracy and Δ_4 degeneracy. The increase of the energy splitting reduces the intervalley scattering and increases the electron mobility. Electrons are favorable to occupy the lower Δ_2 degeneracy, thus resulting in great enhancement of the electron mobility. With germanium fraction of about 25–30%, the optimum electron mobility enhancement in tensile strained Si can be observed (Jung, 2009). Electron mobility measurement on a large area of long channel devices by (Rim et al., 2003) is ~70% higher than for the conventional bulk Si devices. What is remarkable is that in early 2001, Rim (2001) reported an enhanced mobility at vertical field as ~1.5 MV/cm for sub-70 nm tensile strained Si. The comparison study of electron mobility between surface and buried n-channel strained Si MOSFETs was performed by Welser et al. (1994). It was found that the electron mobility for both the surface and buried n-channel MOSFETs are larger than that of the conventional bulk Si.

However, hole mobility in the tensile strained Si shows only little enhancement. Furthermore, the hole mobility enhancement diminishes at a higher electric field. As a result, this architecture benefits only the electron confinement and consequently is widely fabricated only for n-channel MOSFETs.

13.5 ADVANCES IN STRAINED Sɪ/SɪGᴇ MOSFETS TECHNOLOGY

The potentiality of strained $Si/Si_{1-x}Ge_x$ MOSFETs is now fully recognized by both academic researchers and industries. Significant interests have been raised toward the study and understanding of strained technology. Another important strained technology is the development of the dual channel heterostructure. Dual channel heterostructure consists of tensile strained Si on a compressively strained $Si_{1-x}Ge_x$ on a relaxed $Si_{1-y}Ge_y$, with $x > y$. The cross section and schematic diagram of dual channel heterostructure is illustrated in Figure 13.8. Higher hole mobility can be obtained with higher Ge content, x (Lee and Fitzgerald, 2005). The main advantage of this heterostructure is that both n-channel and p-channel MOSFETs can be

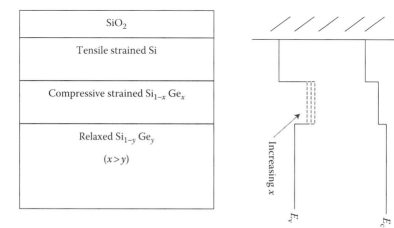

FIGURE 13.8 Schematic band alignment of dual channel heterostructure consisting of compressive strained $Si_{1-x}Ge_x$ on top of relaxed $Si_{1-y}Ge_y$ ($x > y$), capped with tensile strained Si.

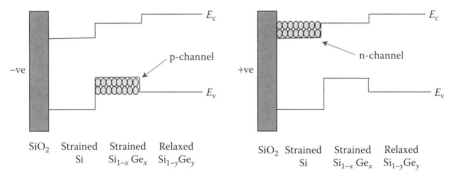

SiO$_2$ Strained   Strained   Relaxed   SiO$_2$ Strained   Strained   Relaxed
  Si Si$_{1-x}$Ge$_x$ Si$_{1-y}$Ge$_y$ Si Si$_{1-x}$Ge$_x$ Si$_{1-y}$Ge$_y$

FIGURE 13.9 Carrier confinement in the energy band diagram when negative and positive gate bias are applied.

fabricated using the same structure. This heterostructure achieves greater hole confinement in the dual channel compressive strained layer compared to single channel compressive strained. The high Ge content in the strained layer provided enhanced p-channel performance, material stability, and adaptable design (Badcock et al., 2002). From Figure 13.8, it is observed that the type I band alignment at the strained Si$_{1-x}$Ge$_x$/relaxed Si$_{1-y}$Ge$_y$ heterojunction is suitable for the large hole confinement. Additionally, type II band alignment at the heterojunction of strained Si/strained Si$_{1-x}$Ge$_x$ is suitable for large electron confinement.

In strained Si/strained Si$_{1-x}$Ge$_x$/relaxed Si$_{1-y}$Ge$_y$ MOSFETs, the type of channel depends on the applied gate voltage. When a positive gate voltage is applied, the tensile n-channel is induced within the tensile strained Si layer. On the other hand, for a negative gate voltage, p-channel is induced first in the compressive strained Si$_{1-x}$Ge$_x$ and then at the strained Si layer (Olsen et al., 2001). The energy band diagrams showing the carrier confinement at different gate voltages is illustrated in Figure 13.9.

Experimental results for dual channel heterostructure were reported by Olsen et al. (2005). The results show an ~70% electron mobility enhancement of single channel and ~62% electron mobility enhancement of dual channel heterostructure compared to the conventional bulk Si. Dual channel architecture observed a large hole mobility enhancement owing to the splitting of the valence band degeneracy. Besides, it is also caused by the type I band alignment between the strained Si$_{1-x}$Ge$_x$ and the relaxed Si$_{1-y}$Ge$_y$, which form the deep quantum well for holes in the compressive strained Si$_{1-x}$Ge$_x$.

13.6 QUANTUM MECHANICS APPROACH

Quantum mechanics is a fundamental theory that explains the inaccuracy of classical mechanics at the atomic level. It precisely describes many physical phenomena, where classical mechanics fails to agree with the system behavior.

In MOSFETs, shrinking the size of the transistor to a smaller dimension causes the correlation effects to become more dominant. One of the significant phenomena is the

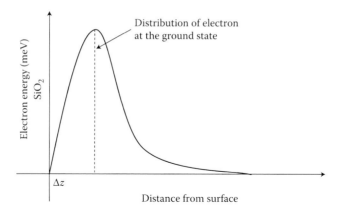

FIGURE 13.10 Schematic diagram illustrates the peak electron distribution is shifted away from the SiO_2 by a constant of Δz.

presence of QMEs. In the conventional MOSFETs, the potential well forms exactly at the interface between the silicon conduction band and oxide barrier (Ohkura, 1990). Accordingly, the confining carriers are adjacent to the silicon surface (Bratati et al., 2008). However, at an extremely small scale, the charge carrier concentration is shifted away from the Si–SiO interface by an amount of ΔZ, where ΔZ is the average distance of charges from the Si–SiO interface. The shift of the charge carrier concentration contributes to the increased threshold voltage. Moreover, in state-of-the-art nanoscale MOSFETs, at a high electric field, the width of the well is small enough so that the motion of the carriers is only in the direction perpendicular to the Si–SiO_2 interface (Ma et al., 2000; Stern, 1972). In contradiction to classical mechanics, the carriers are considered as 2D standing waves, and thus must be treated quantum mechanically. Due to the carrier quantization and the changes in the band structure, the maximum inversion layer charge density is attained at some distance away from the interface, as demonstrated in Figure 13.10. QMEs lead to an increase in the surface potential as higher gate voltage is necessary to produce the same level of inversion (Bratati et al., 2008).

13.7 QUANTUM-BASED THRESHOLD VOLTAGE MODEL IN STRAINED Si

Pseudomorphic growth of a thin Si film on relaxed $Si_{1-x}Ge_x$ introduces the strain into the channel, which alters the bandgap, electron affinity, flatband voltage, and heterointerface band offset at the strained Si and relaxed $Si_{1-x}Ge_x$. As discussed earlier, the energy splitting attributed to the stress increases the carrier mobility and also reduces the threshold voltage. Threshold voltage is an important parameter, particularly affecting the AC and DC characteristics, subthreshold effects, and so on (Qin et al., 2010). It can be defined as the required gate voltage to generate a strong inversion at the interface between the oxide and the substrate. In simple words, it is a minimum voltage to trigger the gate to operate, either in positive or negative voltage.

A reduction in the native threshold voltage of tensile strained Si has been reported by several researchers when compared to the conventional bulk Si (Goo, 2003; Rim et al., 2000). Threshold voltage reduction is also reported by Olsen (2005) in a dual channel heterostructure nMOSFET with 15% germanium content. The results showed that dual channel heterostructure exhibits lower threshold voltage compared to the tensile strained nMOSFETs attributed to the changes in band alignment and increased oxide trap density. In the studies by Kumar et al. (2007), an analytical threshold voltage model was developed for nanoscale strained Si/SiGe, investigating the effects of strain, strained Si thin film thickness, and gate work function. Although the models were carried out for sub-50 nm channel length, the QMEs were taken into considered for the analytical calculations. There is still a lack of analytical models and experimental studies concerning the quantum induced threshold voltage of strained silicon MOSFET.

Figure 13.11 shows the cross section of a nanoscale strained Si/Si$_{1-x}$Ge$_x$ MOSFET. Both source body and drain body depth (x_{dl}) contribute to the nonuniformity of the depletion region along the channel (Kumar, 2006). To obtain a uniform charge density N_A and depletion thickness x_d, the transformation of the structure to a box type–approximation is applied (Venkataraman et al., 2007). The average uniform depletion depth x_d can be approximated as follows:

$$x_d = \frac{x_{dl}(4r_j + \pi x_{dl}) + 2x_{dV}(L - 2x_{dl})}{2L} \tag{13.5}$$

where

x_{dV} represents the depletion region width due to the gate bias
x_{dl} is the depletion region depth due to source–drain voltage as in Equations 13.6 and 13.7
r_j is the source–drain junction depth

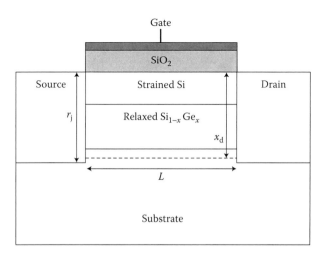

FIGURE 13.11 Cross section of nanoscale strained Si/Si$_{1-x}$Ge$_x$ MOSFETs after box approximation.

V_{sub} is the substrate bias

L is the gate length

In the following equations, ε_{SiGe} is the permittivity of SiGe, V_{sub} is the substrate bias, N_A is the effective channel doping concentration, $V_{bi,SiGe}$ is the built-in potential in the relaxed $Si_{1-x}Ge_x$, q is the carrier charge, and φ_{th} is defined as the surface potential where the charge density of electron in the strained Si is identical to conventional bulk Si at threshold (Zhang and Fossum, 2005).

$$x_{dv} = \sqrt{\frac{2\varepsilon_{SiGe}(\varphi_{th} - V_{sub})}{qN_A}} \qquad (13.6)$$

$$x_{dl} = \sqrt{\frac{2\varepsilon_{SiGe}V_{bi,SiGe}}{qN_A}} \qquad (13.7)$$

For sub-50 nm gate length in sizes, the classical mechanism no longer precisely models the device characteristics, particularly the threshold voltage. In conjunction to gain, an accurate value of threshold voltage V_{th}, the quantum effects on the carrier energy quantization, and the carrier distribution must be taken into consideration. In this section, the QMEs have been incorporated by considering two components: effective oxide thickness $T_{ox,eff}$ and effective flatband voltage $V_{fb,eff}$. With the introduction of the QMEs, the expressions for the effective oxide thickness $T_{ox,eff}$ and effective flatband voltage $V_{fb,eff}$ are modified as listed in Pirovano et al. (2002) as the following:

$$T_{ox,eff} = T_{ox} + \frac{\varepsilon_{ox}}{\varepsilon_{Si}}\Delta z \qquad (13.8)$$

$$V_{fb,eff} = V_{fb} + qN_A\left(\frac{\Delta z}{2\varepsilon_{Si}} + \frac{T_{ox}}{\varepsilon_{ox}}\right)\Delta z \qquad (13.9)$$

where

ε_{ox} and ε_{Si} are, respectively, the permittivity of SiO_2 and silicon

Δz is the quantum correlation constant derived from the difference of the average layer depth with a constant value of about 1.2 nm (Ohkura, 1990)

13.7.1 FORMULATION OF SURFACE POTENTIAL

2D Poisson equation for the area under the gate of the transformed structure reveals the QMEs that change the carrier concentration. The 2D Poisson equation can be described as (Kumar, 2007)

- Strained Si layer

$$\frac{d^2\varphi_1(x,y)}{dx^2} + \frac{d^2\varphi_1(x,y)}{dy^2} = \frac{qN_A}{\varepsilon_{Si}} \qquad (13.10)$$

- Relaxed $Si_{1-x}Ge_x$ layer

$$\frac{d^2\varphi_2(x,Y)}{dx^2} + \frac{d^2\varphi_2(x,Y)}{dy^2} = \frac{qN_A}{\varepsilon_{SiGe}} \tag{13.11}$$

where $\varphi_1(x,y)$ and $\varphi_2(x,y)$ are the electric potential in the active device, respectively, in the horizontal x-direction and the vertical y-direction for both the strained Si layer and the relaxed $Si_{1-x}Ge_x$ layer.

Potential distribution along the channel of the strained Si and relaxed $Si_{1-x}Ge_x$ substrates is approximated as parabolic equations as shown in Saxena et al. (2002):

- Strained Si layer

$$\varphi_1\left(x,y\right) = M_1\left(x\right)y + M_2\left(x\right)y^2 + \varphi_{s1}(x) \tag{13.12}$$

- Relaxed $Si_{1-x}Ge_x$ layer

$$\varphi_2\left(x,y\right) = N_1\left(x\right)Y + N_2\left(x\right)Y^2 + V_{sub} \tag{13.13}$$

where
$\varphi_{s1}(x)$ refers to the surface potential which is only the function of x
M_1, M_2, N_1, and N_2 are the unknown coefficients in only x-direction

Equations 13.14–13.16 represent the boundary condition derived for the gate electric field, and Equations 13.17–13.19 represent the boundary condition for the potential:

- Continuous electric flux from the gate oxide to strained Si thin film interface satisfies the equation

$$\frac{d\varphi_1\left(x,0\right)}{dy} = \frac{\varepsilon_{ox}}{\varepsilon_{si}}\frac{\left(\varphi_{s1} - V_{GS}'\right)}{T_{ox,eff}} \tag{13.14}$$

- Continuous electric flux from the strained Si thin film interface to $Si_{1-x}Ge_x$ substrate satisfies the equation

$$\frac{d\varphi_2\left(x,T_{Si}\right)}{dy} = \frac{\varepsilon_{SiGe}}{\varepsilon_{Si}}\frac{d\varphi_4\left(x,T_{SiGe}\right)}{dY} \tag{13.15}$$

- Electric flux at the depletion edge is equal to zero as shown in the following:

$$\frac{d\varphi_2\left(x,0\right)}{dY} = 0 \tag{13.16}$$

- Continuous potential distribution from the strained Si to relaxed $Si_{1-x}Ge_x$ substrate is expressed as

$$\varphi_1\left(x, T_{Si}\right) = \varphi_3\left(x, T_{SiGe}\right) \qquad (13.17)$$

- Surface potential at the drain and source edge is expressed as

$$\varphi_s(0) = V_{sub} + V_{bi} \qquad (13.18)$$

and

$$\varphi_s(L) = V_{sub} + V_{bi} + V_{DS} \qquad (13.19)$$

In these equations, $V'_{GS} = V_{GS} - V_{fb,eff}$, where $V_{fb,eff}$ is the effective flatband voltage incorporated into the QMEs for the strained Si MOSFETs. T_{Si} is the strained Si thickness, while T_{SiGe} is the thickness of $Si_{1-x}Ge_x$ substrate.

The unknown coefficients M_1, M_2, N_1, and N_2 can be figured out using the boundary condition. With the coefficients obtained, the surface potential distribution is resolved as

$$\varphi_s(x) = -\frac{\beta}{\alpha} + e^{x\sqrt{\alpha}}A + e^{-x\sqrt{\alpha}}B \qquad (13.20)$$

where the derivation for the parameters α, β, A, and B are calculated as shown in the appendix.

The surface potential along the channel for 30 nm gate length of both conventional strained Si and quantum strained Si is plotted as shown in Figure 13.12. It can be observed from the figure that the minimum surface potential increases when the drain bias is increased. The potentials at the source and drain end remain the same for both quantum and conventional strained Si models. The only difference is the required potential for the minimum point of inversion. The quantum strained Si model demonstrated a lower minimum point for the inversion to happen due to the changes in flatband voltage as well as the effective oxide thickness, resulting in a higher required gate voltage overdrive to reach the minimum point of inversion.

13.7.2 FORMULATION OF THRESHOLD VOLTAGE

To calculate the threshold voltage, the minimum surface potential is required. The minimum point of surface potential $\varphi_{s,min}$ can be obtained by differentiating Equation 13.20 with respect to x and then equating it to zero as in Equation 13.21. The minimum of surface potential, $\varphi_{s,min}$, is the point where the strong inversion starts to accumulate.

$$\frac{d\varphi_s(x)}{dx}\bigg|_{x=\varphi_{s,min}} = 0 \qquad (13.21)$$

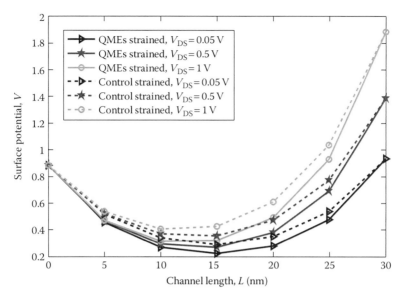

FIGURE 13.12 **(See color insert.)** Surface potential variation along the channel for 30 nm channel length with 20% germanium content, respectively, for conventional and quantum tensile strained Si. The solid curves obtained by the quantum approach and the dashed represent the conventional strained Si.

It is certain that the maximum barrier that carriers need to overcome when traveling from the source to drain is located at the strained Si thin film channel. Thus, the threshold voltage here is determined as the required gate voltage to achieve the strong inversion. The threshold voltage point in this model, in turn, is defined as the condition when the surface potential is given as

$$\varphi_{s,min} = 2\varphi_{F,Si} - \frac{E_{g,Si}}{q} + V_T \ln\left(\frac{N_{V,si}}{N_{V,ss}}\right) \tag{13.22}$$

where $\varphi_{F,Si}$ is the difference between intrinsic Fermi energy and Fermi energy level. Substituting the value of $\varphi_{s,min}$ into Equation 13.22, the threshold voltage for strained Si thin film can be determined as explained in the appendix.

Figure 13.13 illustrates the threshold voltage variation with change in the channel length, for germanium content of 0.2 and 0.4. The threshold voltage for bulk strained Si is extracted from a 2D ATLAS simulator. It is demonstrated that the threshold voltage experiences a significant fall with decreasing the channel length. The sharp fall below the channel length of 60 nm is mainly caused by the punch-through current that flows from the source to drain through the silicon substrate. At a very small channel length, in our case smaller than 50 nm, the threshold variation for quantum and bulk strained Si model demonstrated significant differences. For channel length larger than 50 nm, there are no discernible changes in the threshold voltage between bulk strained and quantum

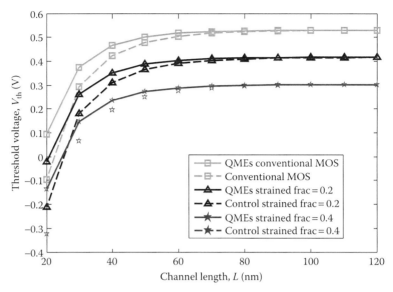

FIGURE 13.13   (**See color insert.**) Threshold voltage variation along the channel, comparing conventional MOSFETs, and conventional and quantum strained Si model, for germanium content 0.2 and 0.4.

strained Si models. This strongly indicates that QMEs are only dominant at small channel length and must be taken into account. As discussed earlier for a nanoscale device, the inversion layer charge concentration value maximizes at a distance away from the Si–SiO interface by a value of ΔZ. The shift of the peak carrier concentration away from the interface thereby degrades the oxide capacitance with greater effective oxide thickness. Thus, it requires a larger gate voltage to reach the threshold inversion point.

Figure 13.14 demonstrates the threshold voltage variation with the germanium content in $Si_{1-x}Ge_x$ substrate for both the conventional and quantum strained Si models. In this part, several stress situations induced underneath the strained Si, by changing the germanium content in $Si_{1-x}Ge_x$, are examined. Both trends demonstrate that the threshold voltage will fall linearly with increasing the germanium content. This reduction is attributed to the decrease of flatband voltage, and the earlier offset of inversion is due to the strain. In terms of the quantum mechanical approach, we observe that there are significant differences between the threshold voltage conventional and quantum strained Si models, particularly in lower channel length. For a higher channel length, the threshold variation has remained the same for both the threshold voltage models. It is again evident that the QMEs only take place in short channel regime instead of long channel.

Threshold voltage performance with strained Si thin film thickness variation is illustrated in Figure 13.15. As can be seen from the trend, the quantum strained Si required higher threshold voltage than the conventional strained Si, and the discrepancy between them becomes larger as the channel length decreases.

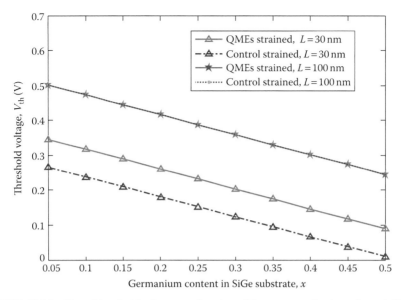

FIGURE 13.14 Plot of threshold voltage as a function of Ge content, x for short channel (30 nm) and long channel length strained Si MOSFETs.

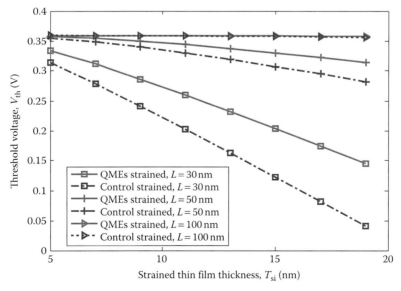

FIGURE 13.15 (**See color insert.**) Plot of threshold voltage performance as the function of strained Si thickness for $L = 30$, 50, and 100 nm for 30% germanium content in $Si_{1-x}Ge_x$ substrate.

The differences between both models become significant once the dimension of channel length is decreased. Again, this proves the existence of quantum confinement effect in shorter channel length. On the other hand, it has been found that when the thickness of strained Si thin film is reduced to less than 10 nm in size, the threshold voltage is increased due to the quantum confinement of the carrier in the thin silicon film (Omura et al., 1993). In the model, the strained Si thin film thickness was kept low enough to maintain the earlier elastic strain to avoid misfit dislocation.

13.8 CONCLUSION

SCEs resulting from the aggressive downscaling of the conventional CMOS devices accelerate the requirement to alternative new materials and new architecture technologies. Si/SiGe material, particularly the strained Si technology, emerged as an attractive approach offering a significant enhancement over the conventional MOSFETs, namely the carrier mobility and the drive current. Nowadays, advanced designs in strained Si technology below 32 nm gate length started appearing. In this sub-50 nm channel length, the impact of QMEs becomes extremely dominant in determining the system properties. This chapter includes a brief introduction to the strained Si technology as well as the QMEs on the threshold voltage of tensile strained Si MOSFETs, considering the impact of channel, germanium content, and strained thin film thickness.

APPENDIX

Using the boundary condition, the coefficients M_1, M_2, N_1, and N_2 are calculated as

$$M_1 = \frac{\varepsilon_{ox}[\varphi_s(x) - V_{GS}']}{T_{ox}\varepsilon_{Si}}$$ (13.A.1)

$$M_2 = -(-2T_{ox}V_{sub}\varepsilon_{Si}\varepsilon_{SiGe} + T_{SiGe}\varepsilon_{ox}\varepsilon_{Si}\varphi_s(x) + 2T_{Si}\varepsilon_{ox}\varepsilon_{SiGe}\varphi_s(x)$$
$$+ 2T_{ox}\varepsilon_{si}\varepsilon_{SiGe}\varphi_s(x) - T_{SiGe}\varepsilon_{ox}\varepsilon_{Si}V_{GS}' - 2T_{Si}\varepsilon_{ox}\varepsilon_{SiGe}V_{GS}')/$$
$$(2T_{ox}T_{Si}\varepsilon_{Si}(T_{SiGe}\varepsilon_{Si} + T_{Si}\varepsilon_{SiGe}))$$ (13.A.2)

$$N_1 = 0$$ (13.A.3)

$$N_2 = -\frac{2T_{ox}V_{sub}\varepsilon_{Si} - T_{Si}\varepsilon_{ox}\varphi_s(x) - 2T_{ox}\varepsilon_{Si}\varphi_s(x) + T_{Si}\varepsilon_{ox}V_{GS}'}{2T_{ox}T_{SiGe}(T_{SiGe}\varepsilon_{Si} + T_{Si}\varepsilon_{SiGe})}$$ (13.A.4)

Surface potential distribution is resolved as

$$\varphi_s(x) = -\frac{\beta}{\alpha} + e^{x\sqrt{\alpha}}A + e^{-x\sqrt{\alpha}}B$$ (13.A.5)

where

$$\alpha = \frac{T_{SiGe}\varepsilon_{ox}\varepsilon_{Si} + 2(T_{Si}\varepsilon_{ox} + T_{ox,eff}\varepsilon_{Si})\varepsilon_{SiGe}}{T_{ox,eff}T_{Si}\varepsilon_{Si}(T_{SiGe}\varepsilon_{Si} + T_{Si}\varepsilon_{SiGe})} \tag{13.A.6}$$

$$\beta = \frac{qN_A}{\varepsilon_{Si}} - \frac{\varepsilon_{ox}V_{GS}'(2T_{Si}\varepsilon_{SiGe} + T_{SiGe}\varepsilon_{Si})}{T_{ox,eff}T_{Si}\varepsilon_{Si}(T_{Si}\varepsilon_{SiGe} + T_{SiGe}\varepsilon_{Si})} - \frac{2V_{Sub}\varepsilon_{SiGe}}{T_{Si}(T_{Si}\varepsilon_{SiGe} + T_{SiGe}\varepsilon_{Si})} \tag{13.A.7}$$

Using boundary conditions, parameters A and B in Equation 13.A.5 can be expressed as

$$A = \frac{[\coth(L\sqrt{\alpha} - 1)][e^{L\sqrt{\alpha}}(\alpha(V_{bi} + V_{DS} + V_{sub}) + \beta) + \alpha(-V_{bi} - V_{sub})] - \beta]}{2\alpha} \tag{13.A.8}$$

$$B = \frac{[\coth(L\sqrt{\alpha}) - 1][e^{2L\sqrt{\alpha}}(\alpha(V_{bi} + V_{sub}) + \beta) - e^{L\sqrt{\alpha}}(\alpha(V_{bi} + V_{DS} + V_{sub}) + \beta)]}{2\alpha} \tag{13.A.9}$$

Threshold voltage for strained Si with QMEs finally can be determined as

$$V_{th} = V_{fb} + {} - 4\alpha^2(A_1B_2 + A_2B_2) + 2vw + 2\alpha\varphi_{s,min}w$$

$$+ \sqrt{4(4A_2B_2\alpha^2 - w^2)[(v + \alpha\varphi_{s,min})^2 - 4A_1B_1\alpha^2}$$

$$\frac{+ 4\alpha^2(A_1B_2 + A_2B_1) - 2w(v + \alpha\varphi_{s,min})]^2}{2(4A_2B_2\alpha^2 - w^2)} \tag{13.A.10}$$

$$A_1 = \frac{-v + e^{L\sqrt{\alpha}}v - V_{bi}\alpha + e^{L\sqrt{\alpha}}V_{bi}\alpha + e^{L\sqrt{\alpha}}V_{DS}\alpha - V_{sub}\alpha + e^{L\sqrt{\alpha}}V_{sub}\alpha}{(-1 + e^{2L\sqrt{\alpha}})\alpha} \tag{13.A.11}$$

$$B_1 = -\left(\frac{-V_{bi} + e^{L\sqrt{\alpha}}V_{bi} - V_{ds} - V_{sub} + e^{L\sqrt{\alpha}}V_{sub} - \dfrac{v}{\alpha} + \dfrac{e^{L\sqrt{\alpha}}v}{\alpha}}{e^{-L\sqrt{\alpha}} - e^{L\sqrt{\alpha}}}\right) \tag{13.A.12}$$

$$A_2 = \frac{(-w + e^{L\sqrt{\alpha}}w)}{(-1 + e^{2L\sqrt{\alpha}})\alpha} \tag{13.A.13}$$

$$B_2 = -\left(\frac{-\dfrac{w}{\alpha} + \dfrac{e^{L\sqrt{\alpha}}w}{\alpha}}{e^{-L\sqrt{\alpha}} - e^{L\sqrt{\alpha}}}\right) \tag{13.A.14}$$

REFERENCES

S. G. Badcock, A. G. O'Neill, and E. G. Chester. 2002. Device and circuit performance of SiGe/Si MOSFETs. *Solid State Electron.* 46, 11, 1925–1932.

R. Braunstein, A. R. Moore, and F. Herman. 1958. Intrinsic optical absorption germanium silicon alloys. *Physical Review.* 109, 3, 695–710.

M. Bratati, A. Biswas, P. K. Basu, G. Eneman, P. Verheyen, E. Simoen, and C. Claeys. 2008. Modeling of the threshold voltage and subthreshold slope of strained Si MOSFET including quantum mechanical effects. *Semiconductors Science Technology.* 23, 095017.

A. Chaudhry, J. N. Roy, and G. Joshi. 2010. Nanoscale strained Si MOSFETs physics and modeling approaches: A review. *Journal of Semiconductor.* 31, 10, 400-1–400-6.

J. S. Goo. 2003. Band offset induced threshold voltage variation in strained Si nMOSFETs. *IEEE Transaction on Electron Device Letters.* 24, 9, 568–570.

J. C. Hensel and G. Feher. 1963. Cyclotron resonance experiments in uniaxially stressed silicon: Valence band inverse mass parameters and deformation potentials. *Physical Review.* 129, 1041–1062.

J. L. Hoyt, H. M. Nayfeh, S. Eguchi, I. Aberg, G. Xia, T. Drake, E. A. Fitzgerald, and D. A. Antoniadis. 2002. Strained silicon MOSFET technology. *IEDM Technical Digest.* San Francisco, Paper No. 2.1, pp. 23–26.

K. Ismail, S. F. Nelson, J. O. Chu, and B. S. Meyerson. 1993. Electron transport properties of Si/SiGe heterostructures: Measurement and device implication. *Applied Physics Letters.* 63, 660–662.

J. W. Jung. 2009. Hole mobility and device characteristics of SiGe dual structure. *Current Applied Physics.* 9, 47–50.

S. Kaya, Y. P. Zhao, J. R. Watling, A. Asenov, J. R. Barker, G. Ansaripour, G. Braithwaite, T. E. Whall, and E. H. C. Parker. 2000. Indication of velocity overshoot in strained $Si_{0.8}Ge_{0.2}$ p-channel MOSFETs. *Semiconductors Science Technology.* 15, 573–578.

M. J. Kumar. 2006. A simple analytical threshold voltage model of nanoscale single layer fully depleted strained siliocn on insulator MOSFETs. *IEEE Transactions on Electron Devices.* 53(10), 2500–2506.

M. J. Kumar, V. Venkataraman, and S. Nawal. 2007. Impact of strain or Ge content on the threshold voltage of nanoscale strained Si/SiGe bulk MOSFETs. *IEEE Transaction on Device and Materials Reliability.* 7(1), 181–187.

D. V. Lang, R. People, J. C. Bean, and A. M. Sergent. 1985. Measurement of the bandgap of $Si_{1-x}Ge_x$/Si strained layer heterostructures. *Applied Physics Letters.* 47, 12, 1333–1335.

M. J. Lee and E. A. Fitzgerald. 2005. Strained Si, SiGe, and Ge channels for high-mobility metal-oxide-semiconductor field effect transistor. *Journal of Applied Physics.* 97, 011101.

F. Lime, F. Andrieu, J. Derx, G. Ghibaudo, F. Boeuf, and T. Skotnicki. 2006. Low temperature characterization of effective mobility in unixially and biaxially strained nMOSFETs. *Solid State Electronics.* 50, 4, 644–649.

Y. T. Ma, L. T. Liu, L. L. Tian, Z. J. Li, and Z. P. Yu. 2000. A new charge model including quantum mechanical effects in MOS structure inversion layer. *Solid State Electronics.* 44, 1697–1702.

S. H. Olsen, A. G. O'Neill, S. Chattopadhyay, L. S. Driscoll, K. S. K. Kwa, D. J. Norris, A. G. Cullis, and D. J. Paul. 2001. Study of single and dual channel design for high performance strained Si-SiGe nMOSFETs. *IEEE Transaction on Electron Devices.* 51(7), 1245–1253.

S. H. Olsen, P. Dobrosz, E. E. Cousin, S. J. Bill, and A. G. O'Neill. 2005. Mobility limiting mechanisms in single and dual channel strained Si/SiGe MOSFETs. *Materials Science and Engineering B.* 124–125, 107–112.

Y. Omura, S. Horiguchi, M. Tabe, and K. Kishi. 1993. Quantum mechanical effects on the threshold voltage of ultrathin SOI nMOSFETs. *IEEE Electron Device Letters*. 14, 12, 569–571.

Y. Ohkura. 1990. Quantum effects in Si n-MOS inversion layer at high substrate concentration. *Solid-State Electron.* 33, 1581–1585.

R. People and J. C. Bean. 1986. Band alignment of coherantly strained $Si_{1-x}Ge_x/Si$ heterostructures on <001> $Si_{1-y}Ge_y$ substrates. *Applied Physics Letters.* 48, 8, 530–540.

A. Pirovano, A. L. Lacaita, and A. S. Spinelli. 2002. Two dimensional quantum effects in nanoscale MOSFETs. *Transactions on Electron Devices.* 49(1), 25–31.

F. H. Pollak and M. Cardona. 1968. Piezo electrorefectance in Ge, GaAs, and Si. *Physical Review.* 172, 3, 816–837.

S. S. Qin, H. M. Zhang, H. Y. Hu, X. Y. Dai, R. X. Xuan, and B. Shu. 2010. An analytical threshold voltage model for dual channel PMOSFET. *Chinese Physics B.* 19(11), 117309.

K. Rim. 2001. Strained Si nMOSFETs for high performance CMOS technology. *Symposium on VLSI Technology Digest of Technical Paper.* 12–14 June, pp. 59–60.

K. Rim, R. Anderson, D. Boyd, F. Cardone, K. Chan, H. Chen, S. Christansen, J. Chu, K. Jenkins, T. Kanarsky, S. Koester, B.H. Lee, K. Lee, V. Mazzeo, A. Mocuta, D. Mocuta, P. M. Mooney, P. Oldiges, J. Ott, P. Ronshem, R. Roy, A. Steegen, M. Yang, H. Zhu, M. Ieong, and H.-S.P. Wong. 2003. Strained Si CMOS (SS CMOS) technology: Opportunities and challenges. *Solid State Electronics.* 47, pp. 1133–1139.

K. K. Rim, J. L. Hoyt, and J. F. Gibbons. 2000. Fabrication and analysis of deep sub-micron strained Si n-MOSFETs. *IEEE Transactions Electron Devices.* 47(7), 1406–1415.

M. Saxena, S. Haldar, M. Gupta, and R. S. Gupta. 2002. Physics based analytical modeling of potential and electrical field distribution in dual material gate (DMG)-MOSFET for improved hot electron effect and carrier transport efficiency. *IEEE Transactions on Electron Devices.* 49(11), 1928–1938.

M. Shima. 2003. <100> Strained SiGe channel pMOSFETs with enhanced hole mobility and lower parasitic resistance. *Fujitsu Science and Technical Journal.* 39(1), 78–83.

F. Stern. 1972. Self consistent results for n-type Si inversion layers. *Physics Review B. Condensed Matter.* 5(12), 4891–4899.

V. Venkataraman, S. Nawal, and M. J. Kumar. 2007. Compact analytical threshold voltage model of nanoscale fully depleted strained Si on silicon-germanium-on-insulator (SGOI) MOSFETs. *IEEE Transaction on Electron Devices.* 54(3), 554–562.

J. Welser, J. L. Hoyt, and J, F, Gibbons. 1994. Electron mobility enhancement in strained Si n-type metal oxide semiconductor field effects transistors. *IEEE Electron Device Letters.* 15, 100–102.

W. Zhang and J. G. Fossum. 2005. On the threshold voltage strained Si/MOSFETs. *IEEE Transactions on Electron Devices.* 52, 2, pp. 263–268.

14 Nanoelectronics Research and Commercialization in the United States

Sohail Anwar

CONTENTS

14.1 INTRODUCTION

Nanoelectronics is defined as the ability to manipulate matter on a scale of less than 100 nm to create structures with many useful electronic properties. Nanoelectronics is a rapidly developing technological field with potential impact across a broad industry range. At present, significant research efforts focusing on graphene-based electronics, molecular organic-based electronics, single electron devices, nano-tubes, and nanowires are underway. These efforts are needed to continue the ever-increasing miniaturization of semiconductor devices. Researchers are developing and using innovative approaches based on nanoscale science and engineering to change the very nature of electronics.

Shrinking device dimensions are important in order to increase device density and processing speed, reduce switching energy, enhance system functionality, and decrease manufacturing cost. But as the dimensions of the critical elements of devices approach atomic size, quantum tunneling and other quant effects cause deterioration of conventional device operation. Therefore, new innovative approaches are being developed to overcome the limitations imposed by the fundamental laws of physics. The ultimate goal of all these efforts is to advance the state of the art in sensor technologies, information processing, communication infrastructure, and computation.

As listed in NSTC (2010), the five thrust areas for the NNI (National Nanotechnology Initiative) signature initiative "Nanoelectronics for 2020 and Beyond" are

- Exploring new or alternative architectures and modes of operation for computing
- Merger of nanoelectronics with nanophotonics
- Enhancement of research activities associated with carbon-based nanoelectronics
- Investigating nanoscale processes and phenomena for the development of quantum information systems
- Development of University-Based National Nanoelectronics Research and Manufacturing Infrastructure Network

Though nanoelectronics has numerous applications, ranging from solar cells to ambulatory monitoring, it faces significant technical and economic challenges. A major hurdle in the nanoelectronics pathway is the long time period between researching a concept and creating a commercial product (Dhillon et al. 2008). The serious gap between research and commercialization must be addressed by industry, government, and academia. The lack of a coherent policy on technology transfer from academia to industry constitutes another hurdle (McNeil et al. 2007). In addition, there is a shortage of proper infrastructure needed to develop cutting-edge nanoelectronics research.

Measures that can be adopted to lower the barriers to nanoelectronics commercialization are described in Anwar et al. (2010). One of the proposed measures calls for early market research to close the concept to product gap. Potential markets and products must be identified early during the research and development phase. This should be followed by periodic reviews as innovations evolve into prototypes that later transform into marketable products.

14.2 NANOELECTRONICS RESEARCH INITIATIVES IN THE UNITED STATES

As explained in Etzkowitz (2011), a new mode of production based on the linkages among government, industry, and academia has been gaining popularity. Universities, government institutions, and industrial organizations are increasingly developing research and development collaborations with a spiral pattern of linkages emerging at various stages of the innovation process. In the United States, several collaborative initiatives based on the triple helix model are currently involved in advancing the forefront of nanoelectronics.

The National Nanotechnology Infrastructure Network (NNIN) is a prime example of triple helix model. It is an integrated network of user facilities, supported by the National Science Foundation (NSF) serving the resource needs of users in academia, industry, and government by providing open access to cutting-edge nano tools, instrumentation, and capabilities for fabrication, synthesis, characterization design, simulation, and integration. The NNIN technical research and development areas include nanoscale physics, electronics, optics and optoelectronics, materials science, nanotechnology applications in biology and life sciences, MEMs and microfluidics, and chemistry. Since NNIN is not a funding or research organization, no research is directly funded by this integrated network. It serves as a research facilitator providing state-of-the-art equipment resources and human expertise to promote high-quality nanotechnology research. NNIN user facilities exist at 14 universities, which include Cornell University; Stanford University; Georgia Institute of Technology; University of Washington; University of Michigan; University of Minnesota; The Pennsylvania State University; University of California, Santa Barbara; University of Texas at Austin; Arizona State University; University of Colorado at Boulder; Harvard University; Howard University; and Washington University at St. Louis. More detailed information regarding NNIN can be obtained from the NNIN website: http://www.nnin.org/

The Nanoelectronics Research Initiative (NRI), a consortium of companies in the Semiconductor Industry Association (SIA), seeks to accelerate research and development in nanoelectronics for the benefit of the technology industry. The Initiative is concentrated in more than 30 universities within the United States. The research projects are organized into multiuniversity centers (MIND, WIN, INDEX, SWAN) and at the NSF nanoscience centers (NRI–NSF). The NRI consortium promotes industry–academia–government research collaboration. Major semiconductor corporations such as IBM, Intel, Global Foundries, Micron, and Texas Instruments contribute millions of dollars to this major research effort. The government organizations supporting the NRI consortium include NSF, National Institute of Standards & Technology (NIST), the state governments of California, Indiana, New York, and Texas, and the City of South Bend. The participating universities include several campuses of University of California, Portland State University, University of Nebraska, Notre Dame University, Purdue University, The Pennsylvania State University, Harvard University, University of Texas at Austin, Rice University, University of Maryland, North Carolina State University, Columbia University, MIT, Stanford University, University of Alabama, Brown, Northwestern University, University of Virginia, and Old Dominion University. The NRI consortium is managed by the Semiconductor Research Corporation (SRC). More information regarding NRI can be obtained from the NRI website: http://www.src.org/program/nri/

The Network for Computational Nanotechnology (NCN) is a research network of seven universities with the mission to transform nanoscience into nanotechnology. The network achieves this mission by designing and operating a cyber-resource for nanotechnology theory, modeling, and simulation. The cyber-resource is termed *nanoHUB.org*. The NCN focus areas include nanoelectronics, nanophotonics, nanoelectromagnetics, and nanodevices for biology and medicine. The NCN partners include Purdue University, University of California at Berkeley, University of Illinois at Urbana–Champaign, Molecular Foundry at Lawrence Berkeley National Laboratory, Norfolk State University, Northwestern University, MIT, and University of Texas at El Paso. The *nanoHUB.org* is an operational cyber-infrastructure with global use in research and education. This infrastructure has effectively developed a virtual community among many different disciplines and organizations involved in nanoscience and nanotechnology. In addition, *nanoHUB.org* has a significant impact on education. The use of remote and freely available software tools on *nanoHUB.org* helps in developing a solid understanding of nanoscience and nanotechnology concepts. More detailed information regarding *nanoHUB.org* can be obtained from the following website: http://nanohub.org/

More information regarding the NCN consortium is available at http://www.ncn.purdue.edu/

14.3 NANOELECTRONICS RESEARCH THRUST AREAS

The five thrust areas for the NNI signature initiative "Nanoelectronics for 2020 and Beyond" are now described in detail. The focus areas and their descriptions are as follows:

- Exploring alternative architectures and modes of operation, other than change transfer by conventional transistors, for computing. Such alternatives include electron spin devices, magnetic devices, quantum cellular automata, and computations on biological substrates.
- Merging nanophotonics with nanoelectronics. Technological approaches based on the merger of these two technologies may lead to the realization of new light sources, detectors, sensors, frequency converters, modulators, quantum optical, and ultrafast optical devices and components. To achieve this goal, significant research activities in the following topical areas will be needed.
  - Novel material processing and fabrication techniques to develop hybrid metallic/semiconducting materials and nanostructures.
  - CMOS—compatible fabrication techniques to be applied toward the development of hybrid nanophotonic systems.
  - Nanoscale plasmonic sources and detectors.
  - Plasmonic cavities doped with semiconductor gain materials.
  - Plasmonic structures that incorporate nonlinear materials for ultracompact optical modulators or frequency converters.
  - Doped nanoscale optical antennas for directed and smart sources.
  - Computational and analytical design approaches for hybrid nanophotonic devices.
- Exploring carbon-based nanoelectronics. This thrust area is characterized by research in the following topical areas:
  - Enhancement of research efforts to understand the fundamental science underlying the unusual physical phenomena in carbon nanomaterials, carbon nanotubes, and graphene.
  - Improvement of materials quality associated with large-area graphene synthesis, low-cost and high-throughput carbon nanotube sorting, and integration of carbon with CMOS technology.
  - Exploring novel functionalities in carbon nanomaterials.
  - Exploring innovative device construction approaches that go beyond the limits imposed by the field effect transistors (FETs).

- Investigating nanoscale processes and phenomena for developing quantum information systems. The research thrust areas include the following:
  - Control of the quantum state of nanoscale devices at the atomic level.
  - Simulation of quantum phenomena in the nanoscale regime.
  - Connecting quantum and nanoscale phenomena across macroscopic length scales and time scales.
  - Creating hybrid quantum systems to connect nanoscale quantum systems with other quantum systems.

14.4   COMMERCIALIZATION

The path to nanoelectronics commercialization is not linear. Several barriers exist in the transition from research to commercial ventures. Thus, there is a need to bridge the gap between science and commercialization. Target markets and future products need to be identified early during the research and development phase. Support from government and industry will be necessary to continue building industry–academia–government collaborations aimed at promoting nanoelectronics research and commercialization.

The key elements of an effective strategy for nanoelectronics commercialization are as follows (Anwar et al. 2010):

1. Proposing an idea—Usually driven by a specific need.
2. Design, modeling, and simulation—The knowledge and design tools must be available to perform this function.
3. Protoyping—Protoyping, rapid prototyping, and preproduction prototyping are needed for volume manufacturing.
4. Packaging—Needed to protect and provide an interface to the macroworld.
5. Testing and reliability—Reliability plans must identify and prevent failures before a product moves up to the manufacturing phase.
6. Product realization and marketing—It is important to have an effective marketing plan in place before the end-product phase is reached.

14.5   CASE STUDIES DEMONSTRATING EFFECTIVE INDUSTRY–ACADEMIA–GOVERNMENT RESEARCH PARTNERSHIPS

In this section, various examples of several university–industry–government partnerships based on triple helix model are described. All of these partnerships are involved in advancing the forefront of nanoelectronics research and commercialization.

- College of Nanoscale Science and Engineering (CNSE), University at Albany—State University of New York. The College of Nanoscale Science and Engineering (CNSE) and the Albany Nanotech Complex were established in 2001 at the University of New York. The CNSE's Albany Nanotech Complex, through partnerships of industry, academia, and government, has created a triple helix model to support an accelerated nanotechnology commercialization in pursuit of economic development and job creation in the state of New York (Dhillon et al. 2008). The Nanotech Complex is a world class research, development, prototyping, and educational facility with onsite corporate partners including IBM, Intel, Global Foundries, SEMATECH, Samsung, Toshiba, Applied Materials, and Tokyo Electron. It receives significant funding from industry, which is matched by the state of New York. In 2011, the total facility space occupied by the CNSE and the Nanotech Complex was 800,000 ft^2. At that time, expansion was underway to add 500,000 ft^2 of additional space.

The research constellations within the CNSE include the following:

- Nanoscience Constellation
- Nanoengineering Constellation
- Nanoeconomics Constellation

The key research centers and programs within the CNSE and the Albany Nanotech Complex include the following:

- Center for Excellence in Nanoelectronics and Nanotechnology
- National Institute for Sustainable Energy
- Center for Advanced Technology in Nanomaterials and Nanoelectronics
- Center for Nanoscale Lithography
- Applied Materials R&D Center
- Center for Intelligent Power

The $12 billion Albany Nanotech Complex attracts over 300 global corporate partners. It provides technical, marketing, and business development assistance to its corporate partners through technology incubation, pilot prototyping, and test-bed integration support. All the partners have access to the state-of-the art laboratories and an array of scientific centers serving their long- and short-term technology development needs. The partners are able to collaborate, establish joint ventures, or form alliances within a technically and financially competitive environment. More information regarding CNSE and the Albany Nanotech Complex may be obtained from the following website: http://cnse.albany.edu

- The Pennsylvania State University Nanofabrication Facility
The Pennsylvania State University Nanofabrication Facility (Nanofab) is an open access NNIN user facility. Academic and industry users can perform research on-site using facility equipment, training, and technical staff support. The Nanofab facility provides use of state-of-the-art micro- and nanofabrication equipment housed in Class 1 and Class 10 clean rooms.

The Nanofab facility was established to enable advanced research and development in semiconductor electronics, micro- and nano-electromechanical systems (MEMS/NEMS), materials, and molecular-scale technologies. The Nanofab has an established record of identifying nontraditional users of nanotechnology and linking nanoscience and nanoengineering to new disciplinary fields. The spin-out companies established by The Pennsylvania State University have brought several nanotechnologies to the marketplace. A prime example of such spin-offs is the Nano Horizons company. It was founded in 2002 by a team of scientists from The Pennsylvania State University. This company develops, designs, and manufactures advanced nanoscale silver additives that add antimicrobial and performance-enhancing characteristics to consumer, commercial, and industrial products.

The Pennsylvania State University provides education and training to develop nanoscience and nanotechnology workforce through its Center for Nanotechnology Education and Utilization (CNEU). The CNEU focuses on incorporating nanotechnology into secondary education, postsecondary education, and industrial applications. The CNEU is a partner in the Pennsylvania Nanofabrication Manufacturing Technology (NMT) Partnership. The NMT Partnership is a higher education collaborative actively engaged in creating and updating a workforce trained in the discipline of nanotechnology. All the activities and programs of the NMT Partnership are coordinated by the CNEU. The NMT educational partners include the following:

- The Pennsylvania Commission for Community Colleges
- The Pennsylvania State System of Higher Education
- Pennsylvania College of Technology
- The Pennsylvania State University

More information regarding the research, development, and workforce training programs conducted by CNEU is available at the following website: http://www.cneu.psu.edu

Additional information regarding the Penn State Nanofabrication Facility is available at the following website: http://www.mri,psu.edu/facilities/nanofab/

- The Rhode Island Consortium for Nanoscience and Nanotechnology
  The Rhode Island Consortium for Nanoscience and Nanotechnology was established in 2010 by U.S. Congress. The Consortium is a joint partnership between the University of Rhode Island and Brown University. As listed on the website, http://www.uri.edu/nano/, the Consortium strives to attain the following goals:

  - Enhance the state of Rhode Island's position as a center of excellence
  - Develop industry academia collaboration and new technology ventures to accelerate economic growth and job creation in the region.
  - Build research teams that exploit the research and development capabilities of URI, Brown University, and industry to solve technology problems.
  - Provide entrepreneurship awards for selected new ventures.
  - Establish the necessary infrastructure to allow user access to state-of-the-art instrumentation for nanoscale research and development.

The focus areas of the above-mentioned alliance include nanomaterials, nanohealth technologies, nanotools, and nanoenergy. The Consortium provides access to state-of-the-art calorimetry, microscopy, and spectroscopy equipment, computational and analysis capabilities, and micro/nano manufacturing facilities to conduct research and development activities. More information regarding the Rhode Island Consortium for Nanoscience and Nanotechnology is available at the following website: http://www.uri.edu/nano

• Virginia Nanoelectronics Center
The University of Virginia, in collaboration with Old Dominion University and the College of William and Mary, has recently launched the Virginia Nanoelectronics Center (ViNC) to conduct research and development activities and programs focusing on next-generation electronics. The ViNC partnership will bring together researchers to explore and develop advanced materials, novel devices, and systems at nanoscale dimensions. The center has been launched in partnership with Micron Technology Inc. The center is supported by the Nanoelectronics Research Initiative (NRI) described earlier in this chapter, Funding for the ViNC has been provided by Micron Technology, Intel, IBM, Texas Instruments, Global Foundries, National Institute of Standards and Technology, Commonwealth of Virginia, National Science Foundation, and the Defense Advanced Research Projects Agency. Additional information can be obtained from the following website: http://chem.virginia.edu/faculty-research/centers-programs/

14.6   CONCLUSION

The unrivaled pace of advances in nanoelectronics research, development, commercialization, and manufacturing is driven by a relentless pursuit of innovation. The ultimate goal of nanoelectronics research is to transform the advanced knowledge into complex systems. The success of nanoelectronics research and development will be measured using the following criteria:

• Ability to reduce power density and/or energy consumption
• Ability to cohost the three basic system functionalities which include computation, storage, and transmission of information
• Controlled device variability
• Cost-effective technological solutions
• The possibility of building analog or mixed signal systems from the new logic devices
• Sufficient reliability and yield for applications

There is a need to bridge the gap between scientific discovery and practical applications. Triple helix models involving universities, industry, and government can help bridge this gap and lower the barriers to nanoelectronics commercialization. In regard to the education and training needs for the future nanoelectronics workforce, the continuing evolution of electronics beyond the scaling limits of Moore's law will require a broad thinking across multiple disciplines. A coordinated effort will be needed to define the knowledge base and competencies needed by future nano-electronics engineers and technologists.

REFERENCES

Anwar, S., M. Y. Raja, S. Qazi, and M. Ilyas. 2010. *Nanotechnology for Telecommunications.* Boca Raton, FL: Taylor & Francis Group/CRC Press.

Dhillon, H., S. Qazi, and S. Anwar. 2008. Mitigation of barriers to commercialization of nano-technology: An overview of two successful university-based initiatives. *Proceedings of the ASEE 2008 Annual Conference and Exposition.* Pittsburg, PA.

Etzkowitz, H. 2011. The triple helix: Science, technology and the entrepreneurial spirit. *Journal of Knowledge-Based Innovation in China.* 3(2): 76–90.

Mc Neil, R. D., M. Lowe, and D. Ferk. 2007. Barriers to nanotechnology commercializa-tion. *Report Prepared for US Department of Commerce, Technology Administration.* Springfield, IL: College of Business and Management, University of Illinois.

National Science and Technology Council (NSTC) Subcommittee on Nanoscale Science, Engineering and Technology. 2010. *National Nanotechnology Initiative Signature Initiative*: *Nanoelectronics for 2020 and Beyond.* NNI Publications and Reports.

Appendix

MATLAB CODES USED TO GENERATE SOME OF THE TEXT FIGURES

Copyright, Computational Nanoelectronic Research Group (CoNE) University Teknologi Malaysia (UTM): All codes included herein.

FIGURE 3.5: CNT DOS

```
%CNT DOS
clc
t=3;
m=200;
a0=0.14;
D=2.*a0.*sqrt(3)./(2.*pi);
Eg=2.*t.*0.14./D;
c=pi.*D;
L=1,D;
nu0=round(2.*m./3);
a=3.*a0./2;
E=linspace(0,0.125,100);
DG=(2*c.*L./(2.*pi*a*a*t*t)).*E;
DN=zeros(1,100);
for nu=nu0-101:nu0+101;
        Ek=((t.*2.*pi./sqrt(3)).*((3.*nu./(2.*m))-1)+(i.*1e-12));
        d=abs(real(E./(sqrt((E.^2)-(Ek.^2)))));
        DN=DN+((2.*L./(pi.*a.*t)).*d);
end
hold on
h1=plot(E,DN,'b',-E,DN,'b');
axis([-0.125 0.125 0 10])
set(h1,'linewidth',[3.0])
xlabel('E(ev)','fontsize',14)
ylabel('DOS(ev.cm)^{-1}','fontsize',14)
```

FIGURE 3.7: Fermi Integral Approximation for 1D

```
clc
eta=[-5:0.1:5];
y=exp(eta);
yd1=fermi(eta,-1/2);
ynd1=2.*sqrt(eta)/sqrt(pi);
% semilogy(eta,y,'-k',eta,yd1,'-g',eta,ynd1,'.r','LineWidth',4)
% axis([-5 5 -5 5])
% xlabel('{\eta}','fontsize',20)
```

```
% ylabel('{\Im}_{-1/2}','fontsize',20)
% H=legend('nondegenerate','{\Im}_{-1/2}','degenrate',25);
semilogx(y,eta,'-k',yd1,eta,'-g',ynd1,eta,'.r','LineWidth',4)
axis([-5 5 -5 5])
ylabel('{\eta}','fontsize',20)
xlabel('{\Im}_{-1/2}','fontsize',20)
H=legend('nondegenerate','{\Im}_{-1/2}','degenrate',25);
```

FIGURE 3.9: CNT Velocity versus Temperature

```
%This program calculates relative velocity for CNT (9,2)CNT m=0.099 m0
%This program calculates relative velocity for CNT (5,3)CNT m=0.189 m0
kb=1.38066*10^-23;
m=0.189*9.1*10^-31;%for (5,3)
%m=0.099*9.1*10^-31;%for (9,2)
hbar=1.0545887*10^-34;
Nc1300=sqrt(2*m*kb*300/(pi*hbar^2));
vt300=sqrt(2*kb*300/m);
T=linspace(0.001,400,400);
vtT=vt300.*sqrt(T/300);
viND=(1/sqrt(pi))*vtT/10^5;
Nc1T=Nc1300.*sqrt(T/300);
 %n3=5*10^8
n3=5*10^8;
u=n3./Nc1T;
for i=1:length(T);
    f=@(x)(u(i)-fermi(x,-1/2));
    eta1(i)=fzero(f,0);
    vrel1(i)=(1/sqrt(pi))*log(exp(eta1(i))+1)/u(i);
    vi8(i)=vrel1(i)*vtT(i)/10^5;
end

%n4=10^6
n4=10^8;
u=n4./Nc1T;
for i=1:length(T);
    f=@(x)(u(i)-fermi(x,-1/2));
    eta1(i)=fzero(f,0);
    vrel1(i)=(1/sqrt(pi))*log(exp(eta1(i))+1)/u(i);
    vi6(i)=vrel1(i)*vtT(i)/10^5;
end
%n4=10^6
n5=10^6;
u=n5./Nc1T;
for i=1:length(T);
    f=@(x)(u(i)-fermi(x,-1/2));
    eta1(i)=fzero(f,0);
    vrel1(i)=(1/sqrt(pi))*log(exp(eta1(i))+1)/u(i);
    vi7(i)=vrel1(i)*vtT(i)/10^5;
end
```

```
%semilogx(ncm,vdeg)
plot(T,vi8,'r-',T,vi6,'b:',T,viND,'k-',T,vi7,':m','Linewidth', 4.0);
legend('n=5*10^{8} m^{-1}','n=10^{8} m^{-1}','n=10^{6} m^{-1}','NONDEG',4);
xlabel('TEMPERATURE (K)', 'Fontsize',24);
ylabel('vELOCITY (10^{5} m/s)', 'Fontsize',24);
```

FIGURE 3.10: CNT Velocity versus Carrier Concentration

```
%This program calculates relative velocity for nanowire (9,2)m=0.099m0
%This program calculates relative velocity for nanowire (5,3)m=0.189m0
kb=1.38066*10^-23;
%m=0.099*9.1*10^-31;%for(9,2)
m=0.189*9.1*10^-31;%for(5,3)
hear=1.0545887*10^-34;
Nc1300=1.*sqrt(2*m*kb*300/(pi*hbar^2));
vt300=sqrt(2*kb*300/m);
n1=logspace(6,9,100);
%T=300K
T=300;
Nc1T=Nc1300*sqrt(T/300);
vtT=vt300*sqrt(T/300);
u=n1./Nc1T;
for i=1:100;
    f=@(x)(u(i)-fermi(x,-1/2));
    eta1(i)=fzero(f,0);
    vrel1(i)=(1/sqrt(pi))*log(exp(eta1(i))+1)/u(i);
end
vi300=vrel1*vtT/10^5;
 %T=77K
T=77;
Nc1T=Nc1300*sqrt(T/300);
vtT=vt300*sqrt(T/300);
u=n1./Nc1T;
for i=1:100;
    f=@(x)(u(i)-fermi(x,-1/2));
    eta1(i)=fzero(f,0);
    vrel1(i)=(1/sqrt(pi))*log(exp(eta1(i))+1)/u(i);
end
vi77=vrel1*vtT/10^5;
 %T=4.2K
T=4.2;
Nc1T=Nc1300*sqrt(T/300);
vtT=vt300*sqrt(T/300);
u=n1./Nc1T;
for i=1:100;
    f=@(x)(u(i)-fermi(x,-1/2));
    eta1(i)=fzero(f,0);
    vrel1(i)=(1/sqrt(pi))*log(exp(eta1(i))+1)/u(i);
end
```

```
vi42=vrel1*vtT/10^5;
vd=hbar*n1*pi./(4*m*10^5);
semilogx(n1,vi300,'r-.',n1,vi77,'g--',n1,vi42,'b-',n1,vd,'k:',
  'Linewidth', 4.0 );
h=legend('T = 300 K','T =  77 K','T =  4.2 K','Degenrate',2);
xlabel('n (m^{-1})', 'Fontsize',24);
ylabel('v_{i} (10^{5} m/s)', 'Fontsize',24);
axis([10^7 9*10^8 0 6])
```

FIGURE 3.11: CNT Diameter Dependence Carrier Drift Velocity

```
%This program calculates diameter dependence Carrier Drift Velocity
%This program calculates diameter dependence Carrier Drift Velocity
kb=1.38066*10^-23;
m=0.099*9.1*10^-31;%for(9,2)
m2=0.189*9.1*10^-31;%for(5,3)
m3=0.255*9.1*10^-31;%for(6,1)
m4=0.408*9.1*10^-31;%for(5,0)
%m=0.099*9.1*10^-31;%for (9,2)
hbar=1.0545887*10^-34;
Nc1300=sqrt(2*m*kb*300/(pi*hbar^2));
vt300=sqrt(2*kb*300/m);
T=linspace(0.001,400,400);
vtT=vt300.*sqrt(T/300);
viND=(1/sqrt(pi))*vtT/10^5;
Nc13002=sqrt(2*m2*kb*300/(pi*hbar^2));
vt3002=sqrt(2*kb*300/m2);
T=linspace(0.001,400,400);
vtT2=vt3002.*sqrt(T/300);
viND2=(1/sqrt(pi))*vtT2/10^5;
Nc13003=sqrt(2*m3*kb*300/(pi*hbar^2));
vt3003=sqrt(2*kb*300/m3);
T=linspace(0.001,400,400);
vtT3=vt3003.*sqrt(T/300);
viND3=(1/sqrt(pi))*vtT3/10^5;

Nc13004=sqrt(2*m4*kb*300/(pi*hbar^2));
vt3004=sqrt(2*kb*300/m4);
T=linspace(0.001,400,400);
vtT4=vt3004.*sqrt(T/300);
viND4=(1/sqrt(pi))*vtT4/10^5;
 plot(T,viND,'r-',T,viND2,'b:',T,viND3,'k-',T,viND4,':m','Linewidth', 4.0);
h=legend('(9,2),d=0.8081nm','(5,3),d=0.4233nm','(6,1),d=0.3137nm','(5,0),
    d=0.1961nm',2);
%legend('n=5*10^{8} m^{-1}','n=10^{8} m^{-1}','n=10^{6} m^{-1}','NONDEG',4);
xlabel('TEMPERATURE (K)', 'Fontsize',24);
ylabel('vELOCITY (10^{5} m/s)', 'Fontsize',24);
```

FIGURE 3.12: CNT Diameter Dependence Carrier Drift Velocity

```
%This program calculates C.4 diameter dependence Carrier Drift Velocity
%This program calculates C.4 diameter dependence Carrier Drift Velocity
kb=1.38066*10^-23;
m=0.099*9.1*10^-31;%for(9,2)
m2=0.189*9.1*10^-31;%for(5,3)
m3=0.255*9.1*10^-31;%for(6,1)
m4=0.408*9.1*10^-31;%for(5,0)
hbar=1.0545887*10^-34;
Nc1300=1.*sqrt(2*m*kb*300/(pi*hbar^2));
vt300=sqrt(2*kb*300/m);
n1=logspace(6,9,100);
vd=hbar*n1*pi./(4*m*10^5);
vd2=hbar*n1*pi./(4*m2*10^5);
vd3=hbar*n1*pi./(4*m3*10^5);
vd4=hbar*n1*pi./(4*m4*10^5);
d=0.08*9.1*10^-31/m
d2=0.08*9.1*10^-31/m2
d3=0.08*9.1*10^-31/m3
d4=0.08*9.1*10^-31/m4
semilogy(vd,n1,'b-.',vd2,n1,'g--',vd3,n1,'r-.',vd4,n1,'k-.','Linewidth', 4.0 );
h=legend('(9,2),d=0.8081nm','(5,3),d=0.4233nm','(6,1),d=0.3137nm','(5,0),
   d=0.1961nm',2);
ylabel('n (m^{-1})', 'Fontsize',24);
xlabel('v_{i} (10^{5} m/s)', 'Fontsize',24);
axis([ 0 6 10^7 9*10^8])
```

FIGURE 3.16: CNT Current–Voltage

```
%I - V charactristic of the CNTFET with experimental data from
   Navab@ime.a-sta.edu.sg paper(2007)
clc
hbar=1.0545887e-34;%j.s
kb=1.380662e-23;%j/k
q=1.6021892e-19;%C
T=300;%K
r=1.7e-9;%m
m=0.099*9.1*10^-31;%for (9,2)
e=20*8.8541878e-12;%hfo
%m=(0.08/1.7)*9.109534e-31%kg
vth1=sqrt(2*kb*T/(pi*m)); %nondegenerate 1D velocity
vth=sqrt(2*kb*T/m);% genral thermal velocity
gv=4;
NcT=gv*sqrt(2*m*kb*T)/sqrt(pi*hbar^2);%1D Effective DOS with gv=4
tox=11*10^-9;%m
L=300*10^-9;%m
```

```
%Threshold voltage VT=-0.2;
VT=0.0037;
%The following component calculates the vsat value for ElecFDrain infinity
    for k=1:4
    VGS(k)=0.2+(k-1)*0.2;
    VGT(k)=VGS(k)-VT;
     Et(k)=(VGS(k)+VT)/(6*tox);
    Eo(k)=(hbar^2/(2*m))^(1/3)*(9*pi*q*Et(k)/8)^(2/3);
    zo(k)=Eo(k)/(q*Et(k));
    zQM(k)=2*zo(k)/3;
    toxeff(k)=tox;
    toxeffp(k)=r-zQM(k);
%CG(k)=2*pi*ehf/(log((toxeff(k)+toxeffp(k)+sqrt((toxeff(k)).^2+
  2.*toxeffp(k).*toxeff(k)))./(toxeffp(k)))));
    vf=8*10^5;
Ci=(2*pi*e)/(log((2*tox)/r));%Ci / L is (F/m)
Cq=2*q/vf;%vahede zarfiiat F/m hast
%CG(k)=Ci*Cq/(Ci+Cq);
CG(k)=1.8*10.^-10;
    uo(k)=0.7397934;
    vsat1(k)=vth1;  % seed value calculated from the
      nondegenrate velocity
          for j=1:10
        Vc(k)=vsat1(k)*L/uo(k);
        VDsat1(k)=Vc(k)*(sqrt(1+(2*VGT(k)/Vc(k)))-1);
        n1(k)=CG(k)*(VGT(k)-VDsat1(k))/q;
        u(k)=n1(k)/NcT;  %find fermi(-1/2) eta1 from general
          formula of carrier statistics
            for i=1:10;
    f=@(x)(u(k)-fermi(x,-1/2));
    eta(k)=fzero(f,0);
        end
    t1=fermi(eta(k),-1/2); %find fermi -0.5 eta1
        vi1(k)=vth1*(t1/u(k));
        vsat1(k)=vi1(k);
        end
     n2cm(k)=n1(k)/1e4;
gmo(k)=CG(k)*vsat1(k);
K(k)=CG(k)*uo(k)/L;
IDsat1(k)=K(k)*VDsat1(k)^2/2;
gch1(k)=(K(k)/2)*(2*(VGT(k)-VDsat1(k))-(VDsat1(k)^2/Vc(k)))/
  (1+VDsat1(k)/Vc(k))^2;
end
%The following component calculates I-V characterisitcs and alpha for
%ElecFDrain finite
for k=1:4
alphanew=1;
```

```
    for i=1:20
    alpha1=alphanew;
    s=sqrt((alpha1+((1-alpha1)*VGT(k)/Vc(k)))^2+2*alpha1*(2*alpha1-1)*
      VGT(k)/Vc(k));
    VDsat(k)=(1/(2*alpha1-1))*((s-alpha1)*Vc(k)-(1-alpha1)*
      VGT(k));
    %alpha is for EDrain=(1/alph)*VDsat/Vc.  Iterative solution
      gives alpha
    alphanew=((1/alpha1)*VDsat(k)/Vc(k))/(1+((1/alpha1)*VDsat(k)/Vc(k)));
    end
alpha(k)=alphanew;
vdrain(k)=alpha(k)*vi1(k);
%alpha not equal to 1
s=sqrt((alpha(k)+((1-alpha(k))*VGT(k)/Vc(k)))^2+2*alpha(k)*
  (2*alpha(k)-1)*VGT(k)/Vc(k));
VDsat=(1/(2*alpha(k)-1))*((s-alpha(k))*Vc(k)-(1-alpha(k))*VGT(k));
IDsat=(alpha(k)/(2*alpha(k)-1))*K(k)*Vc(k)*(alpha(k)*VGT(k)-
  (s-alpha(k))*Vc(k));
VDsatP(k)=VDsat;
IDsatP(k)=IDsat;
end
%This block will use the values calculated before to generate I-V
%characteristics
for k=1:4
    VD=0:0.01:VDsatP(k);
    ID=(K(k)/2)*(2*VGT(k)*VD-VD.^2)./(1+(VD./Vc(k)));
        VDM=VDsatP(k):0.01:0.4;
    alphaprime=((1/alpha(k))*VDM/Vc(k))./(1+((1/alpha(k))*VDM/Vc(k)));
    IDMprime=alphaprime*CG(k)*vsat1(k)*(VGT(k)-VDsatP(k));
        VD1=VDsatP(k):0.01:VDsat1(k);
    ID1=(K(k)/2)*(2*VGT(k)*VD1-VD1.^2)./(1+(VD1./Vc(k)));
        VD1M=VDsat1(k):0.01:0.4;
    ID1M=ones(1,length(VD1M))*IDsat1(k);
%plot(VD,ID/1e-6,'-k',VDM,IDMprime/1e-6,':k',VD1,ID1/1e-6,'-r',
  VD1M,ID1M/1e-6, '--b','Linewidth', 4)
plot(VD,ID/1e-6,'-k',VD1,ID1/1e-6, '-r',VD1M,ID1M/1e-6, '--b',
  'Linewidth', 4)
hold on
end
xlabel('V (v) ')
ylabel('I (\muA)')
axis([0 0.4 0 25])
text(VDsat,IDsat+0.25,'V_{GS}= 0.2','FontSize',10);
text(VDsat,IDsat+0.85,'V_{GS}= 0.4','FontSize',10);
text(VDsat,IDsat+1.57,'V_{GS}= 0.6','FontSize',10);
text(VDsat,IDsat+2.49,'V_{GS}= 0.8','FontSize',10);
%text(VDsat,IDsat+3.4,'V_{GS}= 1','FontSize',10);
```

FIGURE 7.5: Numerical Solution of General Model of BGNs Conductance and Degenerate and Nondegenerate Approximation

```
clc;clear;
syms x;
==========CONSTANT VALUES=========
a=0.142e-9;                    % m
t=3.1*1.6e-19;             % J
tp=0.39*1.6e-19;          % J
vf=3*t*a/2;
kb=1.381e-23;                  % (J/K) Boltzmann's constant
q=1.6e-19;                     % C
h=6.63e-34;                    % (J/s) Planck constant
==========TEMPREATURE=========
T=55;                      % k
==========LENGTH=========
l=7e-6;
==========BIASED VOLTAGE=========
V=0.55;                        % 1volt=1 J/C = 6.25e+018 eV/C
==================
Gn=(-4*q^2)*(kb*T)/(h*l*sqrt(2));
Gnp=(-4*q^2)/(h*l);
alfa=((V/tp^2)*vf^2);
beta=(vf^4)/(V*tp^2);
==================
for vgb=-8:1:5;        % vgb is bottom gate voltage
    vgt=-2.55;              % vgt is top gate voltage
    vg=vgb-vgt;
    eta=((vg-V).*q)./(kb*T)
    Q=alfa^2+4*beta*kb*T.*x;
    a1=(3*beta).*sqrt((alfa+sqrt(Q))./(beta.*Q));
    a2=-alfa./(sqrt(Q.*(alfa+sqrt(Q))/(beta)));
    a3=(1+exp(x-eta));
    a4=(exp(x-eta));
==========MODEL=========
    fn11=(a1./a3);
    q11=Gn.*int(fn11,-vg,0);
    t11=double(vpa(q11,5));
    fn12=(a2./a3);
    q12=Gn.*int(fn12,-vg,0);
    t12=double(vpa(q12,5));
    t1=t12+t11;
    fn21=(a1./a3);
    q21=Gn.*int(fn21,0,vg);
    t21=double(vpa(q21,5));
    fn22=(a2./a3);
```

```
    q22=Gn.*int(fn22,0,vg);
    t22=double(vpa(q22,5));
    t2=t21+t22;
    t=t2-t1;
============DEGENERATE=========
    fn11d=(a1);
    q11d=Gn.*int(fn11d,-vg,0);
    t11d=double(vpa(q11d,10));
    fn12d=(a2);
    q12d=Gn.*int(fn12d,-vg,0);
    t12d=double(vpa(q12d,10));
    t1d=t12d+t11d;
    fn21d=(a1);
    q21d=Gn.*int(fn21d,0,vg);
    t21d=double(vpa(q21d,10));
    fn22d=(a2);
    q22d=Gn.*int(fn22d,0,vg);
    t22d=double(vpa(q22d,10));
    t2d=t21d+t22d;
    td=t2d-t1d;
===========NON-DEGENERATE=========
    fn11nd=(a1./a4);
    q11nd=Gn.*int(fn11nd,-vg,0);
    t11nd=double(vpa(q11nd,10));
    fn12nd=(a2./a4);
    q12nd=Gn.*int(fn12nd,-vg,0);
    t12nd=double(vpa(q12nd,10));
    t1nd=t12nd+t11nd;
    fn21nd=(a1./a4);
    q21nd=Gn.*int(fn21nd,0,vg);
    t21nd=double(vpa(q21nd,10));
    fn22nd=(a2./a4);
    q22nd=Gn.*int(fn22nd,0,vg);
    t22nd=double(vpa(q22nd,10));
    t2nd=t21nd+t22nd;
    tnd=t2nd-t1nd;
  ===========Sketch=========
semilogy (vgb,t,'*r',vgb,td,'*r',vgb,tnd,'*r')
xlabel('vg')
ylabel('G')
hold on
h=legend('Model','Degenrate approximation','Non-Degenrate
  approximation',4);
end
```

FIGURE 7.6: Experimental Data and General Model of Conductance BGN

```
clc;clear;
syms x;
%    ==========CONSTANT VALUES=========
a=0.141e-9;                      % m
t=3.15*1.6e-19;                            %(j)
tp=0.39*1.6e-19;                     % (j)
vf=8*10^5;                               %m/s
kbT=0.025;                          %j
q=1.6e-19;                        %C
h=6.63e-34;                      % (J/s) Planck constant
%    ==========LENGTH=========
l=12e-6;          %m
w=10e-6;        %m
%    ==========BIASED VOLTAGE=========
V=0.6;                            % 1volt=1 J/C = 6.25e+018 eV/C
%    ====================
Gn=(-4*q^2)*(kbT)/(h*l*sqrt(2));
Gnp=(-4*q^2)/(h*l);
alfa=((V/tp^2)*vf^2);
beta=(vf^4)/(V*tp^2);
%    ====================
for vgb=-2.925:0.1:0;        % vgb is bottom gate voltage
    vgt=-2.56;                % vgt is top gate voltage
    vg=vgb-vgt;
    eta=((vg-V).*q)./(kbT);
    Q=alfa^2+4*beta*kbT.*x;
    a1=(3*beta).*sqrt((alfa+sqrt(Q))./(beta.*Q));
    a2=-alfa./(sqrt(Q.*(alfa+sqrt(Q))/(beta)));
    a3=(1+exp(x-eta));
    a4=(exp(x-eta));
%    ==========MODEL=========
    fn11=(a1./a3);
    q11=Gn.*int(fn11,-vg,0);
    t11=double(vpa(q11,5));
    fn12=(a2./a3);
    q12=Gn.*int(fn12,-vg,0);
    t12=double(vpa(q12,5));
    t1=t12+t11;
    fn21=(a1./a3);
    q21=Gn.*int(fn21,0,vg);
    t21=double(vpa(q21,5));
    fn22=(a2./a3);
    q22=Gn.*int(fn22,0,vg);
    t22=double(vpa(q22,5));
    t2=t21+t22;
    t=t2-t1;
```

```
semilogy (vgb,t,'*b',vgb,t,'ob')
xlabel('vg')
ylabel('G')
hold on
end
x=[-3 -2.97581 -2.95968 -2.95161 -2.92742 -2.93548 -2.8871
  -2.83065 ...
-2.81452 -2.81452 -2.79839 -2.77419 -2.74194 -2.75806
  -2.71774 ...
-2.70161 -2.67742 -2.65323 -2.6129 -2.58871 -2.57258 -2.56452
  -2.55645 ...
-2.54839 -2.52419 -2.53226 -2.52419 -2.5 -2.47581 -2.45161
  -2.44355 ...
-2.42742 -2.39516 -2.3629 -2.34677 -2.29032 -2.25 -2.21774
  -2.16935 ...
-2.14516 -2.08871 -1.93548 -2.04032 -1.97581 -2.01613 -1.8871
  -1.84677 ...
-1.79032 -1.74194 -1.69355 -1.64516 -1.59677 -1.52419 -1.45968
  -1.3871 ...
-1.31452 -1.23387 -1.16935 -1.10484 -1.02419 -0.951613
  -0.862903 ...
-0.774194 -0.693548 -0.612903 -0.532258 -0.443548 -0.362903
  -0.266129 ...
-0.185484 -0.120968 -0.0725806 -0.00806452];
yp=[0.0485121 0.0555608 0.0680998 0.0834686 0.12972 0.102306
  0.134195 ...
0.125394 0.16448 0.255622 0.335301 0.292763 0.346868 0.397267
  0.397267 ...
0.50372 0.6174 0.7315 0.837785 1.09893 1.3934 2.02349 3.98742
  93.4417 ...
1.4912 0.959513 0.596812 0.302863 0.371213 0.283 0.194878
  0.134195 ...
0.117171 0.0806851 0.0615116 0.051917 0.0438189 0.033406
  0.029168 ...
0.0246184 0.0237974 0.0148018 0.0222367 0.0158407 0.0187682
  0.012924 ...
0.0116737 0.0112844 0.0109081 0.0101927 0.00985283 0.00889964
  0.00831598 ...
0.00803866 0.00777059 0.00777059 0.00751147 0.00751147
  0.00751147 ...
0.00726098 0.00701884 0.00701884 0.00678478 0.00655853
  0.00655853 ...
0.00645853 0.00635853 0.00635853 0.00625853 0.00615853
  0.00603982 ...
0.00593982 0.00595853];
y=(1./(yp*10^6*l*w));
semilogy (x,y,'*r',x,y,'or')
hold on
```

FIGURE 7.7: BGN Conductance Analytical Model in Degenerate Regime

```
clc;clear;
vgb=-6:0.01:4;
vgt=-2.55;
vg=vgb-vgt;
%    ==========CONSTANT VALUES=========
a=0.142;
t=3.15;
tp=0.39;
vf=3*t*a/2;
kb=(8.62e-5);
q=1.6e-19;
h=6.63e-34;
%    ==========TEMPREATURE=========
T=300;                          % k
%    ==========LENGTH=========
l=12e-6;
%    ==========BIASED VOLTAGE=========
V=0.55;
%    ====================
Gn=-(4*q^2)/(h*l);
alfa=((V/tp^2)*vf^2);
beta=(vf^4)/(V*tp^2);
    P=alfa^2+4*beta*kb*T.*(vg);
    b1=(2*beta)*sqrt(((alfa+sqrt(P))./(2*beta)).^3);
    b2=-(alfa).*sqrt((alfa+sqrt(P))./(2*beta));
    Pp=alfa^2+4*beta*kb*T.*(-vg);
    b1p=(2*beta)*sqrt(((alfa+sqrt(Pp))./(2*beta)).^3);
    b2p=-(alfa).*sqrt((alfa+sqrt(Pp))./(2*beta));
    G1=Gn*(b2+b1);
    G2=Gn*(b1p+b2p);
    G=(G1-G2)*10^4;
semilogy (vgb,G,'*g',-vgb+2*vgt,G,'*g')
hold on
```

FIGURE 8.8: *E–k* Relationship in Trilayer Graphene Nanoribbon

```
clc
t=2.7;       %eV
a=0.141*10^-9;%m
vf=10^6;%m/s
delta=0.3;
tp=0.3;       %eV
T=300;%k
kb=8.62e-5;  %(eV/k)  Boltzmann's constant
B=sqrt(2);
alfa=((vf*delta)./(tp*B));
beta=vf^3/(tp*B*delta);
```

```
x=-0.000001:10^-7.5:0.000001;
y=alfa*x-beta*x.^3;
z=-alfa*x+beta*x.^3;
plot(x,y,':r',x,z,'.-k')
 h=legend('E',3,'-E',3);
xlabel('k (m^{-1})','Fontsize',24)
ylabel('Energy(eV)','Fontsize',24)
hold on
```

FIGURE 8.9: Density of States for Trilayer Graphene Nanoribbon

```
clc
hbar=1.05*10^-34/2*pi;
t=2.7;        %eV
a=0.141*10^-9;%m
vf=3*t*a/2;%m/s
delta=1;%???
tp=0.3;       %eV
T=300;
E=-3:0.01:3;
kb=8.62e-5;  %(eV/k)  Boltzmann's constant
B=sqrt(2);
alfa=((vf*delta)./(tp*B));
beta=vf^3/(tp*B*delta);
A=2*pi*alfa-12*alfa*(1/3)^(1/3)*pi/3^(2/3);
B=6*(2/3)^(2/3)*pi*alfa^2*beta;
C=(6*pi)/(2^(2/3)*3^(4/3)*beta);
D=-9*beta^2;
F=(-12*alfa^3)/(71*beta);
%k=-100000000:1:100000000;
Dos=1./(-
A+(B./(D.*E+sqrt(F+E.^2)).^(2/3))+((C.*(D.*E+sqrt(F+E.^2)).^(2/3))));
plot(E,Dos*10^28,'.b')
xlabel('Energy')
ylabel('Dos')
h=legend('Model','Degenerate approximation',4);
hold on
```

FIGURE 8.10: Carrier Concentration Trilayer Graphene Nanoribbon

```
clc
hbar=1*10^-34/1.6*10^-19;%evs
a=0.142;%nm
t=2.7;%ev
kb=8.62e-5;%ev/k   Boltzmann's constant
T=300;%k
vf=3*t*a/2;   %nm/s
```

```
delta=2;    %V
tp=0.3;      %eV
alfa=((vf*delta)/(tp*sqrt(2)));
beta=(vf^3/(tp*sqrt(2)*delta));
E=8;
Ec=4;
a1=-17.0340;
a2=23.6998;
a3=12.2430;
a4=-0.4522;
a5=-15.0235;
a6= -2.2465e4;
a7= 0.0874;
Eco=Ec/(kb*T);
for y=-5:0.1:15
% a9=(0.0874*(-0.4522*(x+Eco)+sqrt(-2.2465e4+(x+Eco)^2))^(2/3));
% %for a9=-1:-1:-5;
% F=@(x)(8.62e-5*300)./((-17.0340-(23.6998./a9)-(12.2430*a9)).* (1+exp(x-
y)));
%a9=((-1*(x+Eco)+sqrt(-2+(x+Eco)^2))^(2/3));
%a9=((x+3)^2/3)
F=@(x)((8.62e-5*300).*10.^2)./((-17.0340-(23.6998./(0.0874*(-
0.4522*(x+Eco)+sqrt(-2.2465e4+(x+Eco).^2)).^(2/3)))-(12.2430*(0.0874*(-
0.4522*(x+Eco)+sqrt(-2.2465e4+(x+Eco).^2)).^(2/3)))).*(1+exp(x-y)));
Q = quad(F,0,10);
y1=Q;
plot(y,y1,'.r','Linewidth', 2.0);
%end
hold on
%save Q5
end
xlabel('\eta')
ylabel('n (Normalized Carrier Concentration)')
legend('nondegenerate','numerical integral','degenerate')
```

FIGURE 8.10: Carrier Concentration on Degenerate Regime (Trilayer Graphene Nanoribbon)

```
clc
hbar=1*10^-34/1.6*10^-19;%evs
a=0.142;%nm
t=2.7;%ev
kb=8.62e-5;%ev/k   Boltzmann's constant
T=300;%k
vf=3*t*a/2;  %nm/s
delta=2;    %V
tp=0.3;      %eV
alfa=((vf*delta)/(tp*sqrt(2)));
```

```
beta=(vf^3/(tp*sqrt(2)*delta));
E=8;
Ec=4;
a1=-17.0340;
a2=23.6998;
a3=12.2430;
a4=-0.4522;
a5=-15.0235;
a6= -2.2465e4;
a7= 0.0874;
Eco=Ec/(kb*T);
for y=-5:0.1:15
% a9=(0.0874*(-0.4522*(x+Eco)+sqrt(-2.2465e4+(x+Eco)^2))^(2/3));
% %for a9=-1:-1:-5;
% F=@(x)(8.62e-5*300)./((-17.0340-(23.6998./a9)-(12.2430*a9)).*(1+exp(x-
y)));
%a9=((-1*(x+Eco)+sqrt(-2+(x+Eco)^2))^(2/3));
%a9=((x+3)^2/3)
F=@(x)((8.62e-5*300).*10.^2)./((-17.0340-(23.6998./(0.0874*(-
0.4522*(x+Eco)+sqrt(-2.2465e4+(x+Eco).^2)).^(2/3)))-(12.2430*(0.0874*(-
0.4522*(x+Eco)+sqrt(-2.2465e4+(x+Eco).^2)).^(2/3)))));
Q = quad(F,0,10);
y1=Q;
plot(y,y1,'--g','Linewidth', 2.0);
%end
hold on
%save Q5
end
xlabel('\eta')
ylabel('n (Normalized Carrier Concentration)')
legend('nondegenerate','numerical integral','degenerate')
```

FIGURE 8.11: Carrier Concentration on Nondegenerate Regime (Trilayer Graphene Nanoribbon)

```
clc
hbar=1*10^-34/1.6*10^-19;%evs
a=0.142;%nm
t=2.7;%ev
kb=8.62e-5;%ev/k   Boltzmann's constant
T=300;%k
vf=3*t*a/2;  %nm/s
delta=2;   %V
tp=0.3;       %eV
alfa=((vf*delta)/(tp*sqrt(2)));
beta=(vf^3/(tp*sqrt(2)*delta));
E=8;
Ec=4;
a1=-17.0340;
```

```
a2=23.6998;
a3=12.2430;
a4=-0.4522;
a5=-15.0235;
a6= -2.2465e4;
a7= 0.0874;
Eco=Ec/(kb*T);
for y=-5:0.1:2
% a9=(0.0874*(-0.4522*(x+Eco)+sqrt(-2.2465e4+(x+Eco)^2))^(2/3));
%  %for a9=-1:-1:-5;
%  F=@(x)(8.62e-5*300)./((-17.0340-(23.6998./a9)-(12.2430*a9)).*(1+exp(x-
y)));
%a9=((-1*(x+Eco)+sqrt(-2+(x+Eco)^2))^(2/3));
%a9=((x+3)^2/3)
F=@(x)((8.62e-5*300).*(exp(-x+y)).*10.^2)./((-17.0340-(23.6998./(0.0874*(-
0.4522*(x+Eco)+sqrt(-2.2465e4+(x+Eco).^2)).^(2/3)))-(12.2430*(0.0874*(-
0.4522*(x+Eco)+sqrt(-2.2465e4+(x+Eco).^2)).^(2/3)))));
Q = quad(F,0,10);
y1=Q;
plot(y,y1,'.-b','Linewidth', 2.0);
%end
hold on
%save Q5
end
xlabel('\eta')
ylabel('n (Normalized Carrier Concentration)')
legend('nondegenerate','numerical integral','degenerate')
```

FIGURE 8.13: The Effect of Applied Voltage on the Carrier Effective Mass

```
clc
t=2.7;        %eV
a=0.141*10^-9;%m
vf=10^6;%m/s
for delta=0.1:0.02:0.2;
%delta=0.5;
tp=0.3;       %eV
T=300;%k
kb=8.62e-5;  %(eV/k)  Boltzmann's constant
B=sqrt(2);
alfa=((vf*delta)./(tp*B));
beta=vf^3/(tp*B*delta);
x=-0.000001:10^-7.5:0.000001;
y=alfa*x-beta*x.^3;
z=-alfa*x+beta*x.^3;
plot(x,y,':r',x,z,'.-k')
hold on
end
```

```
% plot(x,y,'or',x,z,'*k')
 h=legend('E',3,'-E',3);
xlabel('k')
ylabel('Energy')
hold on
```

Glossary

Band theory: As shown by quantum theory, the energy levels of electrons in single atoms are discrete. However, in a solid the atoms are close together. This causes the discrete energy levels of the individual atoms in the solid to merge into a band structure. Although each band contains many discrete energy levels, they are very close together and in band theory the energy within each band can usually be assumed to vary continuously, just as the energies in classical physics do.

Bilayer graphene: Double layers of graphene are two single-graphene layers lying one on the other with different electrostatic potentials parameterized by voltage and that can be accurately controlled at the nanoscale.

Blackbody radiation: All materials emit electromagnetic radiation over a wide frequency range. A perfect emitter is one that emits the maximum possible amount of radiation at any frequency and the emission is called blackbody radiation.

Brillouin zone: The first Brillouin zone is a uniquely defined primitive cell in reciprocal lattice space. The boundaries of this cell are given by planes related to points on the reciprocal lattice. The importance of the Brillouin zone stems from the Bloch wave description of waves in a periodic medium, in which it is found that the solutions can be completely characterized by their behavior in a single Brillouin zone. Considering surfaces at the same distance from one element of the lattice and its neighbors, the volume included is the first Brillouin zone. There are also second, third, etc. Brillouin zones, corresponding to a sequence of disjoint regions (all with the same volume) at increasing distances from the origin, but these are used less frequently. As a result, the first Brillouin zone is often simply called the Brillouin zone.

Carbon nanotube: Conceptually, a carbon nanotube can be thought of as being made up of a rolled up sheet of graphene with a nanoscale diameter. In practice, this is not the way carbon nanotubes are produced, however. (One way of producing them is by arc discharge of carbon electrodes.)

Carrier concentration: The amount of available carriers in the BLG channel that leads to improve the carrier transport as well as the current in a transistor.

Chiral angle (θ): Determines the amount of "twist" of the carbon nanotube.

Chiral vector: Also known as the roll-up vector, it is defined by the integers n and m, which are the number of steps along the unit vectors.

Classical physics: The physics before relativity and quantum theory that was developed around the beginning of the twentieth century. Newton's laws were central to the ideas in classical physics, but at atomic-size scales they are superseded by quantum theory and, at speeds comparable to the speed of light, by Einstein's theories. Classical physics, therefore, is also known as Newtonian physics.

Conductance: The unimpeded flow of charge or energy carrying particles over relatively long distances in a material.

Conduction band: The uppermost band in a band structure in which conduction takes place. The band is not completely full so that there are empty states for the conduction.

Coulomb potential: The magnitude of the force between two point charges is inversely proportional to their distance apart and to the product of the value of each charge. The direction of the force is along the line that passes through the charges such that like charges (both positive or both negative) repel and unlike charges attract. This is Coulomb's law and the potential associated with the force on a unit point charge in the presence of another charge at a given point is the Coulomb potential. The potential is given by the negative of the work done in moving the unit charge from an infinite distance from the other charge to the given point. The work done is the integral of the force times the distance traveled when moving to the given point.

Crystal lattice: A crystal is a solid made up of a regularly repeating configuration of atoms. This regularly repeating structure is the crystal lattice.

Current–voltage: It is calculated for ballistic BLG FET by employing the "top-of-barrier" approach. The I–V characteristics of BLG FET were then compared to those of GNR and silicon MOSFET (Si MOSFET), where the result showed that the BLG FET has better performance compared to the GNR and Si MOSFET.

de Broglie wavelength: A feature of quantum mechanics is that particles can also behave like waves. The wavelength of these so-called matter waves is the de Broglie wavelength, after Louis de Broglie who first proposed the idea.

Degenerate approximation: In terms of a semiconductor, if the Fermi level is situated at less than $3K_BT$ from the conduction and the valence bands, or located within a band, degenerate approximation will play an important role on carrier statistics study. In the other words, in this regime, $e^{\frac{x-\eta}{K_BT}}$, in comparison with 1, can be neglected because the amount of $(x - \eta)$ is very small.

Density of states: Quantum theory shows that energy values are discrete rather than continuous. The spacing between energy levels is not necessarily continuous. Each energy level corresponds to an energy state and can be considered to occupy a point in state space, and points in state space can occupy a volume. The density of states is the number of states per unit volume (in this space) per unit energy around an energy E.

Dirac fermions: In graphene, electrons in terms of their $E-k$ relationship can behave like photons, in which case they are referred to as Dirac fermions.

Doped semiconductor: A semiconductor with impurities added deliberately to control and improve the electrical conduction properties.

Effective mass: The electron energy–wave number $(E-k)$ relation for band edges can be modeled as parabolic. There is a parallel for electrons in free space: the $E-k$ relation is also parabolic. The shape of the parabolas for the band structure is not the same as in the free space case. But this can be accounted for by introducing an effective mass m^*; the charge carriers can then be treated as though they are in free space but with a mass given by the effective mass rather than the actual mass, thus simplifying the mathematical treatment.

Endohedral doping: Molecules or atoms can be encapsulated inside nanotubes, which involves filling of MWNT cores with Pb or Pb oxide by heating the metallic Pb in air together with tubes.

Energy delay product (EDP): A quality measure for a logic gate. Generally, these two measures determine the average performance of a logic gate from the aspects of power consumption.

Energy levels: A key signature of quantum mechanics is that energy can only take on discrete values unlike the classical case in which energy can be continuously varied. Each discrete energy value (E_0/n^2) is called an energy level. Energy levels are often measured in electron volts.

Excitons: Electron–hole pairs are created when an electron moves from the valence band to the conduction band; each electron and hole can contribute to conduction but the electron–hole pairs are still bound together by Coulomb attraction and can behave something like a hydrogen atom with the nucleus replaced by the hole. These hydrogen-like electron–hole pairs are called excitons.

Extrinsic capacitance: Serves as the gate capacitance for the second inverter, and the intrinsic capacitance consists of gate-to-drain capacitance and drain-to-bulk capacitance in first inverter.

Fermi–Dirac distribution: The probability distribution for indistinguishable particles such that each particle can only occupy a single state (usually electrons) that gives the probability that a state at energy E will be occupied.

Fermi–Dirac distribution function: The probability of occupation $f(E)$ can be discussed in the form of Fermi–Dirac distribution function. For an intrinsic semiconductor, the concentration of electrons in the conduction band is equal to the concentration of holes in the valence band.

Fermi energy: The highest energy to which energy states would be filled to at absolute zero temperature.

Fermi surface: The Fermi surface is an abstract boundary useful for predicting the thermal, electrical, magnetic, and optical properties of metals, semimetals, and doped semiconductors. The shape of the Fermi surface is derived from the periodicity and symmetry of the crystalline lattice and from the occupation of electronic energy bands. The existence of a Fermi surface is a direct consequence of the Pauli exclusion principle, which allows a maximum of one electron per quantum state.

Fullerenes: Spherical shape of graphene sheets.

Graphene: A single layer of graphite (a crystalline allotropic form of carbon): It has a hexagonal lattice structure.

Graphite: A two-dimensional system hexagonal lattice of carbon atoms that can be used in a field effect transistor.

Group velocity: The velocity of the envelope of a wave packet.

Hamiltonian: In classical mechanics, the Hamiltonian of a system is the sum of the kinetic and potential energies of the system. In quantum theory, the Hamiltonian takes the form of an operator and, in conjunction with the wave function, leads to the energy levels of the system.

Heisenberg's uncertainty principle: A consequence of the quantum theory is that there is a fundamental limit to the accuracy with which the position and velocity of a particle can be simultaneously measured. This is Heisenberg's uncertainty principle. It is expressed mathematically as $\Delta p \Delta x \geq \hbar/(4\pi)$, where Δp is the uncertainty in the momentum, Δx is the uncertainty in the position, and $\hbar$ is Planck's constant.

Heterostructure: A layered structure that can be used to form a quantum well.

Hole: When an electron in a semiconductor moves from the valence band to the conduction band, there is an energy state left empty in the valence band that behaves like a positive charge and is called a hole. Charge carriers in semiconductors can thus be positive (holes) or negative (electrons).

Insulator: Materials for which the band structure is such that the conduction and valence bands are separated by an energy band gap which is large compared with the thermal energy of electrons at room temperature. The material will not be able to conduct electrons easily.

Interband transition: An energy transition from one energy band to another.

Intercalation or exohedral doping: The "bundled" form of electronic properties of SWNTs can be modified by doping with donors, accepters, small molecules, and non-carbon atoms existing in the interstitial channels.

Intraband transition: An energy transition between energy levels in a single band.

***K* points:** Points in the plane $E = 0$ of a two-dimensional E–k plot of graphene where bonding (positive energy) and antibonding (negative energy) surfaces meet.

Kronig–Penney model: In this model, the potential seen by charge carriers is assumed to be of a simple form consisting of regularly repeating rectangular wells. The solution to Schrödinger's equation can readily be found in this case. Although the assumption of rectangular wells is not very accurate, several of the main features of conduction in semiconductors such as energy bands are predicted by the Kronig–Penney model.

Lennard-Jones force: Lennard-Jones force is a noncovalent force acting between the layers in a multi-walled carbon nanotube.

Load capacitance: The combination of gate capacitance for the second inverter and the intrinsic capacitance is the so-called load capacitance.

Maxwell–Boltzmann (nondegenerate) approximation: If the Fermi energy lies in the regions above the conduction band edge, the degenerate approximation is appreciated and brought up to the Fermi level where the probability of occupation is equal to one, while it is zero when the approximation is above the Fermi level. Normally, the initiation of strong degeneracy occurs when the Fermi level passes through the conduction band from the forbidden band gap and so the carrier concentration.

Maxwell–Boltzmann statistics: Arise from a statistical analysis of the dynamical behavior of a large number of particles obeying the laws of classical physics for which the quantum effects are insignificant.

Metal: A material, usually a solid, for which the band structure is such that the conduction and valence bands are hardly separated or overlapping. The material will then be a good electrical conductor.

Mobility: An important aspect for MOSFETs in determining the performance of the device and represents the ease of carrier flowing through an area. It is normally expressed as cm^2/V s. The hole and electron mobility depend on various aspects like the doping impurity concentrations of the donors and accepts majority or minority carriers, the temperature, etc.

Moore's law: According to this law by Gordon Moore, the number of transistors on a chip roughly doubles every 2 years with the result that the scale gets smaller and smaller.

MOSFET: Metal–Oxide–Semiconductor Field Effect Transistor.

Nanoelectronics: The technology of electronic devices with dimensions ranging from the atom scale (0.1 nm) to 100 nm.

Newtonian physics: See *Classical physics.*

Ontology: The study of the nature or essence of things.

Pauli exclusion principle: Electrons with the same spin cannot share the same quantum state.

Phase velocity: The velocity of the waves inside a wave packet.

Photoelectric effect: When ultraviolet light (frequency v) is incident on a metal, the electrons released are due to the light behaving as particles called photons of energy $E = \hbar v$. This is independent of the amplitude of the light and is called the photoelectric effect.

Photon: When light shows particle-like properties, the particles are referred to as photons.

Planck's constant: The relationship between the energy E of a photon of light (or any electromagnetic wave) and the frequency v of the light is related by the equation $E = \hbar v$, where $\hbar$ is Planck's constant.

Poisson's equation: The fundamental partial differential equation with broad utility in physics.

Power delay product (PDP): A quality measure for a logic gate. Generally, it determines the average performance of a logic gate from the aspects of propagation delay.

Primitive cell: When describing a crystal structure, a primitive cell is a minimum cell corresponding to a single lattice point of a structure with translational symmetry in two dimensions, three dimensions, or other dimensions. A lattice can be characterized by the geometry of its primitive cell. A crystal can be categorized by its lattice and the atoms that lie in a primitive cell (the basis). A cell will fill all the lattice space without leaving gaps by repetition of crystal translation operations.

Propagation delay: Refers to the amount of time starting from the time when input becomes stable and valid to the time when output becomes stable and valid.

p-Type nanoconductor: Boron and nitrogen substitutional doping within graphene nanocylinders is notable. This method will introduce localized electronic properties in valence or conduction bands. According to the location and concentration of dopants, it will enhance the number of electronic states at the Fermi level (E_f). Substituting boron for carbon within a SWNT, regarding the fact that B has one electron less than C, under the Fermi level (valence band), sharply localized states (three-coordinated B) appear. Considering the existence of holes in the structure, sharp localized states appear in valence band, thus leading to the tubes being assumed as a p-type nanoconductor. p-type nanoconductor can act in response with donor-type molecules.

Pyridine-type N doping SWNT: One type of substitutional doping is pyridine-type N doping SWNT that can be a part of the SWNT lattice. This type of doping makes localized states below and above the Fermi level. Therefore, substitutional N doping in SWNTs leads to n-type conducting behavior.

Quantum dot: Similar to a quantum well except that the confinement now is over a three-dimensional region.

Quantum mechanical effects: Phenomena that arise when the size of channel length is reduced under a certain limit. In terms of MOSFET, it is a branch of physics providing a mathematical description of particle-like and wake-like behavior.

Quantum mechanics: Describes the behavior of things at the microscopic scale such as atoms and electrons. The behavior is radically different from that of macroscopic objects such as billiard balls and is not intuitively obvious. However, a vast body of experimental data is consistent with the predictions of quantum mechanics.

Quantum theory: This is more or less similar to quantum mechanics but places more emphasis, perhaps, on the mathematical theory of quantum mechanics.

Quantum tunneling: A purely quantum effect not encountered in classical physics in which there is a finite probability that a particle with energy less than a potential barrier will be able to pass from one side of the barrier to the other.

Quantum well: A device in which one dimension is small enough to create quantum confinement. For electronic devices, the size scale for this confinement is of the order of the de Broglie wavelength of the electrons or other charge carriers involved in conduction. An example of a quantum well structure is a thin film with a film thickness of the order of the de Broglie wavelength.

Quantum wire: Similar to a quantum well except that the confinement now is over a two-dimensional region.

Quasi-Fermi level: The quasi-Fermi level for electrons and the quasi-Fermi level for holes relate the nonequilibrium electron and hole concentrations, respectively, to the intrinsic carrier concentration and the intrinsic Fermi level.

Reciprocal lattice: A lattice (usually a Bravais lattice) in which the Fourier transform of the spatial function of the original lattice (or direct lattice) is represented. This space is also known as momentum space or less commonly k-space.

Schrödinger's wave equation: The fundamental equation of quantum mechanics that replaces Newton's classical mechanics.

Semiconductor: Solids for which the band structure is such that the conduction and valence bands are separated by an energy band gap which is small enough to be comparable with the thermal energy of electrons at room temperature. The semiconductor will thus be able to conduct due to electrons in the conduction band. The conductivity increases with temperature.

Single-walled carbon nanotube: It can be imagined as a graphene sheet having been rolled into a tube or cylinder with a radius in the range of about 0.4–3 nm. The thickness of the tube's wall, in most of the models presented so far in literature, is considered equal to that of a graphene sheet, which is about 0.34 nm, as carbon atoms in this structure are arranged in a hexagonal array.

Spin: Classically, spin is the rotation of a body about an axis, a generally well-known concept. In quantum mechanics, however, it is a feature of the mathematical theory which is not really an analogue of classical spin. Quantum mechanical spin is intrinsic in the sense that a quantum particle, if it has spin, cannot be separated from the spin. This is quite different from the classical case in which a body can be given spin by causing it to rotate, but it can also have no spin.

Strain silicon: A process that involves physically stretching or compressing the lattice crystal to enhance the carrier transport without having geometry scaling.

Transistor: A basic building block of electronic circuits that facilitates the amplification of signals.

Trilayer graphene nanoribbon: Two well-known forms of trilayer graphene nanoribbon (TGN) with different stacking manners are understood as ABA (Bernal) and ABC (rhombohedral). ABA- and ABC-stacked TGN as two of the most common multilayers of graphene nanoribbon are taken into consideration in most of the studies.

Valence band: The energy band immediately below the conduction band (however, it may overlap with the conduction band in a metal).

Wave packet: A superposition of harmonic waves localized in space.

Wave particle duality: One of the strange features of quantum theory is that whether or not wave- or particle-like behavior is observed is not intrinsic to what is being observed; rather, it is context dependent: depending on the whole situation that includes the experimental observation method and what is being observed, either particle- or wave-like properties are observed. This is often described as wave–particle duality. For example, an electron can behave either as a particle or as a wave depending on how it is observed.

Zigzag graphene nanoribbon (armchair): Interestingly, GNR can be classified into two different nomenclatures based on its edge atom alignment known as armchair GNR (AGNR) and zigzag GNR (ZGNR). Both types of GNRs exhibit semiconducting properties where the band gap increases or decreases proportional to the ribbon width.

Index

K) Ltd, Croydon, CR0 4YY